Bayesian Methods

A Bayesian "posterior distribution" or "predictive distribution" summarizes everything you need to know about an unknown parameter, or future observations. This unique book shows how to use Bayesian statistical techniques in a sound and practically relevant manner. It will guide the reader on inferring scientific, medical, and social conclusions from numerical data. The authors explain the subtle assumptions needed for Bayesian methodology and show how to use them to obtain good-quality conclusions. The methods also perform remarkably well in terms of computer-simulated frequency properties.

The lively introductory chapter on Fisherian methods (the frequency approach), together with a strong overall emphasis on likelihood, makes the text suitable for mainstream statistics courses whose instructors wish to follow mixed or comparative philosophies. A chapter on advances in utility theory, and several sections on time series and forecasting, makes the text also suitable for quantitative economics students. The other chapters contain material on the linear model, categorical data analysis, survival analysis, random effects models, and nonlinear smoothing.

The book contains numerous worked examples, self-study exercises, and practical applications. It provides essential reading for final-year undergraduates, Master's-degree and graduate students, statisticians, and other interdisciplinary researchers wishing to develop good-quality conclusions from their data and to pursue the notion of scientific truth.

Thomas Leonard was appointed to the Chair of Statistics at the University of Edinburgh in 1995. He was previously Professor of Statistics at the University of Wisconsin–Madison. His early career, at the University of Warwick, followed his Ph.D. in statistics, obtained from the University of London in 1973. During the 1980s, Leonard pioneered the introduction of Laplacian methods into Bayesian methodology. He has published numerous papers on applications of statistics and has appeared as statistical expert in numerous U.S. legal cases.

John S. J. Hsu is Associate Professor of Statistics and Applied Probability at the University of California, Santa Barbara. He also holds an honorary appointment at the University of Edinburgh, which cites his contributions to the analysis of log-linear models. He obtained his Ph.D. in statistics from the University of Wisconsin–Madison in 1990 under the supervision of Thomas Leonard and Kam-Wah Tsui. At Santa Barbara, Hsu has worked on applied problems, as Director of the Statistical Laboratory, and has also developed his own Bayesian theoretical research program.

CAMBRIDGE SERIES IN STATISTICAL AND PROBABILISTIC MATHEMATICS

This series of high-quality upper-division textbooks and expository monographs covers all aspects of stochastic applicable mathematics. The topics range from pure and applied statistics to probability theory, operations research, optimization, and mathematical programming. The books contain clear presentations of new developments in the field and also of the state of the art in classical methods. While emphasizing rigorous treatment of theoretical methods, the books also contain applications and discussions of new techniques made possible by advances in computational practice.

Already published

1. Bootstrap Methods and Their Application, by A. C. Davison and D. V. Hinkley
2. Markov Chains, by J. Norris
3. Asymptotic Statistics, by A. W. van der Vaart

Bayesian Methods

An Analysis for Statisticians and Interdisciplinary Researchers

---

THOMAS LEONARD

JOHN S. J. HSU

CAMBRIDGE
UNIVERSITY PRESS

PUBLISHED BY THE PRESS SYNDICATE OF THE UNIVERSITY OF CAMBRIDGE
The Pitt Building, Trumpington Street, Cambridge, United Kingdom

CAMBRIDGE UNIVERSITY PRESS
The Edinburgh Building, Cambridge CB2 2RU, UK http://www.cup.cam.ac.uk
40 West 20th Street, New York, NY 10011–4211, USA http://www.cup.org
10 Stamford Road, Oakleigh, Melbourne 3166, Australia

First published 1999

Printed in the United States of America

Typeface Times Roman 10.5/13 pt. *System* LaTeX 2_ε [RW]

*A catalog record of this book is available from
the British Library.*

Library of Congress Cataloging in Publication data
Leonard, Thomas, 1948–
Bayesian methods: an analysis for statisticians and
interdisciplinary researchers / Thomas Leonard, John S. J. Hsu
p. cm. – (Cambridge series in statistical and probabilistic
mathematics)
Includes bibliographical references and index.
1. Bayesian statistical decision theory. I. Hsu, John S. J., 1955– .
II. Title. III. Series: Cambridge series on statistical and probabilistic
mathematics.
QA279.5.L45 1999
519.5′42 – dc21 98-33722 CIP

ISBN 0 521 59417 0 hardback

To
Sarah-Jane, Helen, James
and
Serene, Justin, Andrew

Contents

Preface

Statistics uses theoretical models and techniques to help applied researchers to extract, and infer, real-life, scientific, medical, and social conclusions from numerical data, which are subject to random uncertainty. For any particular study, it is important to combine theoretical and computational resources, together with applied skills, and an ability to interact with experts with knowledge relating to the background and usefulness of the data.

Many studies and data sets are nonstandard, and it is not always possible to provide a completely convincing analysis based upon preexisting techniques. Therefore, statisticians frequently need to develop new techniques, on line, for a particular practical study. Furthermore, the statistical state of the art is continuously evolving, and it is therefore important for researchers to continue to develop the available statistical methodology. Finally, when existing methodology is available, it is important that this should be applied with specific knowledge of the subtleties of the assumptions involved, together with their consequences.

There are nowadays two main streams of statistical thought. We will refer to these as the "Fisherian" and the "Bayesian" philosophies. The Fisherian philosophy is named after Sir Ronald Fisher and combines the "frequency approach" (unbiased estimators, hypothesis tests, and confidence intervals) with likelihood methods. The Fisherian philosophy also includes the "fiducial approach," an incomplete method, suggested by Fisher, which attempts to achieve some of the advantages of the Bayesian approach (e.g., good conditional inference, given the observed values of the data, combined with appealing frequency properties when repeating the experiment a number of times under identical conditions), but without the assumption of a "prior distribution."

The Bayesian philosophy is named after the Reverend Thomas Bayes and refers to such concepts as "prior and posterior knowledge," "prior, posterior, and predictive distributions," and "Bayes decision rules and estimators." The Bayesian approach possesses many advantages, even when viewed from a Fisherian viewpoint, in particular its inherent long-run frequency properties. In practical terms, this means that if computer simulations are used to compare the mean squared error, prediction error, coverage probability, or power of different procedures, then Bayesian methods can perform remarkably well. This validation is an essential ingredient, when combined with the construction of statistical techniques, and provides just one substantial justification of the Bayesian paradigm. Other advantages are summarized by Berger (1985) and Bernardo and Smith (1994), and in our introductions to Chapters 2, 3, 5, and 6 of the current text.

Chapter 1 describes a number of Fisherian procedures, which comprise important background to the Bayesian approach. It is, for example, essential for the reader to be able to construct and understand likelihood functions before attempting Bayesian techniques. The reader should also understand basic data analysis.

Chapter 2 provides an easy introduction to Bayesian ideas and utilizes easy forms of Bayes' theorem when the parameter space is discrete. These are of particular importance in medical and legal applications.

Chapter 3 develops the Bayesian paradigm when there is a single unknown parameter. In such cases, a univariate probability distribution readily summarizes the posterior information. Frequency properties of related estimators and decision rules are developed.

Chapter 4 provides a break to some of the technicalities and considers the "expected utility hypothesis" and its role in financial decision making. Some extensions to the expected utility hypothesis are considered.

Chapter 5 extends the ideas of Chapter 3 to statistical models with several parameters. Approaches to the linear statistical model, categorical data analysis, and time-series analysis are included.

Chapter 6 provides advanced studies of prior structures, posterior smoothing, and Bayes–Stein estimation. Many of the techniques again achieve appealing frequency properties. Computational techniques, already mentioned in Chapter 5, for approximating or simulating high-dimensional numerical integrations, for example, for providing adequate finite sample size analyses of nonlinear models, are developed further. These include Laplacian methods, importance sampling, and Markov Chain Monte Carlo Methods (MCMC).

The text contains 49 worked examples and 148 self-study exercises, which relate to special cases of methodology more broadly explained in the main body of the text. The reader is thereby provided with layers of knowledge, which can be studied at different levels. The volume progressively develops a number of special themes in a possibly unique manner. A large number of further practical examples are described throughout the text.

The bibliography integrates Bayesian statistics with other statistical methodologies and with interdisciplinary research. While the Bayesian references represent the last four decades of research, they do not provide an exhaustive reference list for the Bayesian literature.

Much of the material in this text has been previously taught to graduate students in statistics, economics, and business attending a Bayesian Decisions course at the University of Wisconsin–Madison, and to graduate students attending a Bayesian Inference course at the University of California at Santa Barbara. The text is also appropriate for the following readerships:

- Students attending a statistics course with a mixture of Fisherian and Bayesian philosophies, at final-year undergraduate or at Master's-degree level. In this case, the instructors should concentrate on the easier parts of Chapter 1, together with Chapters 2 and 3, and the easier parts of Chapter 5. If the course is taught within an economics graduate program, then Chapter 4 and Sections 5.3–5.7 will also be of interest, together with the simulation procedures of Chapter 6.

- Interdisciplinary research specialists wishing to develop statistical models and analyses relating to their own area. We have previously used techniques described in this text for interdisciplinary research in many areas, including geology, psychometrics, medicine, animal science, genetics, biology, archaeology, forensic science, civil engineering, plant science, pathology, and physics. We have been involved in many practical collaborations, as directors of statistical laboratories at the University of Edinburgh and The University of California, with these objectives in mind.
- Doctoral students, and other researchers, in statistics. For example, Chapters 5 and 6 will help you to achieve the research frontiers in Bayesian statistics. Chapters 3, 5, and 6 would provide useful material for an advanced graduate course in statistics.

The first co-author wishes to acknowledge his mentors Anne F. S. Mitchell, Dennis V. Lindley, and A. Philip Dawid for teaching him Bayesian statistics at Imperial College and University College, London. His early Bayesian ideas, also frequently employed in this volume, were further influenced by James M. Dickey, Irving Jack Good, Adrian F. M. Smith, Tony O'Hagan, Jim Q. Smith, Patricia M. E. Altham, and P. Jeffrey Harrison. The second co-author wishes to acknowledge David V. Hinkley and Raisa Feldman for their encouragement. Both co-authors are indebted to Arnold Zellner, George Tiao, and Kam-Wah Tsui for their outstanding help and encouragement. They would also like to thank George E. P. Box, Jeff C. F. Wu, Irwin Guttman, Colin G. Aitken, Grace Wahba, Nan Laird, Michael Newton, Greg Reinsel, Bob Miller, Douglas Bates, and Richard A. Johnson for their previous advice on Bayesian and other related methods contained in this volume. Peter Lee has provided very helpful information in relation to his own writings. Suggestions by Bob Barmisch, Robert McCullough, Peter Wakker, Derek Arthur, and John Searle are indicated in the text. Jerome Klotz has advised us on gambling with roulette. Orestis Papasouliotis collaborated on some of the recent methodological developments and prepared the mathematics and computer program for the graphs in the cover design (these are the posterior densities of the group means in an analysis of covariance model). Geoff McLachlan kindly provided us with a copy of his computer package for multivariate mixtures. Rod Leonard described valuable insights regarding the problem of spurious correlation in the context of the chemical industry.

We should also acknowledge the many graduate students attending our Bayesian courses who have helped or advised us over the years. These include, but are not limited to, Jean Deichtmann, Josep Ginebra-Molins, Robert Tempelman, Taskin Atilgan, Christian Ritter, Tom Chiu, and Jen-Ting Wang.

We would like to thank the following publishers and associations for granting us permission to reproduce previously published material: John Wiley & Sons for Figure 5.2.1 from T. Leonard and J. S. J. Hsu (1994), The Bayesian analysis of categorical data – a selective review, *in* P. R. Freeman and A. F. M. Smith (eds.), *Aspects of Uncertainty: A Tribute to D. V. Lindley*, copyright John Wiley & Sons Limited; the American Educational Research Association for Tables 5.2.4 and 5.2.5 from T. Leonard and M. R. Novick (1986), Bayesian full rank marginalization for two-way contingency tables, *Journal of Educational Statistics* **11**: 33–56; the Royal Statistical Society for Tables

5.2.7 and 5.2.8 from T. Leonard (1975), Bayesian estimation methods for two-way contingency tables, *Journal of Royal Statistical Society*, Ser. B, **37**: 23–37; the American Statistical Association for Table 6.5.1 from L. Sun, J. S. J. Hsu, I. Guttman, and T. Leonard (1996), Bayesian methods for variance component models, *Journal of the American Statistical Association* **91**: 743–52; and the Institute of Statistical Mathematics for Table 6.5.2 from T. Leonard (1984), Some data-analytic modifications to Bayes-Stein estimation, *Annals of the Institute of Statistical Mathematics* **36**: 11–21.

We would also like to thank Gloria Scallissi for her excellent typing of the first draft of the manuscript, Lauren Cowles for her continued encouragement on behalf of the publishers, David Tranah for some helpful suggestions, and Rena S. Wells for her excellent copyediting.

1

Introductory Statistical Concepts

1.0 Preliminaries and Overview

When analyzing numerical data, subject to random uncertainty, which have been collected in some scientific or real-life context, the first "golden rule" is to study the data, for example, using dotplots, bivariate scatterplots, relative frequency histograms, and contingency tables, before applying any formal statistical technique. Complicated data sets deserve several hours, days, or even weeks of study. When studying a data set, you should realize that data are not simply numbers but rather measurements or counts of real entities (e.g., birthweights of babies, numbers of students passing a college test, a measurement of a real chemical). Therefore, any tentative conclusions should be made in the contexts of their meaning in relation to these entities, the real background of the data, and how the data were collected. The same set of numerical data might mean something entirely different in different scientific or real-life contexts.

Sometimes, upon viewing the data, you may discover a particularly distinctive feature that yields a decisive conclusion. In this case, it may not be necessary, or indeed technically feasible, to proceed to a more formal analysis. For example, when investigating the years of service of French generals during the late eighteenth century (see Wetzler, 1983, Appendix), the conclusion was reached, upon viewing a distinctive spike in the scatterplot, that a number of the generals had been rather abruptly dismissed during the French Revolution. As another example, the State of Wisconsin was advised during a court action in 1986, and based upon data for a carefully collected random sample of $n = 120$ nursing homes, that the state was not adequately reimbursing the actual costs (in dollars per patient per day) for nursing homes with costs in excess of \$45. This conclusion was validated by a distinctive blip in an otherwise linear bivariate scatterplot. The State of Wisconsin conceded the case, largely because the state recognized that data based upon a representative sample (a random sample of 120 nursing homes from 600 nursing homes in Wisconsin) had been collected. George and Wecker (1985) also emphasize the importance of using good statistics in legal cases.

For the first of these analyses, a computer package was not used, since formal summary statistics such as the sample mean and variance would not be particularly relevant. For the second analysis, any attempt to fit a regression model without first carefully considering the data could have led to many hours of fruitless analysis. Similarly, you should try to avoid any "black box" data analytic technique that cannot be combined with an interaction between your thought processes and the data. It is particularly

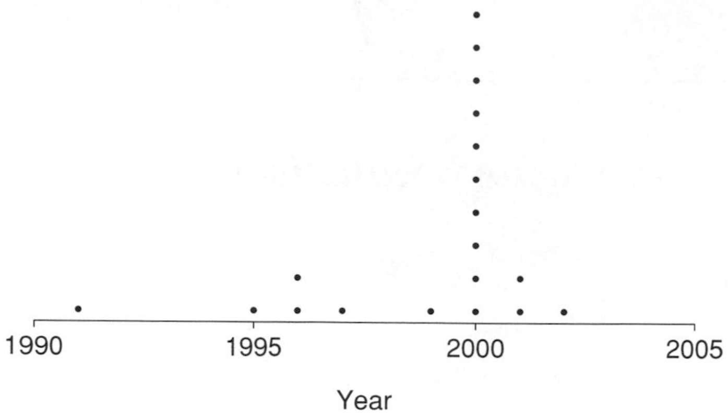

Figure 1.0.1. A dotplot with a spike.

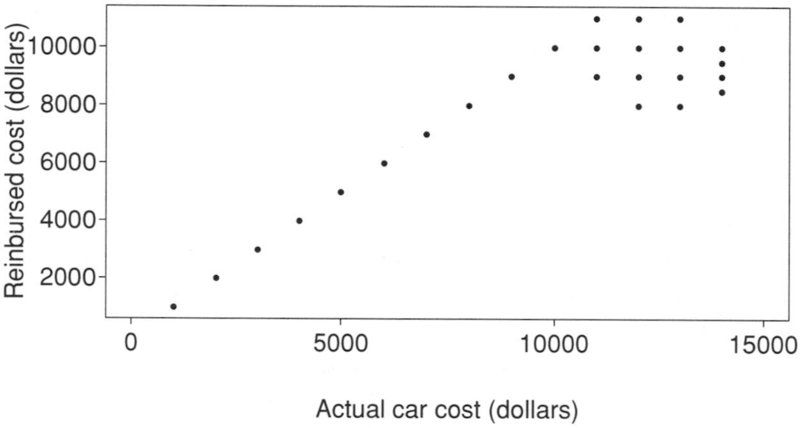

Figure 1.0.2. A bivariate scatterplot with a blip.

important to consider the dotplots and scatterplots. Two further data plots, with a spike and a blip, respectively, are described in Figures 1.0.1 and 1.0.2.

When viewing the data, we should pay careful attention to any outlying observations (see Figure 1.0.3). Outliers are discussed in greater detail by Barnett and Lewis (1978). For example, an outlier can enhance the apparent correlation between two variables that may not otherwise be obviously correlated. Carefully consider the origins and meaning of each outlier and make a careful intuitive decision as to whether or not to include it in the sample. Don't automatically reject outliers, using a "black box" technique,

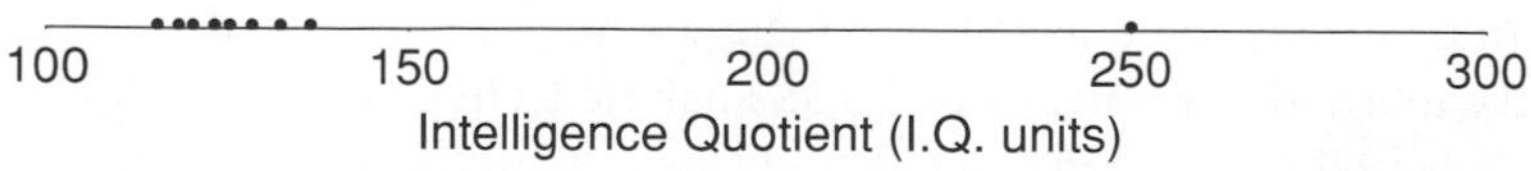

Figure 1.0.3. A dotplot with an outlier.

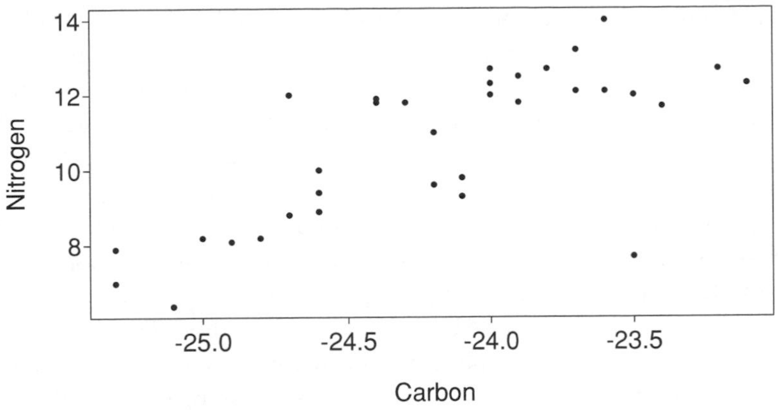

Figure 1.0.4. A scatterplot of the readings on $n = 34$ skeletons.

since they may be quite informative, particularly if they are part of a random sample. Outliers can of course strongly influence any formal analysis, so it is essential to be aware of them. We also regard it as important not to "impute" values for missing data, since the modeling process for the whole data set can then become confused with the imputation process, and the imputed values can exaggerate the information content of the data. Likelihood and Bayesian methods will be able to readily handle missing data problems (just integrate the sampling distribution with respect to the missing observations), without any need to impute the data.

A typical archaeologists' diagram for recording the (transformed) nitrogen and carbon content of skeletons compresses the carbon axis, creating a tendency for the skeletons to be divided into groups, according to nitrogen content alone. In cases where there are two groups, the group of skeletons with the higher nitrogen content is often taken to be Mesolithic, and the group with lower nitrogen content, Neolithic. In Figure 1.0.4, however, we report the entire scatterplot of the readings on $n = 34$ skeletons, at the Lepenski Vir site in the Danube valley, but with the carbon scale substantially broadened when compared with the archaeologists' procedures. Our scatterplot suggests division into several groups rather than two groups. Indeed, McLachlan's package for multivariate mixtures (McLachlan and Basford, 1988) indicates at least six groups. The strong case for at least four or five groups can be confirmed by matching the groups with gender. Further discussions of these data are provided by Bonsall, Lennon, and McSweeney (1997).

For many data sets, it is also of interest either (a) to draw inferences about unknown parameters of interest, for example, the density of a fluid, the recovery rate for patients receiving a particular treatment, or population means, or proportions, or (b) to be able to predict future observations, given the current and previous observations, for example, economic forecasting, the forecasting of the paths of hurricanes, or the prediction of the probability of failure of an engineering design. It is then useful to formalize the random variability or uncertainty in the data, using a mathematical probability model, that is, by taking the numerical observations to be realizations of random variables whose joint distribution comprises the mathematical or "sampling" model. A second "golden rule" is to realize that "true models" are available only in limited situations. In many cases, a variety of different models can be taken to plausibly represent a data set with a finite number n of observations. Which model to use depends partly on statistical technique, but also on the meaning and usefulness of the model in relation to the actual context of the data.

Given a particular sampling model, a key question is, How should the applied statistician use the data to draw inferences about any unknown parameters appearing in the model? In this text, we adhere to the principle, Given the truth of the sampling model, all information in the data is summarized by the likelihood function. The beautiful concept of likelihood links all major philosophies of statistics and provides a cornerstone of the Bayesian paradigm. Its properties and applications are developed in detail throughout this chapter.

It is possible to draw objectively acceptable conclusions from data, when appropriate randomization is performed at the design stage, ideally with further replications of the experiment, to detect unlucky randomizations. For uncontrolled data, an appropriate model can be more difficult to find, and any conclusions are subjective and subject to the effects of "lurking" or "confounding" variables (see Section 1.2 (H)). In general, the conclusions are subject to "shades of subjectivity," depending upon the way the data are collected. For example, Brown et al. (1997) experienced considerable practical difficulties in collecting a random sample while surveying primary-care patients for drug or alcohol abuse. This was mainly the case because the interviewers were under considerable pressure to complete their interviews within time periods agreed upon with the clinics. The conclusions therefore needed to be qualified accordingly. There are also frequently problems with the selective reporting of significant results (see Dawid and Dickey, 1977). Furthermore, the sample size should be chosen with care at the design stage (Donner, 1984).

Both Fisherian and Bayesian statistics depend heavily upon the concept of "probability." What is probability? For a statistical experiment $\mathcal{E}$, with sample space S, mathematicians will tell you that a probability distribution $p(\cdot)$ is a real-valued function defined on all events (strictly speaking, events are constrained to be "measurable subsets") contained in S and satisfying the Kolmogorov axioms (see Exercise 1.1.l). However, philosophically speaking, there are three main types of probabilities:

(A) *Classical probability:* This is defined by an "m over k" rule and is appropriate whenever $S = \{e_1, e_2, \ldots, e_k\}$ possesses k outcomes that are judged to be "equally likely," and when an event A consists of m of these k outcomes. When the equally likely assumption is made objectively, such as when the outcome that occurs has been chosen at random from the k outcomes, or the equally likely assumption has been tested by replicating the experiment numerous times under identical conditions, then the probability $p(A)$ of the event A, defined by $p(A) = m/k$, can be referred to as an "objective classical probability." When the equally likely assumption is made subjectively (e.g., in the absence of evidence to distinguish that any particular outcome is more likely than any other), then $p(A) = m/k$ can be referred to as a "subjective classical probability." See also Exercises 1.1.a and 1.1.b, which tell us that population proportions can be identified with classical probabilities when individuals are chosen at random from the population.

(B) *Frequency probability:* This will be defined by equation (1.1.1). The frequency probability of an event is the long-run proportion of times the event occurs in a large number of replications of the experiment. Objective classical probability provides an example of frequency probability. Therefore, since the objective classical probability that a roulette wheel will give a black number is 9/19, this can also be used to predict the long-run performance of the wheel.

(C) *Subjective probability:* This measures an individual's uncertainty in an event and may vary from individual to individual. You may calibrate your subjective probabilities by judging whether events A are equally likely to events for an objective auxiliary experiment, for example, the spinning pointer of Exercise 1.1.k. In principle, you should assign probabilities to all events $A \subseteq S$ and ensure that your probabilities satisfy the Kolmogorov axioms. An individual who always tries to represent his uncertainty by a subjective probability distribution is referred to as a "Bayesian."

In general, a number between zero and unity can be regarded as a probability only if all other events in the sample space are envisioned, probabilities are assigned to every event, and the laws of probability, as defined by the Kolmogorov axioms, are satisfied by the entire collection of probabilities (referred to as a "probability distribution"). Many "probabilities" quoted in science and the media do not satisfy these conditions. For a sample space with either finitely many outcomes or outcomes that can be arranged in an infinite sequence, it is sufficient to check that the values assigned to the individual outcomes sum to unity.

Consider situations where you possess some information regarding an unknown parameter θ, for values of θ lying in a parametric space Θ. Then a big question is whether or not you can represent this information by a subjective probability distribution on Θ. Some Bayesians say, You should always represent your information by a subjective probability distribution on Θ, since there are some very simple axioms that tell us that if you don't act like this, then you are irrational, incoherent, and moreover, a sure loser! We do not concur with this type of "normative approach," largely because we are unaware of an axiom system that is simple enough, when compared with the Kolmogorov axioms, to justify this viewpoint. Moreover, some information, such as medical knowledge or evidence in a court case, may be too diverse or eclectic to be representable by probabilities. (These views are open to discussion. For a more traditionally Bayesian approach, see Bernardo and Smith, 1994, section 2.3. Many Bayesians believe that the uncertainty in any event is representable by a probability. These aspects are pursued in Exercise 1.1.k and were previously debated by Leonard, 1980, and discussants. In our current chapter, we also debate the likelihood principle. See Section 1.5 (C), Exercises 1.5.c–1.5.f.)

Other topics discussed in the current chapter include Akaike's and Schwarz's information criteria, AIC and BIC, for deciding between different choices of sampling models. These subtract a penalty per parameter from the log-likelihood function (see equations 1.1.5 and 1.1.6). Information criteria are best justified and compared by computer simulation of sets of observations from particular choices of their true sampling model (see Sections 1.2 (F) and (G)). However, it is also important to consider all possible diagnostics, for example, residual analyses for regression models, when comparing models (see Exercise 1.5.l) and also to consider the real-life or scientific reasonability of the candidates.

Any formal statistical procedure, whether for inference about parameters, prediction of future observations, or choice of sampling model, should possess desirable long-run frequency properties (e.g., good mean squared error (MSE) for estimation of parameters, accurate frequency coverage for approximate confidence intervals, high long-run probability of choosing a reasonable model). In situations where these cannot be developed theoretically, computer simulations can produce accurate and meaningful results. Graduate students and research specialists are encouraged to create novel statistical procedures, but then always to check their new ideas by using frequency simulations.

In Section 1.3, procedures are described for obtaining approximate confidence intervals that closely relate to the multivariate normal likelihood approximation (1.3.11) and that refer to the concept of transforming the parameters to achieve possibly better approximate normality. It is particularly important for the research worker to numerically check any theoretical suggestions when using theoretical approximations, since the numerical work may produce some surprises or suggest adjustment terms to the approximations. For example, an "approximately normally distributed random variable X" might not yield values for $p(X < -1.96)$ or $p(X > 1.96)$ that are particularly close to 0.025, as required for an exact result. A variety of practical justifications of the approximations employed in the text are included (e.g., Sections 1.2 (C), 1.4 (A) and (B)). Some key properties of the multivariate normal distribution are developed in Exercise 1.1.n.

A multivariate normal approximation (1.3.11) and related parameter transformations will provide a central theme to a variety of Bayesian ideas developed later in the text, such as the construction of "prior distributions" for several parameters, computational procedures using importance sampling, Laplacian approximations, and rejection sampling. It is more important for the reader to understand the multivariate normal approximation and related approximate confidence intervals than to research the complicated asymptotic theory of maximum likelihood estimators. For any particular model, it is better to check the validity of (1.3.11) computationally and for finite sample sizes.

The works of Sir Ronald Fisher provide excellent background to this chapter. See, for example, Fisher (1925, 1935, 1959) and Bennett (1971–4). Fisher always mixed his techniques with practical common sense.

1.1 Sampling Models and Likelihoods

Numerical data often arise as a result of some statistical experiment $\mathcal{E}$, that is, an occurrence with a random or uncertain outcome. Suppose that on a single repetition of $\mathcal{E}$, you observe n numerical observations $y_1, \ldots, y_n$. Let the sample space S denote the set of all possible realizations of the column vector $\mathbf{y} = (y_1, \ldots, y_n)^T$. Then S is a subset of n-dimensional Euclidean space R^n, and the vector $\mathbf{y}$ consists of the n observations, arranged in a column.

You might be prepared to make the quite strong assumption that $\mathbf{y} = (y_1, \ldots, y_n)^T$ is a numerical realization of a random vector $\mathbf{Y} = (Y_1, \ldots, Y_n)^T$ (i.e., the column vector of the random variables $Y_1, \ldots, Y_n$), which possesses some probability distribution P, defined on events in S. For example, if $\mathcal{E}$ can be repeated a large number of times under identical conditions (i.e., *replicated*), then $P = P(\cdot)$ can be defined in terms of the frequency notion of probability. That is, for any event A contained in S,

$$P(A) = \text{prob}(\mathbf{Y} \in A) = \lim_{m \to \infty} r_m(A), \tag{1.1.1}$$

whenever the limit on the right-hand side exists, where the relative frequency $r_m(A)$ denotes the proportion of times that $\mathbf{y} \in A$, during the first m replications of $\mathcal{E}$. However, if $\mathcal{E}$ can be performed only once, then the assumption that $\mathbf{y}$ is a numerical realization of $\mathbf{Y}$ cannot always be made objectively. In some situations where you use random sampling from a population or in other situations where outcomes of the experiment can be regarded as equally likely, it will still be possible to define P in an objective fashion. However, in many cases, part of the modeling process, that is, the specification of P, will need to be performed subjectively and by reference to the scientific or social background of the data. In many cases, P will be incompletely known, even after a variety of modeling assumptions, and it is therefore frequently necessary to infer reasonable choices of P, based upon the vector $\mathbf{y}$ of observations, for a single repetition of $\mathcal{E}$.

For simplicity, assume that $\mathbf{Y}$ is either a continuous random vector with density $p(\mathbf{y})$, for $\mathbf{y} \in S$, or a discrete random vector with probability mass function $p(\mathbf{y})$, for $\mathbf{y} \in S$. Then, following the tenets of *parametric statistical inference*, you might wish to make an assumption of the form

$$p(\mathbf{y}) = f(\mathbf{y} \mid \boldsymbol{\theta}) \qquad \left(\mathbf{y} \in S,\ \boldsymbol{\theta} \in \Theta \subseteq R^k\right), \tag{1.1.2}$$

where $f(\mathbf{y} \mid \boldsymbol{\theta})$ is specified as a function of both $\mathbf{y} \in S$ and $\boldsymbol{\theta} = (\theta_1, \ldots, \theta_k)^T \in \Theta$. Here $\boldsymbol{\theta}$ is some vector of unknown parameters, and Θ is the parameter space. If $n \geq k$, you can now make inferences about a k-dimensional vector $\boldsymbol{\theta}$ of unknown parameters, rather than an entire function $p(\cdot)$.

Box (1980) distinguishes between this *inference* problem and the problem of statistically *modeling* the choice of functional form for f. Modeling involves both creating an appropriate choice for f in relation to the scientific background and checking the reasonability of this choice against the data. Modeling requires substantial inductive thought, while inference requires deduction, that is, the calculation of mathematical conclusions, given that the functional form of the model is assumed true. This blend of inductive and deductive thought is part of the *inductive synthesis* (Aitken, 1944, p. 3).

Following Birnbaum's (1962) philosophy of "the irrelevance of observations not actually observed" (e.g., why use procedures involving significance probabilities, minimum variance criteria, and confidence statements, which average across the sample space?) and Edwards's famous 1972 treatise on likelihood, it is a reasonable and widely held contention, though without a watertight scientific proof, that the *likelihood function* summarizes all the information about $\boldsymbol{\theta}$ contained in the data, when the functional form of f is assumed true, and you condition on the numerical realization of $\mathbf{y} = (y_1, \ldots, y_n)^T$ actually observed. The likelihood function is a function of $\boldsymbol{\theta}$ and satisfies

$$l(\boldsymbol{\theta} \mid \mathbf{y}) = f(\mathbf{y} \mid \boldsymbol{\theta}) \qquad (\boldsymbol{\theta} \in \Theta), \tag{1.1.3}$$

where $\mathbf{y}$ is the observed $\mathbf{Y}$. Hence (1.1.3) provides a complete solution, at least in principle, for the inference problem. For example, if $k = 1$ or 2, you can just consider a sketch of the likelihood function and draw all your conclusions from this graphical procedure, together with your background experience. Any vector $\hat{\boldsymbol{\theta}}$ maximizing (1.1.3), as a function of $\boldsymbol{\theta} \in \Theta$, with $\mathbf{y}$ fixed, provides a *maximum likelihood estimate* of $\boldsymbol{\theta}$. In intuitive terms, this gives the realization of $\boldsymbol{\theta}$ most likely to have given rise to the current data set, an important finite sample property. Different choices or hypotheses, say, $\boldsymbol{\theta}_1$ and $\boldsymbol{\theta}_2$, for $\boldsymbol{\theta}$ may be compared by choosing the hypothesis that maximizes (1.1.3), with $\mathbf{y}$ fixed. However, areas under the likelihood function are not immediately relevant; they should in particular not be interpreted as probabilities.

Note that the expected log-likelihood

$$E\left[\log l(\boldsymbol{\theta} \mid \mathbf{Y})\right] = \int_S f(\mathbf{y} \mid \boldsymbol{\theta}) \log f(\mathbf{y} \mid \boldsymbol{\theta})\, d\mathbf{y}$$

is the negative of the *entropy* associated with the sampling density $f(\mathbf{y} \mid \boldsymbol{\theta})$. Under broad regularity conditions, $n^{-1} \log l(\hat{\boldsymbol{\theta}} \mid \mathbf{Y})$ will converge, to this expectation, with probability one, as n tends to infinity. This introduces entropy as an information criterion in statistics and tells us that $\hat{\boldsymbol{\theta}}$ is associated with minimizing the disorder or lack of information about the sampling model. The modeling process can be assisted by a closely related criterion. Suppose that you use your knowledge or intuition to constrain f to belong to some family $\mathcal{F}$ of meaningful functional forms, but where the dimension k may vary among members of the family. Then choose $f \in \mathcal{F}$ to maximize

$$\text{GIC} = \text{General Information Criterion} = \log l\big(\hat{\boldsymbol{\theta}} \mid \mathbf{y}\big) - \frac{\alpha k}{2}. \tag{1.1.4}$$

Here $\log l(\hat{\boldsymbol{\theta}} \mid \mathbf{y})$ denotes the supremum of the log-likelihood function, and $\frac{1}{2}\alpha$ provides a *penalty per parameter* in the model. The choices $\alpha = 2$ (Akaike, 1978) and $\alpha = \log(n/2\pi)$, (Schwarz, 1978) have been suggested, with some theoretical justification (e.g., Stone, 1977, 1979), providing

$$\text{AIC} = \text{Akaike's Information Criterion} = \log l\big(\hat{\boldsymbol{\theta}} \mid \mathbf{y}\big) - k \tag{1.1.5}$$

and

$$\text{BIC} = \text{Bayesian Information Criterion} = \log l\big(\hat{\boldsymbol{\theta}} \mid \mathbf{y}\big) - \frac{k}{2} \log \frac{n}{2\pi}. \tag{1.1.6}$$

These criteria help us to find a concise model, with just a few parameters. It is also important to consider the meaning of the model in a scientific context. We should avoid complicated models, or model choice based upon small sample sizes. L. J. Savage (see, Lindley, 1983) felt that "a model should be as big as an elephant," but this can be contrasted with the late Toby Mitchell's philosophy, "the greater the amount of information, the less you know," that is, a complicated model might be difficult to interpret. An overall predictive check of the modeling assumptions proposed by Box (1980, 1983) gives only a partial answer, and it more generally seems to be very difficult to check convincingly any particular model without a particular set of alternatives in mind by reference to the current data alone. In other words, applied scientific judgment is invariably needed. Of course, even if a model is obviously incorrect, it might still help us to extract some conclusions from the data, since a model with k parameters helps us to reduce the dimensionality of the problem from n to k and to make some conditional conclusions. This process can be repeated with different tentative models. We may hence use the model to "telescope" low-dimensional pictures of the data. In other words, some models are objective, others are subjective, and the modeling process helps us to perceive the data.

Three examples are now described to illustrate the algebraic manipulations needed when constructing likelihoods.

Worked Example 1A: *Maximum Likelihood for the Geometric Distribution*

Let $Y_1, Y_2, \ldots, Y_n$ denote a random sample from a geometric distribution, with the probability mass function (p.m.f.)

$$p(Y_i = y_i \mid \theta) = \theta(1-\theta)^{y_i - 1} \qquad (y_i = 1, 2, \ldots). \tag{1.1.7}$$

(a) Find the likelihood of θ.
(b) The maximum likelihood estimate $\hat{\theta}$ of θ maximizes the probability of obtaining the observations actually observed. Find $\hat{\theta}$.
(c) The invariance property of maximum likelihood estimates tells that for any function $\eta = g(\theta)$ of θ, $\hat{\eta} = g(\hat{\theta})$ is the maximum likelihood estimate of $g(\theta)$. Find the maximum likelihood estimate of $\eta = \theta(1-\theta) = p(Y_1 = 1)$.

Model Answer 1A:

(a) The joint p.m.f. of $Y_1, \ldots, Y_n$ is

$$\begin{aligned} &p(Y_1 = y_1, \ldots, Y_n = y_n \mid \theta) \\ &= \prod_{i=1}^{n} p(y_i \mid \theta) = \prod_{i=1}^{n} \theta(1-\theta)^{y_i - 1} = \theta^n \prod_{i=1}^{n} (1-\theta)^{y_i - 1} \\ &= \theta^n (1-\theta)^{\sum_{i=1}^{n}(y_i - 1)} = \theta^n (1-\theta)^{\sum_{i=1}^{n} y_i - n} \\ &= \theta^n (1-\theta)^{n(\bar{y}-1)}, \end{aligned}$$

where $\bar{y} = n^{-1} \sum_{i=1}^{n} y_i$. If $y_1, \ldots, y_n$ are the numerical realizations of $Y_1, Y_2, \ldots, Y_n$, this p.m.f. also provides the likelihood

$$l(\theta \mid \mathbf{y}) = l(\theta \mid y_1, \ldots, y_n) = \theta^n (1-\theta)^{n(\bar{y}-1)} \qquad (0 < \theta < 1),$$

which, as a function of θ, is a beta curve.

(b) It is often technically easier to maximize the likelihood by maximizing the log-likelihood. In our example,

$$\log l(\theta \mid \mathbf{y}) = n \log \theta + n(\bar{y}-1)\log(1-\theta),$$

$$\frac{\partial \log l(\theta \mid \mathbf{y})}{\partial \theta} = \frac{n}{\theta} - \frac{n(\bar{y}-1)}{1-\theta},$$

implying

$$\frac{\partial^2 \log l(\theta \mid \mathbf{y})}{\partial \theta^2} = -\frac{n}{\theta^2} - \frac{n(\bar{y}-1)}{(1-\theta)^2}.$$

Setting the first derivative equal to zero, we find that $\hat{\theta}$ satisfies

$$\frac{n}{\hat{\theta}} = \frac{n(\bar{y}-1)}{1-\hat{\theta}}$$

and

$$1 - \hat{\theta} = (\bar{y}-1)\hat{\theta},$$

so that $\hat{\theta} = 1/\bar{y}$. This single stationary point provides a global maximum, since the second derivative of the log-likelihood is negative, for all values of θ. This implies that the likelihood curve is convex.

(c) $\hat{\eta} = \hat{\theta}(1-\hat{\theta}) = \bar{y}^{-2}(\bar{y}-1)$, since, by the invariance property of maximum likelihood estimates, $\hat{\theta} = 1/\bar{y}$.

Worked Example 1B: Likelihood Methods for the Poisson Distribution

Let $Y_1, Y_2, \ldots, Y_n$ denote a random sample from a Poisson distribution with mean μ and p.m.f.

$$p(y) = \frac{e^{-\mu}\mu^y}{y!} \qquad (y = 0, 1, 2, \ldots; 0 < \mu < \infty).$$

Note that $U = Y_1 + Y_2 + \cdots + Y_n$ possesses a Poisson distribution with mean $n\mu$.

(a) Find the likelihood $l(\mu \mid \mathbf{y})$ of μ, given that $Y_1 = y_1, \ldots, Y_n = y_n$.

(b) Show that $l(\mu \mid \mathbf{y}) = h(\mathbf{y}) l^*(\mu \mid \mathbf{y})$, where $h(\mathbf{y})$ does not depend on μ, and the likelihood kernel $l^*(\mu \mid \mathbf{y})$ depends only upon $y_1, y_2, \ldots, y_n$ via the sample mean $\bar{y}$, when n is specified.

(c) Find the likelihood of μ, given only the observed value $n\bar{y}$ of the total U, and without further information about the individual $y_1, y_2, \ldots, y_n$.

(d) Show that the likelihoods in (a) and (c) possess the same shape, as functions of μ, and that both are maximized when $\mu = \bar{y}$.

Model Answer 1B:

(a) From the p.m.f. of the Poisson distribution,

$$p(Y_i = y_i \mid \mu) = \frac{e^{-\mu}\mu^{y_i}}{y_i!} \qquad (y_i = 0, 1, 2, \ldots).$$

By independence of $Y_1, Y_2, \ldots, Y_n$,

$$\begin{aligned} p(\mathbf{y} \mid \mu) &= p(Y_1 = y_1, Y_2 = y_2, \ldots, Y_n = y_n \mid \mu) \\ &= \prod_{i=1}^{n} p(Y_i = y_i \mid \mu) = \prod_{i=1}^{n} \frac{e^{-\mu}\mu^{y_i}}{y_i!} \\ &= \frac{(e^{-\mu})^n \prod_{i=1}^{n} \mu^{y_i}}{\prod_{i=1}^{n} y_i!} = \frac{e^{-n\mu}\mu^{\sum_{i=1}^{n} y_i}}{\prod_{i=1}^{n} y_i!}. \end{aligned}$$

Therefore, if $y_1, y_2, \ldots, y_n$ are observed, the likelihood of μ is

$$l(\mu \mid \mathbf{y}) = p(\mathbf{y} \mid \mu) = \frac{e^{-n\mu}\mu^{n\bar{y}}}{\prod_{i=1}^{n} y_i!} \qquad (0 < \mu < \infty),$$

which takes the form of a Gamma curve in μ.

(b) Note that

$$l(\mu \mid \mathbf{y}) = h(\mathbf{y})e^{-n\mu}\mu^{n\bar{y}},$$

where $h(\mathbf{y}) = (\prod_{i=1}^{n} y_i!)^{-1}$. Hence, $l^*(\mu \mid \mathbf{y}) = e^{-n\mu}\mu^{n\bar{y}}$, as required.

(c) Since U is Poisson distributed with mean $n\mu$,

$$p(U = u \mid \mu) = \frac{e^{-n\mu}(n\mu)^u}{u!} \qquad (u = 0, 1, \ldots).$$

If we observe $U = n\bar{y}$, then

$$p(U = n\bar{y} \mid \mu) = \frac{e^{-n\mu}(n\mu)^{n\bar{y}}}{(n\bar{y})!},$$

and this is $l(\mu \mid \bar{y})$, the likelihood of μ, when just $\bar{y}$ is observed.

(d) The likelihoods in (a) and (b) are both proportional, as functions of μ, to the likelihood kernel

$$l^*(\mu \mid \mathbf{y}) = e^{-n\mu}\mu^{n\bar{y}},$$

so that they both possess the same shape. Then

$$\log l^*(\mu \mid \mathbf{y}) = -n\mu + n\bar{y}\log\mu.$$

The maximum $\hat{\mu}$ satisfies

$$\left[\frac{\partial \log l^*(\mu \mid \mathbf{y})}{\partial \mu}\right]_{\mu=\hat{\mu}} = 0,$$

implying that

$$-n + \frac{n\bar{y}}{\hat{\mu}} = 0.$$

Hencc $\hat{\mu} = \bar{y}$. This must provide a global maximum of the likelihood function, since

$$\frac{\partial^2 \log l^*(\mu \mid \mathbf{y})}{\partial \mu^2} = -\frac{n\bar{y}}{\mu^2},$$

and this is always negative.

Worked Example 1C: *Maximum Likelihood for the Two-Parameter Exponential Distribution*

Let $Y_1, Y_2, \ldots, Y_n$ denote a random sample from the exponential distribution, with unknown location parameter θ, scale parameter λ, and density

$$p(y \mid \theta, \lambda) = \lambda \exp\{-\lambda(y-\theta)\} \qquad (\theta < y < \infty),$$

where $-\infty < \theta < \infty$ and $0 < \lambda < \infty$.

The common mean and variance of the Y_i are $\mu = \theta + \lambda^{-1}$ and $\sigma^2 = \lambda^{-2}$. Find the likelihood of θ and λ and the maximum likelihood estimates of μ and σ^2, in situations where the observed values $y_1, y_2, \ldots, y_n$ of $Y_1, Y_2, \ldots, Y_n$ are not all equal.

Model Answer 1C:

The joint density of $Y_1, Y_2, \ldots, Y_n$ is

$$\begin{aligned} p(y_1, y_2, \ldots, y_n \mid \theta, \lambda) &= \prod_{i=1}^{n} p(y_i \mid \theta, \lambda) \\ &= \prod_{i=1}^{n} \lambda \exp\{-\lambda(y_i-\theta)\} I[\theta \le y_i], \end{aligned}$$

where $I[A]$ denotes the indicator function for the set A. This expression equals the likelihood of θ and λ, when $y_1, y_2, \ldots, y_n$ are observed, so that

$$\begin{aligned} l(\theta, \lambda \mid \mathbf{y}) &= \lambda^n \prod_{i=1}^{n} \exp\{-\lambda(y_i - \theta)\} \prod_{i=1}^{n} I[\theta \le y_i] \\ &= \lambda^n \exp\left\{-\lambda \sum_{i=1}^{n} (y_i - \theta)\right\} \prod_{i=1}^{n} I[\theta \le y_i] \\ &= \lambda^n \exp\{-n\lambda(\bar{y} - \theta)\} I[\theta \le z] \qquad (-\infty < \theta < \infty, 0 < \lambda < \infty), \end{aligned}$$

where z denotes the minimum of $y_1, y_2, \ldots, y_n$.

As a function of θ, the likelihood is proportional to $\exp(n\lambda\theta)$ for each $\theta \le z$, and zero otherwise. This is maximized when $\theta = \hat{\theta} = z$. Consider the function

$$g(\lambda) = \lambda^n \exp\{-A\lambda\},$$

whenever $A = n(\bar{y} - \theta) > 0$. Then

$$\log g(\lambda) = n \log \lambda - A\lambda.$$

The maximum $\hat{\lambda}$ satisfies

$$\left[\frac{\partial \log g(\lambda)}{\partial \lambda}\right]_{\lambda=\hat{\lambda}} = \frac{n}{\hat{\lambda}} - A = 0,$$

so that $\hat{\lambda} = n/A$. This solution provides a global maximum, since the second derivative is always negative. Hence, for fixed $\theta < \bar{y}$, the likelihood is maximized when

$$\lambda = \hat{\lambda} = \frac{n}{n(\bar{y} - \theta)} = \frac{1}{\bar{y} - \theta}.$$

Since $z < \bar{y}$ whenever $y_1, y_2, \ldots, y_n$ are not all equal, the likelihood is maximized when

$$\theta = \hat{\theta} = z \quad \text{and} \quad \lambda = \hat{\lambda} = \frac{1}{\bar{y} - z}.$$

Hence, by the invariance property of maximum likelihood estimators, the maximum likelihood estimates of μ and σ^2 are

$$\begin{aligned} \hat{\mu} &= \hat{\theta} + \hat{\lambda}^{-1} \\ &= z + (\bar{y} - z) = \bar{y}, \\ \text{and} \quad \hat{\sigma}^2 &= \hat{\lambda}^{-2} = (\bar{y} - z)^2. \end{aligned}$$

SELF-STUDY EXERCISES

1.1.a Consider a sample space $S = \{e_1, e_2, \ldots, e_k\}$ with k equally likely outcomes $e_1, \ldots, e_k$. Then, for any event in A, the "classical probability of A" is defined to be

$$p(A) = \frac{m}{k},$$

where m denotes the number of outcomes in S.

The town of Lesser Brodhead, Wisconsin, has 2,555 inhabitants, of whom 5 possess the Bombay gene. If a person is chosen at random from the Lesser Brodhead population, then show that the classical probability of this person having the Bombay gene is equal to the population proportion (or gene frequency) of $\theta = 1/511$. Carefully describe how your classical probability can also be a frequency probability in this particular situation.

1.1.b The relative frequency histogram, with interval width one, of intelligence quotients (I.Q.s) of all Californians is found to resemble closely a location scale curve taking the form $f(y) = \lambda^{-\frac{1}{2}}\Psi\{(y-\mu)/\lambda^{\frac{1}{2}}\}$, for $y \in (-\infty, \infty)$, where $\Psi(\cdot)$ is a specified probability density, and $\mu \in (-\infty, \infty)$ and $\lambda \in (0, \infty)$ are unknown parameters. A random sample of size n is taken, with replacement, from the Californian population, and n intelligence quotients $y_1, \ldots, y_n$ are observed. Use the concept of classical probability (Exercise 1.1.a) to relate the relative frequencies to areas under the location scale curve, and hence to prove that up to a close approximation, the joint density of $y_1, y_2, \ldots, y_n$ is

$$p(\mathbf{y}) = \lambda^{-\frac{1}{2}n} \prod_{i=1}^{n} \Psi\left(\frac{y_i - \mu}{\lambda^{\frac{1}{2}}}\right). \tag{1.1.8}$$

In particular, you should prove, from first principles, that the observations are independent.

1.1.c In Exercise 1.1.b, take Ψ to denote a standard normal density

$$\Psi(z) = (2\pi)^{-\frac{1}{2}} \exp\left\{-\frac{1}{2}z^2\right\} \qquad (-\infty < z < \infty),$$

so that $f(y)$ denotes a normal $N(\mu, \lambda)$ density, with mean μ and variance λ. Show that the likelihood of μ and λ is

$$l(\mu, \lambda \mid \mathbf{y}) = (2\pi)^{-\frac{n}{2}}\lambda^{-\frac{n}{2}} \exp\left\{-\frac{1}{2}\lambda^{-1}s^2 - \frac{n}{2}\lambda^{-1}(\mu - \bar{y})^2\right\} \tag{1.1.9}$$

for $\infty < \mu < \infty$ and $0 < \lambda < \infty$, where $\bar{y}$ denotes the sample mean, and $s^2 = \sum_{i=1}^{n}(y_i - \bar{y})^2$. Describe the likelihood of μ, when λ is fixed, together with the likelihood of λ, when μ is fixed. Discuss why (1.1.9) provides a reasonable approximation to the likelihood function of μ and λ, when the random sampling is instead without replacement and the population size N is much larger than the sample size n.

Find an unbiased estimate of λ, which is a function of s^2 and n alone, together with the standard error of your estimator. Show that a sample size $n = 801$ gives a standard error equal to $0.05\,\lambda$. (Note that $\lambda^{-1}s^2$ is the realization of a chi-squared variate with $n-1$ degrees of freedom (d.f.).)

Show that the likelihood of $\alpha = \log\lambda$ replaces only λ in (1.1.9) by e^{α} and provide a succinct algebraic form for this likelihood.

Note: Let U possess a chi-squared distribution with ν degrees of freedom. Then U possesses density

$$f(u) = \frac{1}{\Gamma\left(\frac{\nu}{2}\right)2^{\frac{\nu}{2}}} u^{\frac{\nu}{2}-1} \exp\left\{-\frac{u}{2}\right\} \qquad (0 < u < \infty). \tag{1.1.10}$$

A random variable X possesses Gamma $G(\alpha, \beta)$ distribution, with parameters α and β, if X possesses density

$$f(x) = \frac{\beta^\alpha}{\Gamma(\alpha)} x^{\alpha-1} \exp\{-\beta x\} \qquad (0 < x < \infty), \tag{1.1.11}$$

in which case, X has mean α/β and variance α/β^2. Hence, our chi-squared variate U has a Gamma $G(\frac{1}{2}\nu, \frac{1}{2})$ distribution, mean ν, and variance 2ν. In shorthand notation, $U \sim \chi^2_\nu \sim G(\frac{1}{2}\nu, \frac{1}{2})$. Using this notation, also note that if $X \sim G(\alpha, \beta)$, then $2\beta X \sim \chi^2_{2\alpha}$. The Gamma function $\Gamma(\alpha)$ is defined, for any $\alpha > 0$, by

$$\Gamma(\alpha) = \int_0^\infty u^{\alpha-1} e^{-u} du,$$

so that $\Gamma(\alpha + 1) = \alpha\Gamma(\alpha)$. Furthermore, $\Gamma(n + 1) = n!$ for any nonnegative integer n, and $\Gamma(\frac{1}{2}) = \sqrt{\pi}$.

1.1.d Consider the linear statistical model, where an $n \times 1$ random vector $\mathbf{Y}$ possesses a multivariate normal distribution with mean vector $\mathbf{X}\boldsymbol{\beta}$ and covariance matrix $\sigma^2\mathbf{I}_n$, where $\mathbf{X}$ is a specified $n \times p$ design matrix, $\mathbf{X}^T\mathbf{X}$ is nonsingular, $\mathbf{I}_n$ is the $n \times n$ identity matrix, and $\boldsymbol{\beta}$ and σ^2 are unknown parameters. Consider the choice of design matrix $\mathbf{X}$, which maximizes Akaike's information criterion (1.1.5) for an appropriate class $\mathcal{X}$ of design matrices. Show that $\mathbf{X}$ minimizes $n \log S_R^2 + 2(p+1)$, where S_R^2 denotes the usual residual sum of squares. Show how this procedure can be used to fit a polynomial regression function or to decide whether or not to include an extra variable, in multiple regression. Show that the Bayesian information criterion (1.1.6) instead corresponds to minimization of $nS_R^2 + \{\log(n/2\pi)\}(p + 1)$.

Hint: A random vector $\mathbf{Y}$ is said to possess a multivariate normal $N(\boldsymbol{\mu}, \mathbf{C})$ distribution, with mean vector $\boldsymbol{\mu}$ and covariance matrix $\mathbf{C}$ if $\mathbf{Y}$ possesses density

$$p(y) = (2\pi)^{-\frac{1}{2}p}|\mathbf{C}|^{-\frac{1}{2}} \exp\left\{ -\tfrac{1}{2}(\mathbf{y} - \boldsymbol{\mu})^T\mathbf{C}^{-1}(\mathbf{y} - \boldsymbol{\mu})\right\} \qquad (\mathbf{y} \in R^p),$$

where $\boldsymbol{\mu} \in R^p$, and $\mathbf{C}$ is positive definite. More properties of this distribution will be developed in Exercise 1.1.n.

1.1.e In Exercise 1.1.a, assume that the number of inhabitants of Lesser Brodhead, who possess the Bombay gene is, instead, unknown. You take a random sample, with replacement, of size $n = 100$, from the population of Lesser Brodhead. Consider the binary responses $y_1, \ldots, y_n$, where, for $i = 1, \ldots, n$, $y_i = 1$ if the ith person in the sample possesses the Bombay gene, and $y_i = 0$ otherwise. By reference to the probability mass function

$$p(y_i) = \theta^{y_i}(1 - \theta)^{1-y_i} \quad \text{for } y_i = 0, 1,$$

prove that the likelihood of θ, the unknown population proportion of inhabitants with the Bombay gene, given $y_1, y_2, \ldots, y_n$, is

$$l(\theta \mid \mathbf{y}) = \theta^t(1 - \theta)^{n-t} \qquad (0 < \theta < 1), \tag{1.1.12}$$

where $t = \sum_{i=1}^{n} y_i$. Show that the maximum likelihood estimate $\hat{\theta} = t/n$ is the numerical realization of an unbiased estimator of θ, with estimated standard error $s_e = \sqrt{\hat{\theta}(1-\hat{\theta})/n}$. Sketch the likelihood curve and calculate the estimated standard error when $t = 2$ and $n = 100$. Do you think that this is a large enough sample size for a reasonable evaluation of θ? How large a sample size would you prefer? Justify your answers by likelihood and standard error calculations. Is $n = 10{,}000$ large enough? Is this likely to give a reasonable value for the estimated coefficient of variation $s_e/\hat{\theta}$?

A couple has $t = 5$ sons, out of $n = 5$ children. What can they infer about the probability that their sixth child will be a girl? Can likelihood methods handle this problem?

1.1.f In Exercise 1.1.e, show that the likelihood of θ, given only t, is

$$l(\theta \mid t) = K(t)\theta^t(1-\theta)^{n-t} \qquad (0 \le \theta \le 1), \tag{1.1.13}$$

where the constant $K(t)$ does not depend upon θ. Use the hypergeometric distribution (e.g., Lee, 1997, p. 285) to find the exact likelihood of θ, given t, when the random sampling is instead without replacement. Calculate the estimated standard error of $\hat{\theta} = t/n$, in this case, and when $t = 18$, $n = 100$, and the population size is $N = 2{,}555$. Suppose instead that no random sampling occurs at all, but that the members of the sample are chosen haphazardly from the local softball crowd. Is it possible to use $\hat{\theta}$ to draw reasonable inferences about θ, in this case?

Albert (1983) discusses the problem of nonresponse. How might nonresponse affect the reasonability of your inferences?

1.1.g In the State of Fredonia, U.S.A. (population size five million), the population gene frequencies are evaluated by reference to a haphazard sample of 5,500 white males who have attended blood-testing clinics in Fredericksville, mainly for purposes of parentage testing, together with a historical record of about 2,000 patients who have attended the campus hospital at the University of Fredonia at Pardeeville. By reference to your answers to Exercises 1.1.e and 1.1.f, discuss whether or not these data are of high enough quality to accurately investigate population proportions.

1.1.h The average mileages per gallon $y_1, \ldots, y_n$ are observed from a random sample of n cars in New York City, and the observations all lie in the finite range $(a, b] = (10, 40]$. It is required to estimate the common density $f(y)$, concentrated on $(a, b]$, of the observations, based upon a relative frequency histogram, with m equal intervals. Consider, therefore, the estimator

$$f^*(y) = m^{-1}(b-a)\sum_{j=1}^{m} \theta_j I\left[y \in I_j\right],$$

where $\boldsymbol{\theta} = (\theta_1, \ldots, \theta_m)^T$ is constrained to the unit simplex

$$S_U = \left\{\boldsymbol{\theta} : \theta_j \ge 0, \text{ for } j = 1, \ldots, m \text{ and } \theta_1 + \theta_2 + \cdots + \theta_m = 1\right\},$$

$I[A]$ is the indicator for the set A, and $I_j = (a + m^{-1}(j-1)(b-a), a + m^{-1}j(b-a)]$. Show that if m is fixed and f assumes precisely the form in (1.1.12), then the likelihood of $\theta_1, \ldots, \theta_m$, given the data, is

$$l_y(\boldsymbol{\theta}) = m^{-n}(b-a)^n \prod_{j=1}^{m} \theta_j^{n_j} \qquad (\boldsymbol{\theta} \in S_U), \tag{1.1.14}$$

where $n_1, \ldots, n_m$ are the numbers of observations in the respective intervals $I_1, I_2, \ldots,$

I_m. Hence show that AIC in (1.1.5) leads to the value of m maximizing

$$\mathrm{AIC}^* = \sum_{j=1}^{m} n_j \log n_j - n \log m - m, \tag{1.1.15}$$

and show how (1.1.5) thereby suggests a procedure for sensibly estimating the model f. (This method is reported by Atilgan, 1983, 1990.)

1.1.i *Rasch's Model* (e.g., Lord and Novick, 1968, Ch. 17): Let n candidates attempt a test with s items. For $i = 1, \ldots, n$ and $j = 1, \ldots, s$, let d_{ij} equal unity if the ith candidate correctly answers the jth item, and zero otherwise. Let θ_{ij} denote the probability that the ith candidate correctly answers the jth item. Assume that the responses d_{ij} are all independent, given the θ_{ij}. Assume, furthermore, that

$$\theta_{ij} = \frac{e^{\alpha_i - \beta_j}}{1 + e^{\alpha_i - \beta_j}} \qquad (i = 1, \ldots, n;\ j = 1, \ldots, s), \tag{1.1.16}$$

where α_i is the "ability" for the ith candidate, and β_j is the "difficulty" of the jth test.

Show that the likelihood of the α_i and β_j, given the d_{ij}, is

$$l(\boldsymbol{\alpha}, \boldsymbol{\beta} \mid \mathbf{d}) = \exp\left\{\sum_{i=1}^{n} t_i^A \alpha_i - \sum_{j=1}^{s} t_j^B \beta_j\right\} \prod_{i=1}^{n} \prod_{j=1}^{s} \left(1 + e^{\alpha_i - \beta_j}\right)^{-1}, \tag{1.1.17}$$

where t_i^A is the number of items correctly answered by the ith candidate, and t_j^B is the number of candidates correctly answering the jth item. Show that the maximum likelihood estimates are not uniquely defined and discuss how this problem might be resolved.

Discuss which elements of the preceding model are subjective and which, if any, are objective. Note that this model is generalized by Birnbaum (1969) to include parameters a_j measuring the "discriminatory powers" of the tests. See also Leonard and Novick (1986) and Kim et al. (1994). The addition of "guessing parameters" γ_j then leads to the "three-parameter" model

$$\theta_{ij} = \gamma_j + (1 - \gamma_j)\frac{e^{a_j(\alpha_i - \beta_j)}}{1 + e^{a_j(\alpha_i - \beta_j)}}. \tag{1.1.18}$$

However, the guessing parameters depend upon the test rather than the candidate. Note that if $\gamma_j \equiv 0$, then the a_j and b_j in the "two-parameter model" can become more fully identified by assuming that the α_i constitute a random sample from a standard normal distribution. See also Bock and Aitkin (1981). Under a probit, rather than logit, transformation, it is possible to explicitly express the likelihood of the item parameters in terms of the cumulative distribution functions (c.d.f.) of the multivariate normal distributions. Can you do this?

1.1.j *Subjective Classical Probability:* The procecution in a parentage case assumed that before taking a blood test, the alleged father possessed probability 50% of being the actual father. After considering the results of a blood test, this probability was revised to 99.99%. On the witness stand, the mother stated that there were at least ten possible fathers, rather than two, as originally stated. The expert witness for the defense was recalled and stated that if there are ten possible fathers, then in the assumed absence of any further evidence, the only logical probabilities to assign to the ten candidates, before viewing the results of the blood test, are 1/10 to each candidate. Do you think that this is a correct application of the definition of classical probability in Exercise 1.1.a? Do you think that the expert's probabilities of 1/10 are subjective or objective (i.e., consistent with the frequency definition of probability in (1.1.1))?

(These ideas also relate to Le Marquis Pierre Simon Laplace's principle of insufficient reason. See Laplace, 1812, and Press, 1989, p. 25.)

1.1.k *Subjective Probability – The Spinning Pointer:* You wish to assign personal probabilities $p(A)$ to all events contained in a sample space S. To calibrate your probabilities, you can consider an objective auxiliary experiment, such as a spinning pointer or random number generator, which generates a real number X that is uniformly distributed on the interval $[0, 1)$. You then define your $p(A)$ to be the smallest real number a^* such that you would judge the event A to be equally likely to the event that $X \leq a^*$. By reference to Chapter 6 of DeGroot (1970), describe and fully interpret a set of conditions such that this process will lead to a uniquely specified probability distribution.

DeGroot's Axiom 5 effectively states that events in S should be compared with events of the form $\{X \leq a^*\}$ in a manner constrained by his first four axioms. Note that his "random variable" X is a "measurable function" from S to the unit interval (see Ash, 1972, Ch. 1). His Section 6.5 confirms that the full implications of Axiom 5 are needed to verify the properties of a probability distribution. (This logical definition of subjective probability is quite free from notions of betting or utility. See O'Hagan, 1988, Ch. 1, for an alternative definition. The practical relevance of the axiom system underlying the spinning pointer is debated by Leonard, 1980, and discussants.)

1.1.l For a sample space S (or a parameter space Θ), a probability distribution on S must, by definition, satisfy the Kolmogorov axioms (laws of probability), namely,

I. $p(A) \geq 0$, for all $A \subseteq S$.
II. $p(S) = 1$.
III. (The Countable Additivity Property.) For any sequence $A_1, A_2, \ldots$ of disjoint events contained in S,

$$p(A_1 \cup A_1 \cup \cdots) = p(A_1) + p(A_2) + \cdots .$$

Do you think that DeGroot's axiom system (DeGroot, 1970) justifies the philosophy (e.g., Lindley, 1985) that subject to a very simple set of coherency axioms, you must always either represent your information about a parameter $\theta \in \Theta$ by a probability distribution on Θ or act as if you have done this? How would you compare this with the philosophy that quite apart from any detailed underlying foundation, it is simply common sense to suggest that the statistician should always try to represent the current information about θ by a probability distribution on Θ satisfying the usual laws of probability including Kolmogorov's countably additivity property?

1.1.m Show that any probability associated with an experiment $\mathcal{E}$ that cannot be replicated under identical conditions a large number of times and with which you cannot associate a sample space S with equally likely outcomes must be subjective, that is, most probabilities are either subjective or computer simulated.

1.1.n A $p \times 1$ random vector $\mathbf{Y} = (Y_1, \ldots, Y_p)^T$ possesses a multivariate normal distribution, with mean vector $\boldsymbol{\mu} = (\mu_1, \ldots, \mu_p)^T$ and covariance matrix $\mathbf{C}$, that is, $\mathbf{Y}$ is $N(\boldsymbol{\mu}, \mathbf{C})$ distributed, if $\mathbf{Y}$ possesses joint density indicated in Exercise 1.1.d. There are many applications of this distribution in multivariate analysis (e.g., Anderson, 1974).

(a) Show that the moment-generating function of $\mathbf{Y}$ is

$$E\left(e^{\mathbf{t}^T\mathbf{Y}}\right) = \exp\left\{\mathbf{t}^T\boldsymbol{\mu} + \tfrac{1}{2}\mathbf{t}^T\mathbf{C}\mathbf{t}\right\}.$$

(b) Hence show that the marginal distribution of any subvector $\mathbf{Y}_1$ of $\mathbf{Y}$ is $N(\boldsymbol{\mu}_1, \mathbf{C}_1)$, where $\boldsymbol{\mu}_1$ and $\mathbf{C}_1$ are appropriate submatrices of $\boldsymbol{\mu}$ and $\mathbf{C}$.

(c) Let $\mathbf{U} = \mathbf{AY}$, where $\mathbf{A}$ is an $l \times p$ matrix of constants, with $l \leq p$. Show that $\mathbf{U}$ is $N(\mathbf{A}\boldsymbol{\mu}, \mathbf{ACA}^T)$ distributed.

(d) Let $\mathbf{Y}_1$ and $\mathbf{Y}_2$ be $p_1 \times 1$ and $p_2 \times 1$ subvectors, which partition $\mathbf{Y}$. Use $p(\mathbf{y}_2 \mid \mathbf{y}_1) = p(\mathbf{y})/p(\mathbf{y}_1)$ to show that the conditional distribution of $\mathbf{Y}_2$, given that $\mathbf{Y}_1 = \mathbf{y}_1$, is multivariate normal, with mean vector taking the form

$$\boldsymbol{\mu}_2 + \mathbf{C}_{21}\mathbf{C}_{11}^{-1}(\mathbf{y}_1 - \boldsymbol{\mu}_1),$$

and covariance matrix $\mathbf{C}_{22} - \mathbf{C}_{21}\mathbf{C}_{11}^{-1}\mathbf{C}_{12}$, and describe $\mathbf{C}_{11}$, $\mathbf{C}_{21}$, and $\mathbf{C}_{22}$.

(v) Show that the multivariate normal distribution can be used to model linear, rather than nonlinear, relationships between several variables.

1.2 Practical Examples

(A) *The Likelihood Function for a Population Proportion*

Consider first the likelihood function (1.1.12) and Exercise 1.1.e, but when $t = 8$ and $n = 400$. This is described by curve (a) in Figure 1.2.1. For the purpose of comparison of curves with different sample sizes, all five curves in Figure 1.2.1 have been divided by appropriate constants to ensure that the area under each curve is unity. Curve (a) possesses a maximum when $\theta = \hat{\theta} = 8/400 = 0.02$. The (normalized) likelihoods for different values of θ may be obtained from the corresponding heights of the likelihood curve. Note that when $n = 400$, the curve suggests that the data are not overinformative regarding θ. For example, values of $\theta = 0.01$, or 0.03, are still quite likely, given the data. (Remember that when considering any particular likelihood curve, you may compare heights, for example, to obtain a preference ordering for different values of the parameter, but you are not formally justified in comparing areas. Also the shape of the likelihood curve may change substantially under nonlinear transformations of the parameters.) Estimates as small as $\hat{\theta} = 0.02$ might be obtained when investigating the proportion θ of people in a population with a rare genotype, or in DNA testing when investigating the proportion of people in the population whose allele lengths (on a particular DNA probe) lie between two critical values. Notice, from curves (b), (c), (d), and (e) in Figure 1.2.1, that as n increases, with $\hat{\theta} = 0.02$ fixed, the likelihood curves suggest that the data become much more informative. For example, curve (e) describes the likelihood when $n = 100{,}000$. In practice, $n = 100{,}000$ would seem preferable to, say, $n = 7{,}500$ (curve c) if an accurate evaluation of θ is required.

(B) *Gambling at Roulette*

Consider a fair American roulette wheel, with $k = 38$ possible outcomes, including two zeros. Then all $k = 38$ outcomes are equally likely. Let A denote the event that

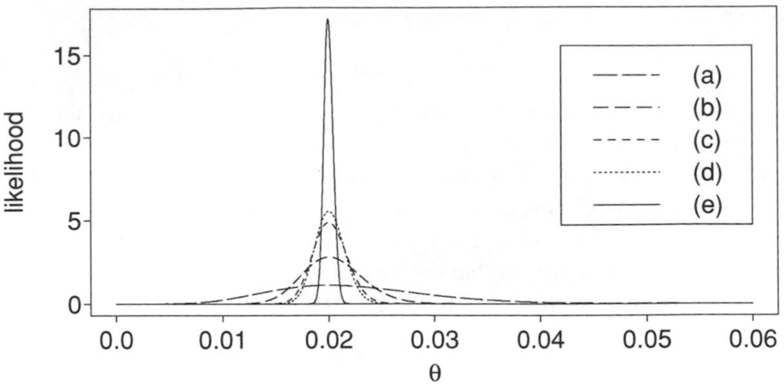

Figure 1.2.1. The likelihood function for a population proportion: (a) $n = 400$; (b) $n = 2{,}500$; (c) $n = 7{,}500$; (d) $n = 10{,}000$; (e) $n = 100{,}000$.

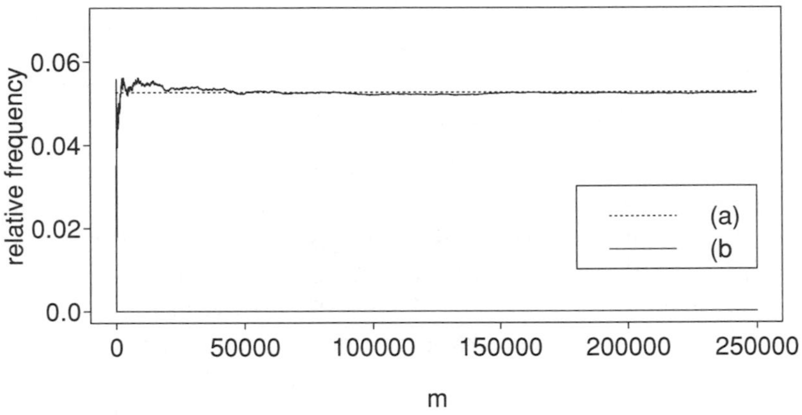

Figure 1.2.2. Frequency simulations for the probability of a zero at American roulette: (a) relative frequency; (b) the limiting value, or frequency probability of 1/19.

a zero occurs on any particular spin of the roulette wheel. From Exercise 1.1.a, the classical probability of A is $p(A) = 2/38 = 1/19 = 0.05263$. This, however, is also the frequency probability of A, as defined by equation (1.1.1).

To illustrate this, we simulated m replications, that is, m independent spins of the roulette wheel, and calculated the relative frequency $r_m(A)$, for $m = 1, 2, \ldots, 250{,}000$. The values of the relative frequencies are described by curve (b) in Figure 1.2.2. Note that $r_m(A)$ ultimately converges to (a), the limiting relative frequency $p(A) = 1/19$. It does, however, take many replications (at least 50,000) to accurately stabilize towards this value.

For gamblers who do not bet on the zeros, the casino always takes all money bet whenever a zero occurs. Indeed, in the long run, the casino will always profit by exactly 1/19 of all money bet if the roulette wheel is fair. An individual with finite resources cannot avoid a corresponding long-run loss, even with the most elaborate strategy.

Some strategies (for example, relating to "double or quit") will win in the long run, but only if the punter possesses infinite resources. If the roulette wheel is sufficiently unfair, it can of course be possible to win in the long run if strategies are devised to detect the biases. Therefore, the best strategy for casinos is to make their wheels as fair as possible.

(C) *Analyzing the Breaking Strengths of 50 Samples of Linen Thread*

Hald (1952, p. 133) reports the breaking strengths, in kilograms, of $n = 50$ samples of linear thread. These yielded $\bar{y} = 2.299$ and $s^2 = \sum(y_i - \bar{y})^2 = 8.274$. Hald (p. 132) reports the sample cumulative distribution function, and this suggests that it is reasonable to take the observations to be independent and normally distributed. The likelihood surface described by equation (1.1.9) is therefore appropriate for the mean μ and variance λ of the observations. In Figure 1.2.3, we report likelihood contours for the corresponding two-dimensional likelihood surface of μ and $\alpha = \log\lambda$ (the contours are calculated for nine equally spaced heights of the likelihood surface).

The likelihood of μ and α is maximized when $\mu = \hat{\mu} = \bar{y} = 2.299$ and when $\alpha = \hat{\alpha} = \log(s^2/n) = -1.799$. For any fixed α, the likelihood is roughly symmetric in μ. However, for any fixed μ, the likelihood is slightly skewed towards negative values of λ. In Figure 1.2.4, we report the corresponding contours of the bivariate normal likelihood approximation

$$l^*(\mu, \alpha \mid \mathbf{y}) = l(\hat{\mu}, \hat{\alpha} \mid \mathbf{y}) \exp\left\{ -\tfrac{1}{2}e^{-\hat{\alpha}}(\mu - \hat{\mu})^2 - \tfrac{1}{4}n(\alpha - \hat{\alpha})^2 \right\} \tag{1.2.1}$$

for $-\infty < \mu < \infty$ and $-\infty < \alpha < \infty$.

The approximation (1.2.1) is intended to motivate the general recommendation (1.3.11), which will be justified in Section 1.3. Note that the approximation (1.2.1) is remarkably accurate, even though the sample size is only 50. It will follow, in

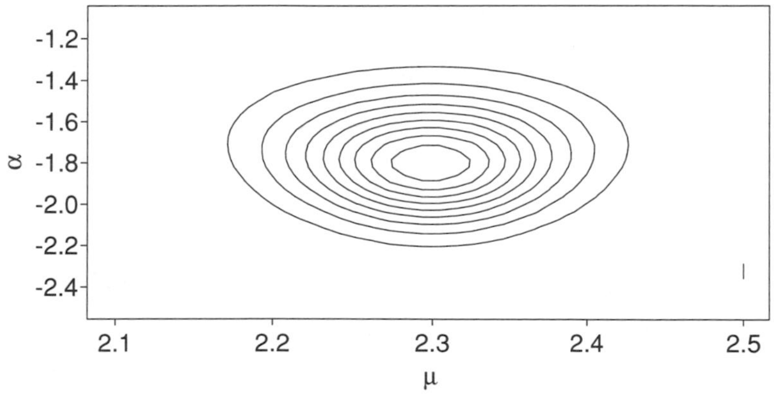

Figure 1.2.3. Likelihood contours for a normal mean and log-variance.

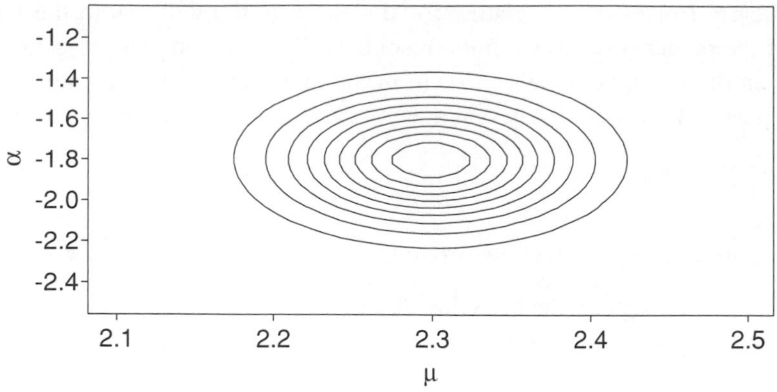

Figure 1.2.4. Approximate bivariate normal contour for a normal mean and log-variance.

Section 1.3, that an approximate 95% confidence intervalfor α is

$$\left(\hat{\alpha} - 1.96\sqrt{\frac{2}{n}}, \hat{\alpha} + 1.96\sqrt{\frac{2}{n}}\right)$$

$$= -1.799 \pm 0.392$$

$$= (-2.191, -1.407), \tag{1.2.2}$$

so that an approximate 95% confidence interval for $\lambda = e^{\alpha}$ is $(e^{-2.191}, e^{-1.407}) = (0.112, 0.245)$. This may be compared with the exact equal-tailed 95% confidence interval (0.118, 0.262) for λ, based upon the percentage points of the chi-squared distribution with 49 degrees of freedom. The approximation may be improved by replacing $\hat{\alpha}$ and n in (1.2.2) by $\log(s^2/(n-1))$ and $n-1$, respectively, yielding the interval (0.114, 0.251). See also Bartlett and Kendall (1946).

(D) *Random Sampling from a t-Density*

In situations where the sampling density is thought to possess thicker tails than the normal, but is to remain symmetric, it is convenient to take the observations $y_1, \ldots, y_n$ to constitute a random sample from a (generalized) t-density, with ν degrees of freedom, location μ, and precision λ^{-1}, that is, assuming the same form as the sampling density in Exercise 1.1.b, but with Ψ replaced by the (standard) t-density with ν degrees of freedom,

$$p_\nu(t) = C(\nu)\left(1 + \frac{t^2}{\nu}\right)^{-\frac{1}{2}(\nu+1)} \qquad (-\infty < t < \infty), \tag{1.2.3}$$

where

$$C(\nu) = \frac{\Gamma\left(\frac{\nu+1}{2}\right)}{\Gamma\left(\frac{\nu}{2}\right)\sqrt{\nu\pi}}, \tag{1.2.4}$$

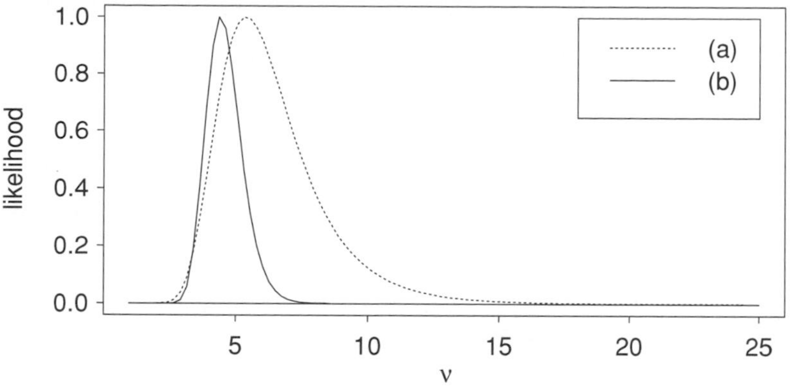

Figure 1.2.5. Likelihood function for the degrees of freedom: (a) $n = 200$; (b) $n = 500$.

and $\Gamma(\alpha)$ denotes the Gamma function (see Note to Exercise 1.1.c). Then the likelihood function of the three unknown parameters ν, μ, and λ is

$$l(\nu, \mu, \lambda \mid \mathbf{y}) = \lambda^{-\frac{1}{2}n} \prod_{i=1}^{n} p_\nu\left(\frac{y_i - \mu}{\lambda^{\frac{1}{2}}}\right) \qquad (0 < \nu, -\infty < \mu < \infty, 0 < \lambda). \tag{1.2.5}$$

Samples of $n = 200$ and $n = 500$ observations were simulated from a generalized t-distribution, with known $\mu = 0$, $\lambda = 1$, and $\nu = 4$. Curve (a) in Figure 1.2.5 describes the likelihood of ν when μ and λ are fixed to equal their true values. This curve suggests that $\nu = \infty$ (normal random sampling) is untrue. Curve (b) in Figure 1.2.5 corresponds to $n = 500$, $\mu = 0$, and $\lambda = 1$. Its increased precision suggests that the data are somewhat more informative than when $n = 200$. The true value ($\nu = 4$) is not refuted.

Curves (a) and (b) in Figure 1.2.6 describe the likelihood of μ when $n = 200$ or 500, $\nu = 4$, and $\lambda = 1$. Note that both curves are skew to the right, a property that cannot occur for the location parameter under a normal random sampling assumption. Curves (a) and (b) in Figure 1.2.7 describe the likelihood of λ when $n = 200$ or 400, $\nu = 4$, and $\mu = 0$. The true value ($\lambda = 1$) is again not refuted by this data set.

(E) *Polynomial Regression; The Paddy-Field Example*

As an application of Exercise 1.1.d, consider the data in Figure 1.2.8, which were reported in raw form by Devore (1991). The yield y of paddy, a grain farmed in India, is regressed against x, the rate of harvesting. For reasons to be discussed in Section 1.9, least squares regression should not be unreservedly applied to data of this type. However, for illustrative purposes, we follow Devore (1991, p. 518) by applying the least squares model of Exercise 1.1.d, for polynomial regression $y = \beta_0 + \beta_1 x + \beta_2 x^2 + \cdots + \beta_q x^q$. The number of unknown parameters in the model is then $m = p+1 = q+2$. The values

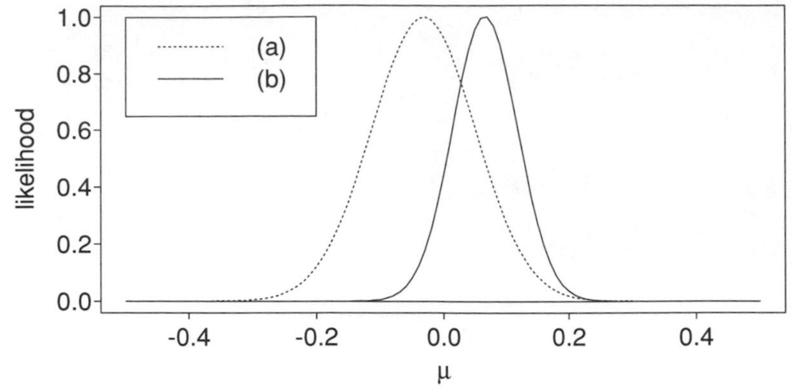

Figure 1.2.6. Likelihood function for the location parameter: (a) $n = 200$; (b) $n = 500$.

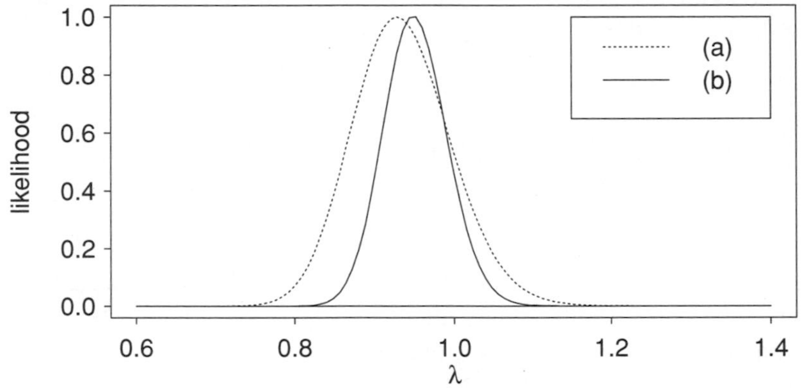

Figure 1.2.7. Likelihood function for the scale parameter: (a) $n = 200$; (b) $n = 500$.

of the log-likelihood L_m, and AIC, for $m = 2, 3, \ldots, 9$ were

m	2	3	4	5	6	7	8	9
L_m	−118.77	−118.12	−106.12	−106.02	−106.01	−105.70	−105.173	−103.716
AIC	−120.77	−121.12	−110.15	−111.02	−112.01	−112.70	−113.173	−112.716

The log-likelihoods L_m are plotted in Figure 1.2.9. The dotted lines in this figure denote the upper convex boundary of the log-likelihoods. The choices $m = 4$ and $q = 2$ are intuitively appealing and lie on the upper convex boundary of the L_m. Furthermore, these are the choices that minimize AIC. In Figure 1.2.8, the quadratic regression curve is fitted to the bivariate scatterplot. We do not report BIC, since the sample size n is small. In general, BIC is not recommended if $\log(n/2\pi) < 2$, that is, $n \leq 46$.

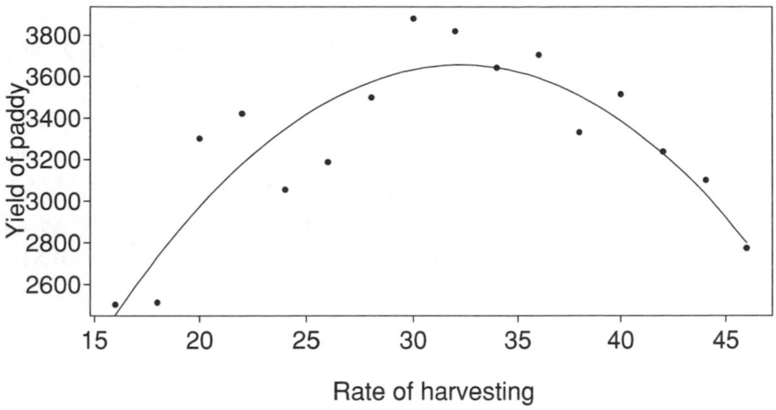

Figure 1.2.8. Fitting a quadratic regression curve.

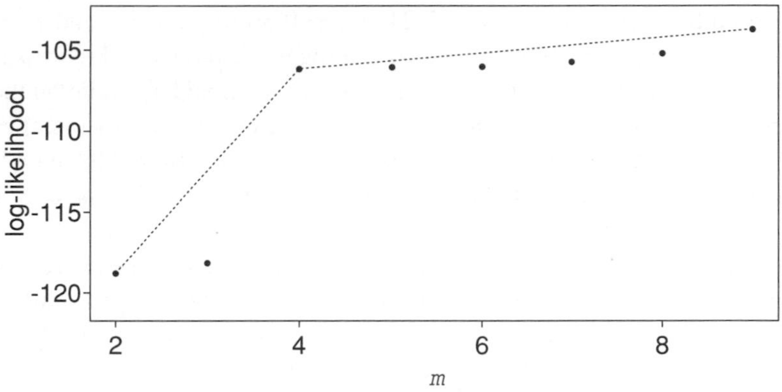

Figure 1.2.9. Log-likelihood plot for polynomial regression.

(F) *Frequency Simulations for Polynomial Regression*

One thousand simulated samples were generated from a quadratic regression model, taking the form $y = \beta_0 + \beta_1 x + \beta_2 x^2$, with $\beta_0 = \beta_1 = \beta_2 = 1$. The choice $\sigma^2 = 1$ was made for the sampling variance. On each simulation, 10 replications were taken at each of the design points $x = 1, 2, \ldots, 10$, giving $n = 100$. Each data set was fitted by a polynomial of degree p, for $p = 1, 2, \ldots, 7$ (i.e., $m = 3, 4, \ldots, 9$ parameters). Out of the 1,000 simulations, the values of m minimizing AIC and BIC varied between 3 and 9 according to the following relative frequencies:

m	3	4	5	6	7	8	9
AIC	0	0.705	0.120	0.078	0.044	0.030	0.023
BIC	0	0.841	0.080	0.046	0.019	0.009	0.005

Note that BIC is correct ($m = 4$) a higher percentage of the time, when compared with AIC. On each simulation, the fitted regression curve was also used to predict the value of a future observation, when $x = 11$. The average mean squared error of prediction, when using AIC, was 13.395, and this can be compared with the value of 3.639, when using BIC. More generally, the criterion with larger penalty (BIC when n is large enough) can perform better, out of the two criteria, unless the true model assumes a very complicated form. This conclusion, regarding the possible superiority of BIC, is substantiated by more detailed comparisons by Koehler and Murphree (1988) in the context of selecting model order for time series, and by Katz (1981) for estimating the order of a Markov chain, though Katz employed a modification to BIC, more appropriate for his particular models.

(G) *Frequency Simulations for Histogram Smoothing*

As an application of Exercise 1.1.h, we generated $n = 100$ observations on the unit interval, from a "true" histogram with $m = 5$ equal intervals and cell probabilities $\theta_1 = 0.1, \theta_2 = 0.2, \theta_3 = 0.3, \theta_4 = 0.2, \theta_5 = 0.2$. The raw observations provided the observed proportions $p_1 = 0.07$, $p_2 = 0.21$, $p_3 = 0.36$, $p_4 = 0.19$, and $p_5 = 0.17$. The log-likelihoods $L_m = n \sum_{k=1}^{m} p_j \log p_j + n \log m$ are plotted in Figure 1.2.10 for the choices $m = 1, 2, \ldots, 30$. A reasonable choice of m should again lie on the upper convex boundary (dotted lines) and at the beginning of the "ridge," that is, the part of the upper convex boundary that most obviously "flattens out." Both AIC in (1.1.5) and BIC from (1.1.6) were optimized, on this particular simulation, when $m = 5$. Further graphical techniques were proposed by Atilgan (1983).

Next, a total of $M = 10{,}000$ samples were generated, each with $n = 100$ observations, from the preceding true distribution ($m = 5$). In the second and third columns (Case 1) of Table 1.2.1, we describe the proportions of times, out of M, that AIC and BIC were optimized, by values of m in the range $m = 1, 2, \ldots, 30$. Note that while BIC is correct 33.50% of the time compared with 35.99% for AIC, BIC undershoots the true value, of $m = 5$, a much larger percentage of the time. However, using these information criteria to evaluate m, and standard histogram formulas to estimate the expectation ($\mu = 0.54$) and variance ($\sigma^2 = 0.624$) of the true distribution,

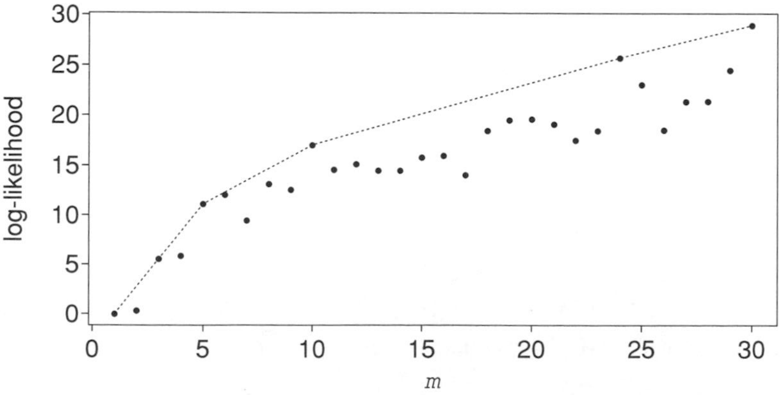

Figure 1.2.10. Log-likelihood plot for histogram smoothing.

Table 1.2.1. *Performance of AIC and BIC ($M = 10{,}000$ simulations).*

	Case 1		Case 2	
m	AIC	BIC	AIC	BIC
1	0.0658	0.1831	0.0000	0.0002
2	0.0261	0.0458	0.0058	0.0277
3	0.1613	0.2247	0.0298	0.1026
4	0.1118	0.1262	0.0519	0.1331
5	0.3599	0.3350	0.0736	0.1510
6	0.0384	0.0237	0.0696	0.1160
7	0.0499	0.0257	0.0758	0.1104
8	0.0254	0.0090	0.0659	0.0781
9	0.0221	0.0076	0.0710	0.0676
10	0.0286	0.0091	0.0594	0.0490
11	0.0169	0.0038	0.0587	0.0388
12	0.0106	0.0014	0.0447	0.0261
13	0.0149	0.0018	0.0502	0.0273
14	0.0092	0.0007	0.0394	0.0162
15	0.0110	0.0010	0.0325	0.0122
16	0.0068	0.0002	0.0324	0.0086
17	0.0065	0.0005	0.0299	0.0093
18	0.0037	0.0000	0.0249	0.0054
19	0.0057	0.0004	0.0272	0.0057
20	0.0039	0.0000	0.0200	0.0029
21	0.0032	0.0001	0.0209	0.0038
22	0.0032	0.0001	0.0183	0.0023
23	0.0027	0.0000	0.0189	0.0019
24	0.0017	0.0000	0.0120	0.0009
25	0.0031	0.0000	0.0143	0.0011
26	0.0012	0.0000	0.0145	0.0008
27	0.0024	0.0001	0.0110	0.0004
28	0.0016	0.0000	0.0095	0.0001
29	0.0013	0.0000	0.0099	0.0003
30	0.0011	0.0000	0.0080	0.0002

BIC performed better than AIC when estimating μ (MSE $= 0.0898$ compared with 0.1247), and roughly the same when estimating σ^2 (MSE $= 0.0053$ compared with 0.0046).

A further $M = 10{,}000$ simulations, each giving $n = 100$ observations, were generated from a beta distribution with mean $\mu = 1/3$ and variance $\sigma^2 = 1/12$. In the fourth and fifth columns (Case 2) of Table 1.2.1, we describe, for this situation, the proportions of times that AIC and BIC were optimized, by values of m in the range $m = 1, 2, \ldots, 30$. In this case, BIC was better than AIC for estimating both μ (MSE $=$ 0.0694 compared with 0.0835) and σ^2 (MSE $= 0.0016$ compared with 0.0028).

(H) *The Death Penalty Data, Simpson's Paradox, and Lurking Variables*

The data in the following 2×2 contingency table were reported by Radelet (1981) and relate to 326 defendants, each found guilty of murder.

	Death penalty	Others	TOTAL
White defendant	19	141	160
Black defendant	17	149	166
TOTAL	36	290	326

This table tells us that 11.88% out of the 160 white defendants in the sample received the death penalty, whereas 10.24% out of the black defendants received the death penalty. Therefore there is negligible evidence of discrimination in the overall table. Simpson's paradox occurs when we split the table according to a third variable, color of victim, which will be referred to as a "lurking variable." It is also referred to as a "confounding variable" or "confounder." The row and column variables (color of defendant and "death or not") are our first and second variables.

	White victim		
	Death penalty	Others	TOTAL
White defendant	19	132	151
Black defendant	11	52	63
TOTAL	30	184	214

	Black victim		
	Death penalty	Others	TOTAL
White defendant	0	9	9
Black defendant	6	97	103
TOTAL	6	106	112

Out of defendants with white victims, 12.58% of white defendants received the death penalty, whereas 17.46% of black defendants received the death penalty. Moreover, out of defendants with black victims, no white defendant received the death penalty, whereas 5.8% of black defendants received the death penalty. This is Simpson's paradox (Simpson, 1951), which at first sight appears to be illogical. The apparent conclusions based upon the overall table can be reversed in each of the two subtables, that is, the lurking variable "color of victim" reverses the previous conclusion regarding the association between the two main variables. Therefore the apparent association between the first two variables is a "spurious association."

Simpson's paradox can frequently occur for uncontrolled or nonrandomized data. If you can identify possible lurking variables in advance, then you can also collect data relating to these variables and analyze a multiway contingency table. However, there are many possible lurking variables, for example, socioeconomic status of victim, degree of violence of the crime, testimony of relatives, and variables that might not be contemplated before collecting the data. Basically, any conclusion from a nonrandomized data set is subject to refutation due to the possible presence of a lurking variable. For example, it is not possible to provide strong statistical evidence for a meaningful association between smoking and lung cancer without random sampling individuals who are required to smoke and individuals who are required not to smoke.

In our example, it is possible to show, using the theory of conditional probability, that the paradox cannot occur in situations where the proportion of defendants with white victims who receive the death penalty ($30/214 = 0.140$) is instead equal to the proportion of defendants with black victims who received the death penalty ($6/112 = 0.054$). If the judges had instead assigned the death penalties at random, but with a fixed probability (e.g., $36/326$), then this would tend to equate these two proportions and would hence reduce the chances of our lurking variable (color of victim) creating Simpson's paradox. It would also reduce the chances of any other lurking variable creating Simpson's paradox, whether or not this variable had been contemplated in advance. In an ideal situation, the experiment could be replicated in order to detect against an unlucky randomization. While these safeguards are precluded by practical constraints in the context of the current sample, it is possible, for example, to assign patients at random to two different treatments, at least at an early stage when it is not obvious which treatment is superior. It is therefore possible to use randomization to reduce the problems of lurking variables during clinical trials (e.g., Byar et al., 1976).

The paradox can be explained by more detailed analysis, providing the second and third stages of a "three-directional approach" (Leonard and Hsu, 1994). At the second stage, consider the possible association between the first variable and the lurking variable.

	White victim	Black victim	TOTAL
White defendant	151	9	160
Black defendant	63	103	166
TOTAL	214	112	326

This table tells us that 94.38% of white defendants had murdered white victims, whereas only 62.05% of black defendants had murdered black victims. Therefore black defendants seem more likely to murder a person of different color than white defendants. This apparent association between the first variable and the lurking variable is not made spurious by the second variable (death or not).

	Death penalty		
	White victim	Black victim	TOTAL
White defendant	19	0	19
Black defendant	11	6	17
TOTAL	30	6	36

	Others		
	White victim	Black victim	TOTAL
White defendant	132	9	141
Black defendant	52	97	149
TOTAL	184	106	290

For defendants receiving the death penalty, 100% of white defendants murdered white victims, and 35.29% of black defendants murdered black victims. For defendants not receiving the death penalty, 93.6% of white defendants murdered white victims, whereas 65% of black defendants murdered white victims. Since 100% > 93.6% and 35.29% < 65%, this confirms that (as far as these three variables alone are concerned) there is a meaningful association between the first two variables, but that the bias is against black defendants.

As a third stage of our analysis, consider the possible association between the lurking variable (color of victim) and the second variable (death or not).

	Death penalty	Others	TOTAL
White victim	30	184	214
Black victim	6	106	112
TOTAL	36	190	326

Note that 14.02% of defendants with white victims received the death penalty, when compared with only 5.36% of defendants with black victims. This apparent association, suggesting a bias based upon color of the victim, is not made spurious by the first variable.

	White defendant		
	Death penalty	Others	TOTAL
White victim	19	132	151
Black victim	0	9	9
TOTAL	19	141	160

	Black defendant		
	Death penalty	Others	TOTAL
White victim	11	52	63
Black victim	6	97	103
TOTAL	17	149	166

Out of white defendants, 12.59% of these with white victims received the death penalty, whereas none of those with black victims received the death penalty. Out of black defendants, 17.46% of those with white victims received the death penalty, whereas only 95.8% of these with black victims received the death penalty.

Simpson's paradox can be explained via the possible association between the first variable and the lurking variable, and between the second variable and the lurking variable. Defendants with white victims are more likely to receive the death penalty, and white defendants are more likely to have white victims. This creates the initial false impression that white defendants are more likely to receive the death penalty. However, our interpretation is itself conditional, since you should also try to think of all other lurking variables, such as the socioeconomic status of the victim, that might influence the conclusions. This process, of listing possible lurking variables, is an important ingredient of the inductive synthesis.

An understanding of Simpson's paradox should change how you think about conclusions from uncontrolled data, since the paradox would initially appear to defy honest common sense. This paradox can also occur in employment discrimination cases, for example, apparent discrimination by an organization based upon gender may be found to be a spurious conclusion when other variables (e.g., rank or merit) are taken into account. The Berkeley Admissions data (e.g., Fienberg, 1987, p. 54) is frequently cited in discrimination cases. Here, six different graduate schools appear to have each tried to avoid discrimination based upon gender, but to have nevertheless given an overall strong impression of discrimination.

Worked Example 1D: *Likelihood Methods for the Multinomial Distribution*

Suppose that $n = 100$ particles are assigned independently and at random to four locations L_1, L_2, L_3, and L_4. Let θ_k denote the probability that any particular particle is assigned to location L_k, $k = 1, 2, 3, 4$, where $\theta_1 + \theta_2 + \theta_3 + \theta_4 = 1$. Let Y_1, Y_2, Y_3, Y_4 be random variables representing the numbers of particles respectively observed at L_1, L_2, L_3, and L_4.

(a) Show that the probability mass function of Y_1, Y_2, Y_3, and Y_4 is

$$p(Y_1 = y_1, Y_2 = y_2, Y_3 = y_3, Y_4 = y_4) = N^* \theta_1^{y_1} \theta_2^{y_2} \theta_3^{y_3} \theta_4^{y_4}$$
$$(y_i = 0, 1, \ldots n, \text{ for } i = 1, 2, 3, 4;\ y_1 + y_2 + y_3 + y_4 = n),$$

where N^* denotes the number of ways in which $n = 100$ particles can be assigned to the four locations, with $Y_1 = y_1$, $Y_2 = y_2$, $Y_3 = y_3$, $Y_4 = y_4$ particles at the respective locations L_1, L_2, L_3, and L_4. Note, without proof, that the *multinomial coefficient* N^* is equal to

$$\frac{n!}{y_1! y_2! y_3! y_4!}.$$

(b) If $\theta_2 = \theta_3 = \theta_4$, then find the likelihood of θ_1, when $y_1 = 41$, $y_2 = 32$, $y_3 = 16$, and $y_4 = 11$.

(c) Show for general $\theta_1, \theta_2, \theta_3$, and θ_4 that $Y = Y_1+Y_3$ possesses the binomial distribution, with probability mass function

$$p(Y = y \mid \phi) = \frac{n!}{y!(n-y)!}\phi^y(1-\phi)^{n-y} \qquad (y = 0, 1, 2, \ldots, n),$$

with $\phi = \theta_1 + \theta_3$.

(d) Suggest a likelihood procedure for drawing inferences regarding $\phi = \theta_1 + \theta_3$.

Model Answer 1D:

(a) By independence, the probability that y_1 particular particles are assigned to L_1 is $\theta_1^{y_1}$, with similar results for the other three locations. Therefore,

$$\theta_1^{y_1}\theta_2^{y_2}\theta_3^{y_3}\theta_4^{y_4}$$

denotes the probability that y_i particular particles are assigned to L_i for each $i = 1, 2, 3, 4$. The answer then follows, since we just multiply by N^*. This gives the probability mass function of a multinomial distribution with four cells, sample size n, and cell probabilities $\theta_1, \theta_2, \theta_3, \theta_4$.

(b) Since $\theta_1+\theta_2+\theta_3+\theta_4 = \theta_1+3\theta_2 = 1$, we have $\theta_2 = \theta_3 = \theta_4 = \frac{1}{3}(1-\theta_1)$. The likelihood of $\theta_1, \theta_2, \theta_3$, and θ_4 is, in general,

$$l(\boldsymbol{\theta} \mid \mathbf{y}) = l(\theta_1, \theta_2, \theta_3, \theta_4 \mid y_1, y_2, y_3, y_4) = N^*\theta_1^{y_1}\theta_2^{y_2}\theta_3^{y_3}\theta_4^{y_4}.$$

This now reduces to

$$\begin{aligned} l(\theta_1 \mid \mathbf{y}) &\propto \theta_1^{y_1}\theta_2^{y_2}\theta_3^{y_3}\theta_4^{y_4} \propto \theta_1^{y_1}(1-\theta_1)^{y_2+y_3+y_4} \\ &\propto \theta_1^{y_1}(1-\theta_1)^{n-y_1} \propto \theta_1^{41}(1-\theta_1)^{59}. \end{aligned}$$

(c) The probability mass function in the answer is just the probability mass function for a multinomial distribution for Y and $n - Y$, with two cells, sample size n, and cell probabilities ϕ and $1 - \phi$. This can therefore be developed by a parallel argument to part (a). Note that $n!/y!(n-y)!$ denotes the number of ways of assigning n particles to two locations, with y particles at the first location.

(d) If instead $Y = Y_1 + Y_2$, then Y is binomial with probability ϕ and sample size n, by the same argument in (c). Therefore, the likelihood of ϕ is

$$\begin{aligned} l(\phi \mid y) &\propto \phi^y(1-\phi)^{n-y} \\ &\propto \phi^{73}(1-\phi)^{27}, \end{aligned}$$

and the maximum likelihood estimate is $\hat{\phi} = 0.73$. The likelihood curve can be sketched, as a function of ϕ, in order to draw broader inferences regarding ϕ.

1.3 Large Sample Properties of Likelihood Procedures

The $n \times 1$ random vector $\mathbf{Y}$ is now assumed to possess the sampling density or probability mass function (1.1.2), depending upon a $k \times 1$ vector $\boldsymbol{\theta}$ of unknown parameters, so that $l(\boldsymbol{\theta} \mid \mathbf{y}) = f(\mathbf{y} \mid \boldsymbol{\theta})$ denotes our likelihood for $\boldsymbol{\theta}$, given the single observation vector $\mathbf{y}$. Consider Fisher's (expected) information matrix

$$\mathbf{F}(\boldsymbol{\theta}) = \mathbf{F}_n(\boldsymbol{\theta}) = E\left[\frac{-\partial^2 \log l(\boldsymbol{\theta} \mid \mathbf{Y})}{\partial\left(\boldsymbol{\theta}\boldsymbol{\theta}^T\right)}\right] \qquad (\boldsymbol{\theta} \in \Theta), \tag{1.3.1}$$

where the expectation, of our $k \times k$ matrix of second derivatives, should be taken with respect to the sampling distribution of $\mathbf{Y}$, for each specified $\boldsymbol{\theta}$. Then $\mathbf{F}(\boldsymbol{\theta})$ possesses the (j, l)th element

$$f_{jl}(\boldsymbol{\theta}) = E\left[\frac{-\partial^2 \log l(\boldsymbol{\theta} \mid \mathbf{Y})}{\partial\theta_j\theta_l}\right] \qquad (\boldsymbol{\theta} \in \Theta, j = 1, \ldots, k; l = 1, \ldots, k). \tag{1.3.2}$$

The matrix $F(\boldsymbol{\theta})$ depends upon both n and $\boldsymbol{\theta}$. It should be carefully distinguished from the (observed) likelihood information matrix

$$\mathbf{R} = \left.\frac{-\partial^2 \log l(\boldsymbol{\theta} \mid \mathbf{y})}{\partial\left(\boldsymbol{\theta}\boldsymbol{\theta}^T\right)}\right|_{\boldsymbol{\theta}=\hat{\boldsymbol{\theta}}}, \tag{1.3.3}$$

which is related to the observed curvature of the likelihood surface, within a small neighborhood of $\boldsymbol{\theta} = \hat{\boldsymbol{\theta}}$. In a large number of situations, $\hat{\boldsymbol{\theta}}$ will, for large n, possess a distribution that is approximately multivariate normal with mean vector $\boldsymbol{\theta}$ and covariance matrix $\mathbf{F}_n^{-1}(\boldsymbol{\theta})$. Indeed, the vector $\mathbf{z} = \{\mathbf{F}_n(\boldsymbol{\theta})\}^{\frac{1}{2}}[\hat{\boldsymbol{\theta}} - \boldsymbol{\theta}]$ is said to *converge in distribution*, as $n \to \infty$, with k fixed, to a standard spherical normal distribution (i.e., a multivariate normal distribution $N(\mathbf{0}, \mathbf{I}_k)$ with zero mean vector, and covariance matrix equal to the $k \times k$ identity matrix $\mathbf{I}_k$). The latter possesses density

$$f(\mathbf{z}) = (2\pi)^{-\frac{1}{2}p} \exp\left\{-\tfrac{1}{2}\mathbf{z}^T\mathbf{z}\right\},$$

and the elements of $\mathbf{z}$ are independent $N(0, 1)$ variates.

By this *convergence in distribution* of the $k \times 1$ vector $\mathbf{Z} = (Z_1, Z_2, \ldots, Z_k)^T$, we mean that the joint c.d.f. $F(z_1, z_2, \ldots, z_k)$ of the elements $Z_1, Z_2, \ldots, Z_k$ of $\mathbf{Z}$ converges, for each fixed $z_1, z_2, \ldots, z_n$ and as $n \to \infty$, to the product $\Phi(z_1)\Phi(z_2)\cdots\Phi(z_n)$, where Φ denotes the standard normal c.d.f., that is, the c.d.f. of an $N(0, 1)$ variate. In many such situations, $\hat{\boldsymbol{\theta}}$ is also a "strongly consistent estimator" of $\boldsymbol{\theta}$, that is, $\hat{\boldsymbol{\theta}}$ converges almost surely (i.e., with sampling probability one) to $\boldsymbol{\theta}$, as $n \to \infty$, with k fixed. These results are discussed more fully by Sen and Singer (1993, pp. 202–210). Examples where the results do not hold include Rasch's model (1.1.16), when r is kept fixed and the ability parameters are not taken to be random, since the number of parameters then increases with the number of observations. Other situations where the regularity conditions break down include exchangeable sampling models (see Exercise 1.3.d) and models where the range of the sample space depends upon the unknown parameters (e.g., Worked Example 1C of Section 1.1). However, the preceding strong consistency and asymptotic normality results hold in, but are not limited to, the following three situations:

(1) Let $Y_1, Y_2, \ldots, Y_n$ denote a random sample from a distribution with density or probability mass function $f(y \mid \boldsymbol{\theta})$, $(y \in \mathcal{Y}; \boldsymbol{\theta} \in \Theta)$, so that

$$f(\mathbf{y} \mid \boldsymbol{\theta}) = l(\boldsymbol{\theta} \mid \mathbf{y}) = \prod_{i=1}^{n} f(y_i \mid \boldsymbol{\theta}). \tag{1.3.4}$$

Then it is enough to assume that the distribution of each Y_i is concentrated on a region not depending on $\boldsymbol{\theta}$, that $f(y \mid \boldsymbol{\theta})$ is twice differentiable in $\boldsymbol{\theta}$, for each $y \in Y$, and that the elements of the matrix $F_n(\boldsymbol{\theta})$ in (1.3.1) are finite. More detailed conditions are described by Sen and Singer (1993, p. 205). Subject to these conditions, we have a "regular estimation problem."

(2) Let $Y_1, \ldots, Y_n$ denote independent variates with respective densities or probability mass functions $f_1(y_1; \alpha_1), f_2(y_1; \alpha_1), \ldots, f_n(y_n; \alpha_n)$, which respectively depend upon unknown parameters $\alpha_1, \ldots, \alpha_n$, but are otherwise completely specified. The distribution of each Y_i should be concentrated on some region not depending upon $\alpha_1, \alpha_2, \ldots, \alpha_n$. Consider a generalized linear model (e.g., McCullagh and Nelder, 1989), where

$$\alpha_i = \mathbf{x}_i^T \boldsymbol{\beta} \qquad (i = 1, \ldots, n),$$

with $\mathbf{x}_1, \mathbf{x}_2, \ldots, \mathbf{x}_n$ denoting specified $k \times 1$ design vectors, and $\boldsymbol{\beta}$ denoting a $k \times 1$ vector of unknown parameters, with $k < n$. Then the likelihood of $\boldsymbol{\beta}$, given that $Y_1 = y_1, \ldots, Y_n = y_n$, is denoted by

$$l(\boldsymbol{\beta} \mid \mathbf{y}) = \prod_{i=1}^{n} f_i(y_i; \boldsymbol{\beta}). \tag{1.3.5}$$

In cases where the maximum likelihood vector $\hat{\boldsymbol{\beta}}$ of $\boldsymbol{\beta}$ is uniquely defined, let $\mathbf{F}(\boldsymbol{\beta})$ denote Fisher's information matrix for $\boldsymbol{\beta}$, paralleling (1.3.1). Then $\hat{\boldsymbol{\beta}}$ will be a strongly consistent estimator for $\boldsymbol{\beta}$, as $n \to \infty$ with k fixed, and $\{\mathbf{F}(\boldsymbol{\beta})\}^{\frac{1}{2}}[\hat{\boldsymbol{\beta}} - \boldsymbol{\beta}]$ will converge in distribution to a standard spherical normal distribution, under quite general conditions, for example, if each $f_i(y_i; \alpha_i)$ possesses similarly regular behavior to $f(y \mid \boldsymbol{\theta})$ in situation (1), and the sequence $\mathbf{x}_1, \mathbf{x}_2, \ldots$ is concentrated on a bounded region of R^k. More detailed conditions for this and related regression problems are discussed by Wu (1981). The conditions can often be expressed in terms of the requirement that Fisher's information matrix remains positive definite as $n \to \infty$. See also Chiu, Leonard, and Tsui (1996).

(3) Models where $Y_1, \ldots, Y_n$ are not independent, for example, where $\mathbf{Y} = (Y_1, \ldots, Y_n)^T$ possesses a multivariate normal distribution with specified known covariance matrix $\mathbf{C}$, and each Y_i possesses a common unknown mean θ, but only under particular conditions, for example, for special cases of the $\mathbf{C}$ matrix. (See Hoadley, 1971.)

Efron and Hinkley (1978) pioneered applications of the result that in many similar circumstances,

$$(\mathbf{R}^*)^{\frac{1}{2}}(\hat{\boldsymbol{\theta}} - \boldsymbol{\theta}) \xrightarrow{d} N(\mathbf{0}, \mathbf{I}_k) \qquad (n \to \infty, k \text{ fixed}). \tag{1.3.6}$$

In other words, $(\mathbf{R}^*)^{\frac{1}{2}}(\hat{\boldsymbol{\theta}} - \boldsymbol{\theta})$ typically converges in distribution to a standard spherical normal distribution, where $\mathbf{R}^*$ denotes the observed likelihood matrix equal to $\mathbf{R}$, in (1.3.3), but with $\mathbf{y}$ replaced by its random counterpart $\mathbf{Y}$. The result in (1.3.6) will always be true, because of Slutsky's theorem (see Sen and Singer, 1993, p. 127), whenever

$$\{\mathbf{F}_n(\boldsymbol{\theta})\}^{\frac{1}{2}}(\hat{\boldsymbol{\theta}} - \boldsymbol{\theta}) \xrightarrow{d} N(\mathbf{0}, \mathbf{I}_k) \qquad (n \to \infty, k \text{ fixed}), \tag{1.3.7}$$

and $n^{-1}[\mathbf{R}^* - F_n(\boldsymbol{\theta})]$ converges in probability to a matrix of zeros, as $n \to \infty$.

The result in (1.3.6) is particularly useful for providing approximate confidence intervals for elements of $\boldsymbol{\theta}$, or for an arbitrary linear transformations $\eta = \mathbf{a}^T\boldsymbol{\theta}$, of elements of $\boldsymbol{\theta}$. Whenever the maximum likelihood estimate $\hat{\boldsymbol{\theta}}$ is unique, $\mathbf{R}$ in (1.3.3) is nonsingular, and (1.3.7) holds, an approximate 95% confidence interval, for $\eta = \mathbf{a}^T\boldsymbol{\theta}$, is given by

$$\left(\mathbf{a}^T\hat{\boldsymbol{\theta}} - 1.96\{\mathbf{a}^T\mathbf{R}^{-1}\mathbf{a}\}^{\frac{1}{2}}, \mathbf{a}^T\hat{\boldsymbol{\theta}} + 1.96\{\mathbf{a}^T\mathbf{R}^{-1}\mathbf{a}\}^{\frac{1}{2}}\right). \tag{1.3.8}$$

Sampling from an Exponential Distribution

Consider, for example, the situation where $Y_1, \ldots, Y_n$ are a random sample, from the negative exponential distribution, with density $f(y \mid \theta) = \theta \exp\{-\theta y\}$, $(0 < \theta < \infty;$ $0 < y < \infty)$. Then

$$l(\theta \mid \mathbf{y}) = \theta^n \exp\{-n\theta\bar{y}\} \qquad (0 < \theta < \infty), \tag{1.3.9}$$

with $\bar{y}$ denoting the sample mean. Consequently $\hat{\theta} = 1/\bar{y}$, and the scalar likelihood information is $R = n/\hat{\theta}^2$, so that an approximate $100(1-\epsilon)\%$ confidence interval for θ is given by

$$\left(\hat{\theta}\left\{1 - n^{-\frac{1}{2}}z_{\epsilon/2}\right\}, \hat{\theta}\left\{1 + n^{-\frac{1}{2}}z_{\epsilon/2}\right\}\right), \tag{1.3.10}$$

where $z_{\epsilon/2}$ is the $100(1-\frac{1}{2}\epsilon)$th percentile of the standard normal distribution. The confidence interval (1.3.10) parallels the likelihood approximation (see also Section 1.4 (B)),

$$l^*(\boldsymbol{\theta} \mid \mathbf{y}) = l(\hat{\boldsymbol{\theta}} \mid \mathbf{y}) \exp\left\{-\tfrac{1}{2}(\boldsymbol{\theta} - \hat{\boldsymbol{\theta}})^T\mathbf{R}(\boldsymbol{\theta} - \hat{\boldsymbol{\theta}})\right\} \qquad (\boldsymbol{\theta} \in R^k), \tag{1.3.11}$$

which reduces in the scalar case to

$$l(\theta \mid \mathbf{y}) \propto \exp\left\{-\tfrac{1}{2}R(\theta - \hat{\theta})^2\right\} \qquad (-\infty < \theta < \infty).$$

Improving the Normality of the Approximation

The interval (1.3.10) can be modified, for finite n, by seeking a parameter whose likelihood more closely resembles a normal curve. One possibility is to consider $\alpha = -\log\theta$, which possesses likelihood

$$l(\alpha \mid \mathbf{y}) = \exp\left\{ -n\alpha - ne^{-\alpha}\bar{y}\right\} \qquad (\infty < \alpha < \infty). \tag{1.3.12}$$

Then $\hat{\alpha} = \log\bar{y}$ and $R = n$, so that an approximate $100(1-\epsilon)\%$ confidence interval for α is

$$\left(\log\bar{y} - z_{\epsilon/2}n^{-\frac{1}{2}}, \log\bar{y} + z_{\epsilon/2}n^{-\frac{1}{2}}\right). \tag{1.3.13}$$

Taking negative exponentials of both confidence limits in (1.3.13) tells us that an alternative approximate $100(1-\epsilon)\%$ confidence interval for θ is given by

$$\left(\hat{\theta}\exp\left\{ -z_{\epsilon/2}n^{-\frac{1}{2}}\right\}, \hat{\theta}\exp\left\{z_{\epsilon/2}n^{-\frac{1}{2}}\right\}\right). \tag{1.3.14}$$

The interval in (1.3.14) provides an interesting finite sample alternative to the more obvious interval (1.3.10) and can lead to substantial numerical differences. Note that any interval of the form $(A\hat{\theta}, B\hat{\theta})$, where A and B are constants, has frequency coverage not depending upon θ and is equal to $p(2A \le U_{2n} \le 2B)$, where U_{2n} is a chi-squared variate with $2n$ degrees of freedom. An exact $100(1-\epsilon)\%$ equal-tailed confidence interval for θ can be based on the distributional result that $2n\theta\bar{Y}$ has a chi-squared distribution with $2n$ degrees of freedom, and the intervals in (1.3.10) and (1.3.14) can alternatively be based upon approximations to this distribution.

Lemma 1.3.1: *Let U_ν possess a chi-squared distribution with ν degrees of freedom and let $W = \log U_\nu$. Then the normal density $p^*(w)$ minimizing the entropy distance*

$$-\int \log\frac{p^*(w)}{p(w)}\,p(w)\,dw \tag{1.3.15}$$

has mean $\log\nu - \nu^{-1}$ and variance $2\nu^{-1}$ (the minimum entropy distance is zero up to an application of Stirling's approximation $\Gamma(\nu) \sim \frac{1}{\sqrt{2\pi}}e^{-\nu}\nu^{\nu-\frac{1}{2}}$ for the Gamma function appearing in the density of the chi-squared distribution).

As a consequence of Lemma 1.3.1, the approximate $100(1-\epsilon)\%$ confidence interval (1.3.14) may be replaced by the refinement

$$\left(\hat{\theta}\exp\left\{ -n^{-1} - z_{\epsilon/2}n^{-\frac{1}{2}}\right\}, \hat{\theta}\exp\left\{ -n^{-1} + z_{\epsilon/2}n^{-\frac{1}{2}}\right\}\right). \tag{1.3.16}$$

When $\{\mathbf{F}_n(\boldsymbol{\theta})\}^{\frac{1}{2}}[\hat{\boldsymbol{\theta}} - \boldsymbol{\theta}]$ converges in distribution to a standard spherical normal distribution, $\hat{\boldsymbol{\theta}} = (\hat{\theta}_1, \hat{\theta}_2, \ldots, \hat{\theta}_k)^T$ will, under broad conditions, possess the property of asymptotic efficiency, that is, no other asymptotically unbiased estimator of $\boldsymbol{\theta}$ will possess better variance properties for large enough values of n. Rasch's model (1.1.16), with r fixed, however, gives an example of a situation where the regularity conditions break down.

Worked Example 1E: *Comparing Two Sampling Models*

The time taken for the next 10 cars to arrive at a particular crossing was observed on $n = 1{,}000$ similar occasions. The observations were recorded as $y_1, y_2, \ldots, y_{1{,}000}$, and these satisfy

$$\bar{y} = n^{-1} \sum_{i=1}^{n} y_i ,$$

$$t = n^{-1} \sum_{i=1}^{n} u_i ,$$

and

$$W = n^{-1} \sum_{i=1}^{n} (u_i - t),$$

where $u_i = \log y_i$. Show how BIC, $\bar{y}$, t, and W can be used to choose between the two models:

Model A: The y_i are a random sample from the Gamma $G(m, \lambda)$ distribution (see Exercise 1.1.c, Note), where $m = 10$ and λ is unknown.

Model B: The $u_i = \log y_i$ are a random sample from a normal $N(\theta, \phi)$ distribution, where both θ and ϕ are unknown.

Model Answer 1E:

Under Model A, the likelihood is

$$\begin{aligned} l(\lambda \mid \mathbf{y}) &= \prod_{i=1}^{n} p(y_i \mid \lambda) \\ &= \lambda^{mn} \left(\prod_{i=1}^{n} y_i \right)^{m-1} \exp\left\{ -\lambda \sum_{i=1}^{n} y_i \right\} D_{m,n}^{-1} \\ &= \lambda^{mn} \exp\left\{ n(m-1)t - n\lambda\bar{y} \right\} D_{m,n}^{-1}, \end{aligned}$$

where $D_{m,n} = \{(m-1)!\}^n$ and t is defined above. Consequently,

$$\log l(\lambda \mid \mathbf{y}) = mn \log \lambda + n(m-1)t - n\lambda\bar{y} - \log D_{m,n}.$$

This is maximized when $\lambda = \hat{\lambda} = m/\bar{y}$, so that

$$\begin{aligned} \text{BIC} &= \log l(\hat{\lambda} \mid \mathbf{y}) - \frac{1}{2} \log \frac{n}{2\pi} \\ &= mn \log \frac{\bar{y}}{m} + (m-1)nt - mn - \log D_{m,n} - \frac{1}{2} \log \frac{n}{2\pi}. \end{aligned}$$

Under Model B, the density of y_i is

$$p(y_i \mid \theta, \phi) = (2\pi)^{-\frac{1}{2}} \phi^{-\frac{1}{2}} y_i^{-\frac{1}{2}} \exp\left\{-\frac{1}{2\phi}(\log y_i - \theta)^2\right\},$$

and the likelihood of θ and ϕ is

$$\begin{aligned} l(\theta, \phi \mid \mathbf{y}) &= \prod_{i=1}^{n} p(y_i \mid \theta, \phi) \\ &= (2\pi)^{-\frac{n}{2}} \phi^{-\frac{n}{2}} \exp\left\{-\frac{1}{2}nt - \frac{1}{2\phi}\sum_{i=1}^{n}(\log y_i - \theta)^2\right\}. \end{aligned}$$

This likelihood is maximized when $\theta = \hat{\theta} = t$ and $\phi = \hat{\phi} = W/n$, so that

$$\begin{aligned} \text{BIC} &= \log l\big(\hat{\theta}, \hat{\phi} \mid \mathbf{y}\big) - \log \frac{n}{2\pi} \\ &= \frac{1}{2}n \log(2\pi) - \frac{1}{2}n \log W - \frac{1}{2}n(t+1). \end{aligned}$$

We should then consider employing the model for which BIC is the larger.

Worked Example 1F: *Fisher's Information and Likelihood Information for a Normal Log-Variance*

Observations $Y_1, Y_2, \ldots, Y_n$ are a random sample from the normal distribution with zero mean and unknown variance ϕ.

(a) Find Fisher's information $F_n(\theta)$ for $\theta = \log \phi$. (Note that with $T = \sum_{i=1}^{n} Y_i^2$, the sampling distribution of T/ϕ is chi-squared with n degrees of freedom.)

(b) Find the likelihood information R for θ.

(c) Give full details of an approximate 95% confidence interval for θ of the form $(\hat{\theta} - 1.96R^{-\frac{1}{2}}, \hat{\theta} + 1.96R^{-\frac{1}{2}})$, where $\hat{\theta}$ denotes the maximum likelihood estimate of θ.

(d) Use an entropy-based normal approximation to the sampling distribution of $\hat{\theta} = \log \hat{\phi}$ to obtain a modified approximate 95% confidence interval for θ, and hence for ϕ.

Model Answer 1F:

(a) The likelihood for ϕ is the product of n normal densities, that is,

$$\begin{aligned} l(\phi \mid \mathbf{y}) &= \prod_{i=1}^{n} \frac{1}{\sqrt{2\pi\phi}} \exp\left\{-\frac{y_i^2}{2\phi}\right\}, \\ &= (2\pi)^{-\frac{n}{2}} \phi^{-\frac{n}{2}} \exp\left\{-\frac{t}{2\phi}\right\}, \end{aligned}$$

where $t = \sum_{i=1}^n y_i^2$. Since $\phi = e^\theta$, the likelihood of θ is

$$l(\theta \mid \mathbf{y}) = (2\pi)^{-\frac{n}{2}} \exp\left\{-\frac{1}{2}n\theta - \frac{t}{2}e^{-\theta}\right\}.$$

Consequently,

$$\frac{\partial \log l(\theta \mid \mathbf{y})}{\partial \theta} = -\frac{1}{2}n + \frac{t}{2}e^{-\theta}$$

and

$$-\frac{\partial^2 \log l(\theta \mid \mathbf{y})}{\partial \theta^2} = \frac{t}{2}e^{-\theta}.$$

Since the mean of T/ϕ is n, T has expectation $n\phi = ne^\theta$. Consequently, Fisher's information is

$$\begin{aligned} F_n(\theta) &= E\left(\tfrac{1}{2}e^{-\theta}T\right) \\ &= \tfrac{1}{2}e^{-\theta}ne^\theta = \tfrac{1}{2}n. \end{aligned}$$

(b) The likelihood of ϕ is maximized when $\phi = \hat{\phi} = t/n$. Consequently, the maximum likelihood estimator of θ is $\hat{\theta} = \log \hat{\phi} = \log(t/n)$. The likelihood information is therefore

$$R = \left[\frac{t}{2}e^{-\theta}\right]_{\theta=\hat{\theta}} = \frac{1}{2}n,$$

which in this case turns out to be identical to Fisher's information.

(c) This interval is

$$\left(\log\frac{t}{n} - 1.96\sqrt{\frac{2}{n}}, \log\frac{t}{n} + 1.96\sqrt{\frac{2}{n}}\right).$$

(d) Since T/ϕ has a chi-squared distribution with n degrees of freedom, $\log T$ is approximately $N(\theta + \log n - n^{-1}, 2n^{-1})$ distributed, by Lemma 1.3.1. Consequently, $\hat{\theta} = \log T - \log n$ is approximately normally distributed with mean $\theta - n^{-1}$ and variance $2n^{-1}$. Standard pivotal arguments for confidence intervals tell that an approximate 95% confidence interval for θ is

$$\left(\hat{\theta} + n^{-1} - 1.96\sqrt{\frac{2}{n}}, \hat{\theta} + n^{-1} + 1.96\sqrt{\frac{2}{n}}\right).$$

Taking exponentials of the limits, we find that an approximate 95% confidence interval for ϕ is

$$\left(\hat{\phi}\exp\left\{n^{-1} - 1.96\sqrt{\frac{2}{n}}\right\}, \hat{\phi}\exp\left\{n^{-1} + 1.96\sqrt{\frac{2}{n}}\right\}\right).$$

SELF-STUDY EXERCISES

1.3.a Let $Y_1, \ldots, Y_n$ denote a random sample from a Poisson distribution with mean θ. Show that the maximum likelihood estimator of θ is $\hat{\theta} = \bar{Y}$ and that Fisher's information is $F_n(\theta) = n/\theta$.

Find the moment-generating function (m.g.f.) $G_n(s) = E(e^{sZ})$ of $Z = \{F_n(\theta)\}^{\frac{1}{2}}(\hat{\theta} - \theta)$ by noting that $E(e^{sY_i}) = \exp\{\theta(e^s - 1)\}$. Show that, as $n \to \infty$, $G_n(s)$ converges to $\exp(\frac{1}{2}s^2)$, the m.g.f. of an $N(0, 1)$ variate (this result, in general, is enough to demonstrate convergence in distribution of $\hat{\theta}$ to a standard normal variate).

1.3.b The "useful limit theorem" or "delta method" tells us that if the sequence $\{\mathbf{U}_n\}$ of random variables converges in distribution, as $n \to \infty$, to a multivariate normal distribution with mean vector $\boldsymbol{\mu}$ and covariance matrix $\mathbf{C}$, and $\mathbf{W}_n = \mathbf{g}(\mathbf{U}_n)$, for some specified real-valued vector function $\mathbf{g}$, then the sequence $\{\mathbf{W}_n\}$ converges in distribution to a multivariate normal distribution with mean vector $\mathbf{g}(\boldsymbol{\mu})$ and covariance matrix

$$\frac{\partial \mathbf{g}(\boldsymbol{\mu})}{\partial \boldsymbol{\mu}^T}\mathbf{C}\frac{\partial \mathbf{g}(\boldsymbol{\mu})}{\partial \boldsymbol{\mu}}.$$

In Exercise 1.3.a, let $\alpha = \log\theta$. Show that $\hat{\alpha} = \log\bar{Y}$ and $F_n(\alpha) = ne^{\alpha}$. Demonstrate that for large n, $\hat{\alpha}$ is approximately normally distributed with mean α and variance $n^{-1}\theta^{-1}$.

1.3.c *The Linear Logistic Model:* Let cell frequencies $Y_1, \ldots, Y_m$ possess binomial distributions, with respective probabilities $\theta_1, \ldots, \theta_m$ and sample sizes $n_1, \ldots, n_m$. Consider the logits $\alpha_i = \log\theta_i - \log(1 - \theta_i)$ and assume that

$$\alpha_i = \mathbf{x}_i^T\boldsymbol{\beta} \qquad (i = 1, \ldots, m), \tag{1.3.17}$$

where $\mathbf{x}_1, \ldots, \mathbf{x}_n$ are specified linearly independent design vectors, and $\boldsymbol{\beta}$ is a $k \times 1$ vector of unknown parameters. Show that the likelihood of $\boldsymbol{\beta}$, given that $Y_i = y_i$, for $i = 1, \ldots, m$, satisfies

$$l(\boldsymbol{\beta} \mid \mathbf{y}) \propto \exp\left\{\mathbf{t}^T\boldsymbol{\beta} - \sum_{i=1}^{m} n_i \log\left(1 + e^{\mathbf{x}_i^T\boldsymbol{\beta}}\right)\right\}, \tag{1.3.18}$$

where $\mathbf{t} = \sum_{i=1}^m y_i\mathbf{x}_i$. Show that the maximum likelihood vector $\hat{\boldsymbol{\beta}}$ for $\boldsymbol{\beta}$ satisfies the nonlinear equation

$$\mathbf{t} = \sum_{i=1}^{m} n_i\mathbf{x}_i\theta_i(\hat{\boldsymbol{\beta}}), \tag{1.3.19}$$

where $\theta_i(\boldsymbol{\beta}) = e^{\mathbf{x}_i^T\boldsymbol{\beta}}/(1 + e^{\mathbf{x}_i^T\boldsymbol{\beta}})$, and that Fisher's information matrix is

$$F_m(\boldsymbol{\beta}) = \sum_{i=1}^{m} n_i\boldsymbol{\theta}_i(\boldsymbol{\beta})\big(1 - \boldsymbol{\theta}_i(\boldsymbol{\beta})\big)\mathbf{x}_i\mathbf{x}_i^T. \tag{1.3.20}$$

The asymptotic results of Section 1.3 apply, as $m \to \infty$, with $n_1, n_2, \ldots, n_m$ fixed, certainly whenever the sequence $\mathbf{x}_1, \mathbf{x}_2, \ldots$ remains concentrated on a bounded region. Find a large sample approximation to the distribution of the ith fitted value $\hat{\alpha}_i = \mathbf{x}_i^T\hat{\boldsymbol{\beta}}$.

Show that the approximation (1.3.11) now reduces to

$$l^*(\boldsymbol{\beta} \mid \mathbf{y}) = l(\hat{\boldsymbol{\beta}} \mid \mathbf{y})\exp\left\{-\tfrac{1}{2}(\boldsymbol{\beta} - \hat{\boldsymbol{\beta}})^T\mathbf{R}(\boldsymbol{\beta} - \hat{\boldsymbol{\beta}})\right\}, \tag{1.3.21}$$

where $\mathbf{R} = F_m(\hat{\boldsymbol{\beta}})$.

1.3.d Observations $y_1, \ldots, y_n$ do not result from random sampling and are not necessarily independent. It is however assumed that each observation y_i is the realization of a normally distributed random variable Y_i and that $\mathbf{Y} = (Y_1, \ldots, Y_n)^T$ possesses a multivariate normal distribution. In the absence of information to the contrary, take $Y_1, \ldots, Y_n$ to be *exchangeable* or *permutable* (e.g., Lindley and Novick, 1981), that is, take their joint density to be invariant under any permutation of the suffices. (This involves a concept similar to Laplace's principle of insufficient reason; see Exercise 1.1.j.)

Show that exchangeability implies that each Y_i possesses a common mean μ and common variance σ^2 and that each distinct pair Y_i and Y_j possesses a common correlation ρ lying in the interval $(-1/(n-1), 1)$. Under the preceding multivariate normal assumption, show that the likelihood of μ, σ^2, and ρ is

$$l(\mu, \sigma^2, \rho \mid \mathbf{y}) \propto (\sigma^2)^{-\frac{1}{2}n}(1-\rho)^{-\frac{1}{2}(n-1)}\left[1+(n-1)\rho\right]^{-\frac{1}{2}} \times \exp\left\{-\frac{s^2}{2\sigma^2} - \frac{n(\bar{y}-\mu)^2}{2\sigma^2[1+(n-1)\rho]}\right\},$$

where $\bar{y}$ denotes the sample mean and $s^2 = \sum_{i=1}^n (y_i - \bar{y})^2$.

Show that $\hat{\boldsymbol{\mu}} = \bar{y}$ is the maximum likelihood estimate of μ, but that ρ cannot be sensibly identified from the data when μ and σ^2 are also unknown. Show that $\hat{\boldsymbol{\mu}} = \bar{y}$ is normally distributed, with mean μ and variance $n^{-1}\sigma^2(1+(n-1)\rho)$. Hence, show that as $n \to \infty$ and whenever $\rho > 0$, $\hat{\boldsymbol{\mu}} = \bar{y}$ behaves like a normally distributed random variable with zero mean and variances $\rho\sigma^2$ and hence cannot converge to μ unless $\rho = 0$.

Let $S^2 = \sum_{i=1}^n (Y_i - \bar{Y})^2$. Then $U = S^2/\sigma^2(1-\rho) \sim \chi^2_\nu$ with $\nu = n-1$, independent of $\bar{Y}$, and possesses density (1.1.10). Construct a 95% confidence interval for μ, when σ^2 is also unknown but μ is known, based upon percentage points of the t-distribution, with $n-1$ degrees of freedom. Show that the width of this interval is $(1+(n-1)\rho)^{\frac{1}{2}}/(1-\rho)^{\frac{1}{2}}$ times the width of the standard t-confidence interval.

Do the above results concern you when considering the modeling of nonrandomized data? Consider also central limit theorems for exchangeable random variables (e.g., Weber, 1980). Note that one possible resolution of the problems created by lack of identification of ρ is provided by maximum entropy (see Exercises 3.4.j and 3.4.k). See also Leonard (1996).

Hint: If

$$T = \frac{Z}{\sqrt{U/\nu}},$$

where $Z \sim N(0, 1)$ and $U \sim \chi^2_\nu$ are independent, then $T \sim t_\nu$, that is, T possesses a student's t-distribution, with ν degrees of freedom and density (1.2.3).

1.3.e In Exercises 1.3.a and 1.3.b, use Chebyshev's inequality (e.g., Sen and Singer, 1993, p. 40) to show that for any $\epsilon > 0$,

$$p\left(\left|\hat{\theta} - \theta\right| > \epsilon\right) \leq \frac{\theta}{n\epsilon^2}. \tag{1.3.22}$$

An estimator $\hat{\theta}$ is said to *converge in probability* to θ, as $n \to \infty$, if for any $\epsilon > 0$, the left-hand side of (1.3.22) converges to zero, as $n \to \infty$. Show that $\bar{Y}$ converges in probability to θ, as $n \to \infty$. Also show that

$$\left[\bar{y}\exp\left\{1.96n^{-\frac{1}{2}}(\bar{y})^{-\frac{1}{2}}\right\}, \bar{y}\exp\left\{-1.96n^{-\frac{1}{2}}(\bar{y})^{-\frac{1}{2}}\right\}\right]$$

is an approximate 95% confidence interval for θ when n is large.

1.4 Practical Examples

(A) *Frequency Coverage of Approximate Confidence Intervals*

In this section, we justify (1.3.16) as an approximate equal-tailed $100(1-\epsilon)\%$ confidence interval for an exponential parameter θ. The fourth column of Table 1.4.1 describes the frequency coverage for the lower tail of the approximate 95% confidence interval based upon (1.3.16), that is, interval B, and the fifth column gives the frequency coverage to the left of the upper tail. These reasonably well approximate the true values (0.025 and 0.975) whenever $n \geq 40$. In the second and third columns, we describe similar quantities, but for the more standard approximation (1.3.10), that is, interval A. This interval is too skew to the left and does not give particularly convincing results when $n < 200$. Similar results are available for the equal-tailed 99% confidence interval (see Table 1.4.2).

(B) *The Linear Logistic Model; The Ingot Data*

The data in Table 1.4.3 were reported by Cox and Snell (1981). Ingots were heated for $m = 4$ different heating times x_i, listed for $i = 1, \ldots, 4$ in the last column of Table 1.4.3. The number y_i out of n_i ingots that were ready for rolling was then observed, for $i = 1, \ldots, 4$ (see second and third columns). These data can be analyzed, under the assumptions of Exercise 1.3.c, with $m = 4$ and $k = 2$, by reference to a linear model

Table 1.4.1. *Frequency coverages for approximate 95% equal-tailed intervals.*

Sample size n	Interval A	Interval B	Interval A	Interval B
1	0.00000	0.94818	0.08189	0.98651
2	0.00000	0.95114	0.05877	0.98574
3	0.00000	0.95350	0.05019	0.98483
4	0.00000	0.95527	0.04561	0.98410
5	0.00045	0.95665	0.04272	0.98351
10	0.00582	0.96074	0.03632	0.98172
20	0.01143	0.96425	0.03243	0.98014
30	0.01403	0.96598	0.03087	0.97934
40	0.01557	0.96706	0.02998	0.97884
50	0.01662	0.96782	0.02939	0.97848
100	0.01917	0.96979	0.02800	0.97755
200	0.02094	0.97125	0.02707	0.97685
300	0.02170	0.97191	0.02667	0.97653
400	0.02215	0.97231	0.02644	0.97634
500	0.02246	0.97259	0.02628	0.97620
600	0.02268	0.97281	0.02616	0.97610
800	0.02299	0.97308	0.02602	0.97595
1,000	0.02321	0.97329	0.02589	0.97587

Table 1.4.2. *Frequency coverages for approximate 99% equal-tailed intervals.*

Sample size n	Interval A	Interval B	Interval A	Interval B
1	0.00000	0.97187	0.04532	0.99964
2	0.00000	0.97633	0.02706	0.99928
3	0.00000	0.97900	0.02077	0.99898
4	0.00000	0.98080	0.01756	0.99874
5	0.00000	0.98212	0.01559	0.99854
10	0.00003	0.98567	0.01145	0.99789
20	0.00054	0.98836	0.00910	0.99726
30	0.00105	0.98959	0.00820	0.99693
40	0.00143	0.99032	0.00770	0.99671
50	0.00173	0.99082	0.00738	0.99656
100	0.00257	0.99206	0.00662	0.99614
200	0.00325	0.99292	0.00613	0.99582
300	0.00356	0.99330	0.00592	0.99567
400	0.00375	0.99352	0.00580	0.99557
500	0.00389	0.99367	0.00572	0.99551
600	0.00399	0.99379	0.00566	0.99546
800	0.00413	0.99394	0.00558	0.99539
1,000	0.00422	0.99405	0.00552	0.99535

Table 1.4.3. *The ingot data.*

i	n_i	y_i	x_i
1	7	0	55
2	14	2	157
3	27	7	159
4	51	3	16

of the form

$$\alpha_i = \beta_1 + \beta_2 x_i \qquad (i = 1, \ldots, m), \tag{1.4.1}$$

for the logits α_i. The maximum likelihood estimates $\hat{\beta}_1$ and $\hat{\beta}_2$ of β_1 and β_2 are readily calculable by reference to a standard statistical computing package (e.g., SAS). This gives $\hat{\beta}_1 = -5.4152$ and $\hat{\beta}_2 = 0.0807$. Furthermore, the inverse of the likelihood information matrix $\mathbf{R}$ is given as

$$\mathbf{R}^{-1} = \begin{pmatrix} 0.5293153 & -0.0148034 \\ -0.0148034 & 0.0004998 \end{pmatrix}, \tag{1.4.2}$$

and this provides the dispersion matrix or estimated covariance matrix for $\hat{\beta} = (\hat{\beta}_1, \hat{\beta}_2)^T$. Therefore an approximate 95% confidence interval for the slope β_2 of the

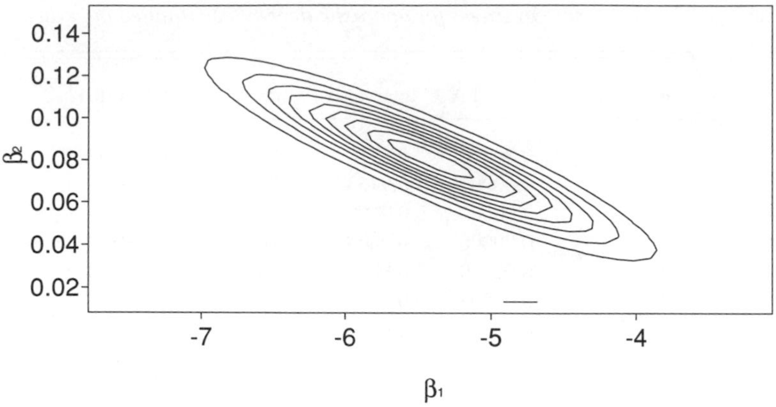

Figure 1.4.1. Likelihood contours for likelihood approximation.

regression line is

$$\begin{aligned} &0.0807 \pm 1.96(0.0004998)^{\frac{1}{2}} \\ &= 0.0807 \pm 1.96 \times 0.02236 \\ &= (0.0369, 0.1245). \end{aligned} \tag{1.4.3}$$

The standard goodness-of-fit statistic was $X^2 = 0.677$, with 2 degrees of freedom. The probability that an ingot, heated for duration $x = 55$, will be ready for rolling may be predicted by

$$\hat{\theta}(55) = \frac{\exp\left(\hat{\beta}_1 + 55\hat{\beta}_2\right)}{1 + \exp\left(\hat{\beta}_1 + 55\hat{\beta}_2\right)} = \frac{e^{-0.9767}}{1 + e^{-0.9767}} = 0.274.$$

However, the predicted logit $\hat{\alpha}(55) = \hat{\beta}_1 + 55\hat{\beta}_2 = -0.9767$ has associated estimated standard error

$$\left(0.5293153 - 2 \times 55 \times 0.0148034 + 55^2 \times 0.0004998\right)^{\frac{1}{2}} = 0.638.$$

Consequently, an approximate 95% prediction interval for the logit $\alpha(55)$ is $(-2.2272, 0.2738)$. Based upon this small amount of data, you can therefore possess only about 95% confidence that $\theta(55)$ lies in the quite wide interval $(0.097, 0.568)$.

The adequacy of the preceding approximations can be indicated by comparing the likelihood approximation (1.3.21) with the exact likelihood (1.3.18). In Figure 1.4.1, the likelihood contours for the bivariate normal approximation (1.3.21) are described, and these are reasonably close to the contours in Figure 1.4.2 for the exact likelihood.

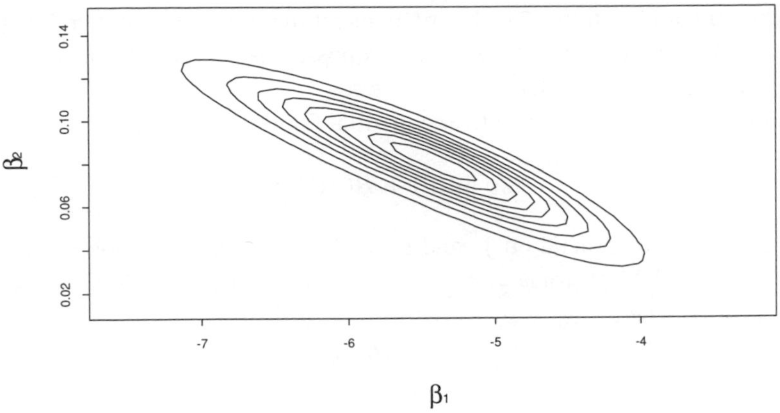

Figure 1.4.2. Likelihood contours for exact likelihood.

1.5 Some Further Properties of Likelihood

(A) *Finite Sample Properties*

It is important to realize that the maximum likelihood vector $\hat{\boldsymbol{\theta}}$ can also possess excellent frequency properties when the sample size n is finite. If, for example, $Y_1, \ldots, Y_n$ are a random sample from the Gamma $G(\theta, 1)$ distribution with density

$$p(y \mid \theta) = \frac{1}{\Gamma(\theta)} y^{\theta-1} e^{-y} \qquad (0 < y < \infty, 0 < \theta < \infty), \tag{1.5.1}$$

then the maximum likelihood estimator $\hat{\theta}$ of θ satisfies

$$\Psi(\hat{\theta}) = t, \tag{1.5.2}$$

where

$$\Psi(\theta) = \frac{\partial \log \Gamma(\theta)}{\partial \theta} = \int_0^\infty (\log u) u^{\theta-1} e^{-u} du$$

denotes the psi or digamma function, and $t = n^{-1} \sum_{i=1}^n \log y_i$. Furthermore, the log-likelihood of $\phi = \Psi(\theta)$ satisfies

$$\frac{\partial \log l(\phi \mid \mathbf{y})}{\partial \phi} = \frac{n\{t - \Psi(\theta)\}}{\partial \Psi(\theta)/\partial \theta}.$$

Owing to this arrangement of terms, Cramer–Rao theory (Wijsman, 1973) tells us that $n^{-1}\partial\Psi(\theta)/\partial\theta$, the reciprocal of the coefficient of $t - \Psi(\theta)$, is also the Cramer–Rao lower bound for the variances of unbiased estimators of $\phi = \Psi(\theta)$. Moreover, $\hat{\phi} = \Psi(\hat{\theta}) = t$ is uniform minimum variance unbiased (UMVU) for $\phi = \Psi(\theta)$ with variance $n^{-1}\partial\Psi(\theta)/\partial\theta$. Consequently, the maximum likelihood estimator of ϕ possesses ideal frequency properties. The maximum likelihood estimator $\hat{\theta}$ of θ, however, is biased.

One possible unbiased estimator is the moment estimator $\theta^* = \bar{y}$. In Section 1.6 (A), we will show that while biased, $\hat{\theta}$ nevertheless possesses superior mean squared error properties when compared with θ^*.

These results can be generalized to exponential families. In the one-parameter case,

$$p(y \mid \theta) = \exp\left\{A(\theta) + B(\theta)t(y) + C(y)\right\},$$

where A and B do not depend on y, and t and C do not depend upon θ. In general, $\hat{\phi} = n^{-1}\sum t(y_i)$ is UMVU for $\phi = \phi(\theta) = -[\partial A(\theta)/\partial\theta]/[\partial B(\theta)/\partial\theta]$ and possesses variance $n^{-1}\tau(\theta)$, where

$$\tau(\theta) = \frac{\partial\phi(\theta)/\partial\theta}{\partial B(\theta)/\partial\theta}.$$

This is again because of the rearrangement of terms

$$\frac{\partial \log l(\phi \mid \mathbf{y})}{\partial\phi} = \frac{n\left(\hat{\phi} - \phi\right)}{\tau(\theta)},$$

which provides a bridge between Cramer–Rao theory and the concepts of likelihood. (See Worked Example 1H for a full statement of this result.) The expectation of this first derivative is always zero within the one-parameter exponential family. It is an immediate consequence of this formulation alone that $\hat{\phi}$ is sufficient, unbiased, and UMVU, for ϕ, and the variance of $\hat{\phi}$ is equal to the Cramer–Rao lower bound $\tau(\theta)/n$.

(B) *A Taylor Series Expansion*

Expanding any twice-differentable log-likelihood function $\log l(\boldsymbol{\theta} \mid \mathbf{y})$ for a $k \times 1$ vector of parameters $\boldsymbol{\theta}$, in a Taylor series about $\boldsymbol{\theta} = \boldsymbol{\theta}_0$, tells us that

$$\begin{aligned}\log l(\boldsymbol{\theta} \mid \mathbf{y}) = \log l(\boldsymbol{\theta}_0 \mid \mathbf{y}) &+ \mathbf{b}^T(\boldsymbol{\theta}_0)(\boldsymbol{\theta} - \boldsymbol{\theta}_0) - \tfrac{1}{2}(\boldsymbol{\theta} - \boldsymbol{\theta}_0)^T\mathbf{B}(\boldsymbol{\theta}_0)(\boldsymbol{\theta} - \boldsymbol{\theta}_0)\\ &+ \text{ cubic and higher terms in } \boldsymbol{\theta} - \boldsymbol{\theta}_0, \end{aligned} \tag{1.5.3}$$

where

$$\mathbf{b}(\boldsymbol{\theta}) = \frac{\partial \log l(\boldsymbol{\theta} \mid \mathbf{y})}{\partial\boldsymbol{\theta}},$$

and

$$\mathbf{B}(\boldsymbol{\theta}) = -\partial^2 \log l(\boldsymbol{\theta} \mid \mathbf{y})/\partial\left(\boldsymbol{\theta}\boldsymbol{\theta}^T\right). \tag{1.5.4}$$

Differentiating (1.5.3) with respect to $\boldsymbol{\theta}$ gives

$$\begin{aligned} b(\boldsymbol{\theta}) = \mathbf{b}(\boldsymbol{\theta}_0) &+ \mathbf{B}(\boldsymbol{\theta}_0)(\boldsymbol{\theta} - \boldsymbol{\theta}_0)\\ &+ \text{ quadratic and higher terms in } \boldsymbol{\theta} - \boldsymbol{\theta}_0. \end{aligned} \tag{1.5.5}$$

Consider (1.5.3) with $\boldsymbol{\theta}_0$ replaced by the maximum likelihood vector $\hat{\boldsymbol{\theta}}$. If the cubic and higher terms can be neglected, then exponentiating (1.5.3) yields the multivariate normal approximation (1.3.11) to the likelihood function, with $\mathbf{R}$ defined in (1.3.3). This approximation will work best when a preliminary transformation of the parameters is chosen to enhance the multivariate normality. For example, (1.3.21) is typically very accurate whenever $\hat{\boldsymbol{\beta}}$ is finite.

The expansion (1.5.3) also provides the basis for the Newton–Raphson procedure for maximizing $l(\boldsymbol{\theta} \mid \mathbf{y})$ with respect to $\boldsymbol{\theta}$ (see Scales, 1985, and Worked Example 1I).

(C) *Sufficiency, the Neyman–Fisher Factorization Theorem, and the Likelihood Principle*

Let $\mathbf{t} = \mathbf{t}(\mathbf{y}) = (t_1(\mathbf{y}), \ldots, t_l(\mathbf{y}))^T$ denote an l-dimensional statistic, consisting of l functions $t_1(\mathbf{y}), \ldots, t_l(\mathbf{y})$ that do not depend upon the unknown $\boldsymbol{\theta}$. Let $\mathbf{T} = t(\mathbf{Y})$.

Definition: The statistic $\mathbf{t} = \mathbf{t}(\mathbf{y})$ is a *sufficient statistic* for $\boldsymbol{\theta}$ if the conditional distribution of $\mathbf{Y}$, given that $\mathbf{T} = \mathbf{t}$, does not depend upon $\boldsymbol{\theta}$. Furthermore, $t = t(\mathbf{y})$ is a *minimal sufficient statistic* if it can be expressed as a function of any other sufficient statistic $t^*(\mathbf{y})$ for $\boldsymbol{\theta}$.

Intuitively speaking, once the realization of $\mathbf{t}$ is recorded, no other quantities calculated from the data give further information about $\boldsymbol{\theta}$ if the sampling model $p(\mathbf{y}) = f(\mathbf{y} \mid \boldsymbol{\theta})$ is assumed true. The full vector $\mathbf{y}$ is always sufficient for $\boldsymbol{\theta}$. A minimal sufficient statistic always possesses as small a dimensionality as possible, among different choices of sufficient statistics. In particular, for the one-parameter exponential family of Section 1.5 (A), the single-valued statistic $t = n^{-1} \sum_{i=1}^{n} t(y_i)$ is minimal sufficient for θ.

The concept of sufficiency is a highlight of the frequency philosophy of statistics, since it involves a manipulation of the sampling distribution of $\mathbf{Y}$, given $\boldsymbol{\theta}$. However, the following theorem links sufficiency to an important genesis for the concept of likelihood and indicates that the shape of the likelihood curve can itself be regarded as summarizing the information content of the data. See Lee (1997, p. 55) for a short proof.

Theorem (Neyman–Fisher factorization theorem): A statistic $\mathbf{t} = \mathbf{t}(\mathbf{y})$ is sufficient for $\boldsymbol{\theta}$ if and only if

$$l(\boldsymbol{\theta} \mid \mathbf{y}) = h(\mathbf{y}) g\{\boldsymbol{\theta}, t(\mathbf{y})\}, \tag{1.5.6}$$

where $h(\mathbf{y})$ does not depend upon $\boldsymbol{\theta}$, and $g(\boldsymbol{\theta}, \mathbf{t})$ does not depend upon $\mathbf{y}$ (i.e., if, as a function of $\boldsymbol{\theta}$, the likelihood curve is proportional to a "likelihood kernel,"

$$l^*(\boldsymbol{\theta} \mid \mathbf{y}) = g\{\boldsymbol{\theta}, t(\mathbf{y})\},$$

which is a function of $\boldsymbol{\theta}$ only, and depends only upon the data via $\mathbf{t} = t(\mathbf{y})$).

Birnbaum (1962) states a theorem proving that a "sufficiency principle" and a "conditionality principle" together imply the "likelihood principle" (i.e., if the likelihoods for $\boldsymbol{\theta}$ based upon two different experiments are proportional, as functions of $\boldsymbol{\theta}$, then the same inference should be drawn about $\boldsymbol{\theta}$ in each case). Note that during the short proof of Birnbaum's theorem (fully reported in Exercise 1.5.e), his "sufficiency principle" is applied to a random mixture of two different experiments. This device can be interpreted in two different ways. Either (a) the application of the sufficiency principle can be regarded as at least as strong as the likelihood principle itself, in which case the theorem can be dismissed as tautologous, or (b) the proof is convincing, since the sufficiency principle is more immediately convincing than the likelihood principle itself. When interpreting this theorem, you should appreciate that the main substantive content is supplied by the Neyman–Fisher factorization theorem, which links the previously mercurial concept of likelihood to the essential frequency concept of sufficiency. If you do accept interpretation (b), standard frequency techniques (for example, UMVU estimation, confidence intervals, and hypothesis testing) are open to some conceptual discussion, since they do not concur with the likelihood principle (e.g., Exercise 1.5.f). Quite apart from Birnbaum's theorem, the likelihood principle is intrinsically appealing. See Exercises 1.5.c–1.5.f, Berger and Wolpert (1984), and Hill (1987).

Consider the normal random sampling Exercise 1.1.c. The expression (1.1.9) for the likelihood of the mean μ and variance λ may be interpreted, via the Neyman–Fisher factorization theorem, as telling us that $\mathbf{t}(\mathbf{y}) = (\bar{y}, s^2)$ is a two-dimensional sufficient statistic for (μ, λ). Therefore, any procedure for drawing inferences about (μ, λ) should depend upon $(\bar{y}, s^2)$, but not further upon $y_1, \ldots, y_n$, if the normal sampling model is correct. While there is no completely convincing theory to definitely prove that we should do so, adherents of the likelihood approach might go one step further and say that any inferential procedure for (μ, λ) should refer to the likelihood surface (1.1.9) and that techniques not relating to this surface must be erroneous. If you accept this quite appealing philosophy, then it makes analysis of data much simpler, since you just need to find a suitable model and sketch or summarize the likelihood function of the unknown parameters.

Worked Example 1G: *Sampling from a Gamma Distribution*

Observations $Y_1, Y_2, \ldots, Y_n$ are a random sample from the Gamma $G(\alpha, \beta)$ distribution, with density (1.1.11).

(a) Express the likelihood of $\boldsymbol{\theta} = (\alpha, \beta)^T$ in terms of the statistics

$$t = n^{-1} \sum_{i=1}^{n} \log y_i$$

and $\bar{y}$.

(b) Describe the matrix $-\partial^2 \log l(\boldsymbol{\theta} \mid \mathbf{y})/\partial(\boldsymbol{\theta}\boldsymbol{\theta}^T)$ of second derivatives, by reference to the digamma function $\Psi(\alpha) = \partial \log \Gamma(\alpha)/\partial\alpha$ and the trigamma function

$$\Psi^{(1)}(\alpha) = \frac{\partial^2 \log \Gamma(\alpha)}{\partial \alpha^2}.$$

Show that this matrix is the same as Fisher's information matrix for $\boldsymbol{\theta}$.

(c) Describe equations for the maximum likelihood estimators $\hat{\alpha}$ and $\hat{\beta}$ of α and β.

(d) Show how $\hat{\alpha}$ and $\hat{\beta}$ can be used to obtain approximate 99% confidence intervals for α and β.

Hint: The inverse of a positive definite 2×2 matrix of the form

$$\begin{pmatrix} a & c \\ c & b \end{pmatrix}$$

is

$$d^{-1}\begin{pmatrix} b & -c \\ -c & a \end{pmatrix},$$

where $d = ab - c^2 > 0$.

Model Answer 1G:

(a)

$$\begin{aligned} l(\boldsymbol{\theta} \mid \mathbf{y}) = l(\alpha, \beta \mid \mathbf{y}) &= \prod_{i=1}^{n} p(y_i \mid \alpha, \beta) \\ &= \frac{\beta^{n\alpha}}{\{\Gamma(\alpha)\}^n}\left(\prod_{i=1}^{n} y_i\right)^{\alpha-1} \exp\left\{-\beta\sum_{i=1}^{n} y_i\right\}. \end{aligned}$$

(b)

$$\begin{aligned} \log l(\boldsymbol{\theta} \mid \mathbf{y}) &= n\alpha \log\beta + (\alpha - 1)\sum_{i=1}^{n} \log y_i - \beta\sum_{i=1}^{n} y_i - n\log\Gamma(\alpha) \\ &= n\alpha\log\beta + n(\alpha - 1)t - n\beta\bar{y} - n\log\Gamma(\alpha). \end{aligned}$$

The first and second derivatives of the log-likelihood, with respect to α and β, are

$$\begin{aligned} \frac{\partial \log l}{\partial\alpha} &= n\log\beta + nt - n\Psi(\alpha), \\ \frac{\partial \log l}{\partial\beta} &= n\frac{\alpha}{\beta} - n\bar{y}, \\ \frac{\partial^2 \log l}{\partial\alpha^2} &= -n\Psi^{(1)}(\alpha), \\ \frac{\partial^2 \log l}{\partial\beta^2} &= -n\frac{\alpha}{\beta^2}, \\ \text{and} \quad \frac{\partial^2 \log l}{\partial\alpha\partial\beta} &= \frac{n}{\beta}. \end{aligned}$$

Consequently,

$$-\frac{\partial^2 \log l}{\partial(\boldsymbol{\theta}\boldsymbol{\theta}^T)} = n \begin{bmatrix} \Psi^{(1)}(\alpha) & -\beta^{-1} \\ -\beta^{-1} & \alpha/\beta^2 \end{bmatrix}.$$

This is also Fisher's information matrix for $\boldsymbol{\theta}$, since it does not depend upon the observations and therefore equals its own expectation.

(c) Setting the first derivatives equal to zero gives the maximum likelihood equations

$$\Psi(\hat{\alpha}) = t - \log \hat{\beta}$$

and

$$\frac{\hat{\alpha}}{\hat{\beta}} = \bar{y},$$

which should be solved iteratively for $\hat{\alpha}$ and $\hat{\beta}$.

(d) The observed information matrix is

$$\mathbf{R} = n \begin{bmatrix} \Psi^{(1)}(\hat{\alpha}) & -\hat{\beta}^{-1} \\ -\hat{\beta}^{-1} & \hat{\alpha}/\hat{\beta}^2 \end{bmatrix},$$

and its inverse is therefore

$$\mathbf{R}^{-1} = n^{-1}d^{-1} \begin{bmatrix} \hat{\alpha}/\hat{\beta}^2 & -\hat{\beta}^{-1} \\ -\hat{\beta}^{-1} & \Psi^{(1)}(\hat{\alpha}) \end{bmatrix},$$

where $d = \{\hat{\alpha}\Psi^{(1)}(\hat{\alpha})/\hat{\beta}^2\} - \hat{\beta}^{-2}$. Consequently an approximate 99% confidence interval for α is

$$\hat{\alpha} \pm 2.58 n^{-\frac{1}{2}} d^{-\frac{1}{2}} \hat{\alpha}^{1/2}/\hat{\beta},$$

and an approximate 99% confidence interval for β is

$$\hat{\beta} \pm 2.58 n^{-\frac{1}{2}} d^{-\frac{1}{2}} \{\Psi^{(1)}(\hat{\alpha})\}^{\frac{1}{2}}.$$

Worked Example 1H: *The Cramer–Rao and Blackwell–Rao Theorems*

The ideas in Section 1.5 (A) can be rephrased by saying that when $Y_1, Y_2, \ldots, Y_n$ comprise a random sample from a model with a single unknown parameter θ, and for the regular estimation problems of Section 1.3, the Cramer–Rao lower bound for the variances of unbiased estimators of θ is equal to $1/F_n(\theta)$, where $F_n(\theta)$ denotes Fisher's information. The lower bound is attained if and only if the first derivative of the log-likelihood $\log l(\theta \mid \mathbf{y})$ satisfies

$$\frac{\partial \log l(\theta \mid \mathbf{y})}{\partial \theta} = c(\theta)\{t(\mathbf{y}) - \theta\},$$

where $c(\theta)$ does not depend upon $\mathbf{y}$ and $t(\mathbf{y})$ does not depend upon θ. In such cases, $\hat{\theta}$ is both maximum likelihood and UMVU for θ, with $\text{var}(t(\mathbf{y})) = 1/c(\theta) = 1/F_n(\theta)$, and $\hat{\theta}$ is said to be *efficient* for θ. Furthermore, if this condition holds, a similar condition cannot hold for any nonlinear function $\phi = g(\theta)$ of θ.

Suppose that $y_1, y_2, \ldots, y_n$ is a random sample from the exponential distribution with parameter λ and density

$$f(y \mid \lambda) = \lambda e^{-\lambda y} \qquad (0 < y < \infty).$$

(a) Show that no bound-attaining (i.e., efficient) estimator for λ exists.

(b) Show that an efficient estimator for $\theta = \lambda^{-1}$ exists and find its variance.

(c) Blackwell–Rao theory (Lehmann, 1991, p. 80) tells us that within the one-parameter exponential family of the form described in Section 1.5 (A), any function of the sufficient statistic $T = n^{-1} \sum_{i=1}^{n} t(y_i)$ is UMVU for its expectation. When $n > 2$, find a UMVU estimator for λ. (Note that $n\bar{Y} \sim G(n, \lambda)$.)

Model Answer 1H:

(a) The likelihood of λ is

$$l(\lambda \mid \mathbf{y}) = \prod_{i=1}^{n} f(y_i \mid \lambda) = \lambda^n e^{-\lambda \sum_{i=1}^{n} y_i} = \lambda^n e^{-n\lambda\bar{y}},$$

so that

$$\log l(\lambda \mid \mathbf{y}) = n \log \lambda - n\lambda\bar{y}$$

and

$$\frac{\partial \log l(\lambda \mid \mathbf{y})}{\partial \lambda} = \frac{n}{\lambda} - n\bar{y},$$

which clearly cannot be arranged in the required form. Therefore no efficient estimator for λ exists.

(b)

$$\frac{\partial \log l}{\partial \theta} = \frac{\partial \lambda}{\partial \theta}\frac{\partial l}{\partial \lambda} = -\frac{1}{\theta^2}\left(\frac{n}{\lambda} - n\bar{y}\right) = \frac{n}{\theta^2}(\bar{y} - \theta).$$

Therefore $\bar{y}$ is efficient and hence UMVU for θ, with variance θ^2/n.

(c) Note that

$$p(y \mid \lambda) = \lambda e^{-\lambda y} = \exp\{\log \lambda - \lambda y\} \qquad (0 < y < \infty).$$

This takes the required form for the one-parameter exponential family, with $t(y) = y$. Hence any function of $T = \bar{Y}$ is UMVU for its expectation. Since $U = n\bar{Y} \sim G(n\lambda)$, the expectation of U^{-1} is

$$\begin{aligned} E\left(U^{-1}\right) &= \int_0^\infty u^{-1}\frac{1}{\Gamma(n)}\lambda^n u^{n-1}\exp\{-\lambda u\}\,du \\ &= \frac{\lambda^n}{\Gamma(n)}\int_0^\infty u^{n-2}\exp\{-\lambda u\}\,du \\ &= \frac{\lambda^n}{\Gamma(n)}\frac{\Gamma(n-1)}{\lambda^{n-1}} = \frac{\lambda}{n-1}. \end{aligned}$$

Consequently,

$$E\left(\frac{1}{n\bar{Y}}\right) = \frac{\lambda}{n-1}.$$

Therefore $T^* = (n-1)/n\bar{Y}$ is unbiased and hence UMVU for λ (though not efficient in the usual sense. Note that an inefficient estimator can be UMVU in situations where the Cramer–Rao lower bound is unattainable).

Worked Example 1I: *Sampling from the Weibull Distribution*

Let $Y_1, Y_2, \ldots, Y_n$ constitute a random sample from the Weibull distribution, with density

$$p(y \mid \alpha, \beta) = \beta y^{\alpha-1}\exp\left\{-\beta y^\alpha\right\} \qquad (0 < y < \infty).$$

(a) Find the likelihood of $\theta_1 = \log\alpha$ and $\theta_2 = \log\beta$, together with the first and second derivatives of the log-likelihood of $\boldsymbol{\theta} = (\theta_1, \theta_2)^T$, with respect to $\boldsymbol{\theta}$.

(b) In terms of the vector $b(\boldsymbol{\theta}) = \partial \log l(\boldsymbol{\theta} \mid \mathbf{y})/\partial\boldsymbol{\theta}$ and the matrix

$$B(\boldsymbol{\theta}) = -\frac{\partial^2 \log l(\boldsymbol{\theta} \mid \mathbf{y})}{\partial\left(\boldsymbol{\theta}\boldsymbol{\theta}^T\right)},$$

develop an iterative scheme for the computation of the maximum likelihood vector $\hat{\boldsymbol{\theta}}$ of $\boldsymbol{\theta}$, by reference to the Taylor series expansion (1.5.3).

(c) Describe, in terms of $\hat{\boldsymbol{\theta}}$, a bivariate normal approximation based upon (1.5.3) to the likelihood of $\boldsymbol{\theta}$.

Model Answer 1I:

(a) The likelihood of α and β is

$$l(\alpha, \beta \mid \mathbf{y}) = \prod_{i=1}^n p(y_i \mid \alpha, \beta) = \beta^n \exp\left\{n(\alpha-1)t - \beta\sum_{i=1}^n y_i^\alpha\right\},$$

where $t = n^{-1} \sum_{i=1}^{n} \log y_i$. Therefore, the likelihood of θ_1 and θ_2 is

$$l(\theta_1, \theta_2 \mid \mathbf{y}) = \exp\left\{ n\theta_2 + n\big(e^{\theta_1} - 1\big)t - e^{\theta_2} \sum_{i=1}^{n} y_i^{e^{\theta_1}} \right\}.$$

Consequently,

$$\log l(\theta_1, \theta_2 \mid \mathbf{y}) = n\theta_2 + n\big(e^{\theta_1} - 1\big)t - e^{\theta_2} \sum_{i=1}^{n} u_i,$$

where u_i takes y_i to the power $\exp(\theta_1)$. Then

$$\begin{aligned}
\frac{\partial \log l}{\partial \theta_1} &= ne^{\theta_1}t - e^{\theta_2} \sum_{i=1}^{n} \frac{\partial}{\partial \theta_1}\left[\exp\left\{ \log y_i e^{\theta_1} \right\}\right] \\
&= ne^{\theta_1}t - e^{\theta_1+\theta_2} \sum_{i=1}^{n} u_i \log y_i, \\
\frac{\partial \log l}{\partial \theta_2} &= n - e^{\theta_2} \sum_{i=1}^{n} u_i, \\
\frac{\partial^2 \log l}{\partial \theta_1^2} &= ne^{\theta_1}t - e^{\theta_1+\theta_2} \sum_{i=1}^{n} u_i \log y_i - e^{2\theta_1+\theta_2} \sum_{i=1}^{n} u_i (\log y_i)^2, \\
\frac{\partial^2 \log l}{\partial \theta_2^2} &= -e^{\theta_2} \sum_{i=1}^{n} u_i,
\end{aligned}$$

and $$\frac{\partial^2 \log l}{\partial \theta_1 \theta_2} = -e^{\theta_1+\theta_2} \sum_{i=1}^{n} u_i \log y_i.$$

(b) Let $\hat{\boldsymbol{\theta}}_n$ be the latest iterate for $\boldsymbol{\theta}$. We seek $\hat{\boldsymbol{\theta}}_{n+1}$. In terms of $\hat{\boldsymbol{\theta}}_n$, the log-likelihood of $\boldsymbol{\theta}$ is approximately (see equation 1.5.3)

$$\log l\big(\hat{\boldsymbol{\theta}}_n \mid \mathbf{y}\big) + \mathbf{b}^T\big(\hat{\boldsymbol{\theta}}_n\big)\big(\boldsymbol{\theta} - \hat{\boldsymbol{\theta}}_n\big) - \tfrac{1}{2}\big(\boldsymbol{\theta} - \hat{\boldsymbol{\theta}}_n\big)^T \mathbf{B}\big(\hat{\boldsymbol{\theta}}_n\big)\big(\boldsymbol{\theta} - \hat{\boldsymbol{\theta}}_n\big),$$

where $\mathbf{B}(\boldsymbol{\theta})$ in (1.5.4) comprises the negatives of the second derivatives developed in part (a). Differentiating with respect to $\boldsymbol{\theta}$ shows that this expression is maximized when

$$\boldsymbol{\theta} = \hat{\boldsymbol{\theta}}_n + \mathbf{B}^{-1}\big(\hat{\boldsymbol{\theta}}_n\big)\mathbf{b}\big(\hat{\boldsymbol{\theta}}_n\big)$$

whenever the inverse exists. Replacing $\boldsymbol{\theta}$ by $\hat{\boldsymbol{\theta}}_n$ gives

$$\hat{\boldsymbol{\theta}}_{n+1} = \hat{\boldsymbol{\theta}}_n + \mathbf{B}^{-1}\big(\hat{\boldsymbol{\theta}}_n\big)\mathbf{b}\big(\hat{\boldsymbol{\theta}}_n\big) \qquad (n = 0, 1, 2, \ldots).$$

Starting with $\hat{\boldsymbol{\theta}}_0$, we therefore have an iterative sequence (Newton–Raphson). Convergence towards the global optimum should be checked computationally.

(c) When $\boldsymbol{\theta}_0$ is replaced by $\hat{\boldsymbol{\theta}}$ and $\mathbf{b}(\hat{\boldsymbol{\theta}})$ is equal to the zero vector, neglecting cubic and higher terms in (1.5.3) gives the approximation

$$\log l(\hat{\boldsymbol{\theta}} \mid \mathbf{y}) - \tfrac{1}{2}(\boldsymbol{\theta} - \hat{\boldsymbol{\theta}})^T \mathbf{B}(\hat{\boldsymbol{\theta}})(\boldsymbol{\theta} - \hat{\boldsymbol{\theta}})$$

to the log-likelihood. Since $R = B(\hat{\boldsymbol{\theta}})$, taking exponentials shows that the likelihood of $\boldsymbol{\theta}$ is approximated by

$$l^*(\boldsymbol{\theta} \mid \mathbf{y}) = l(\hat{\boldsymbol{\theta}} \mid \mathbf{y}) \exp\left\{ -\tfrac{1}{2}(\boldsymbol{\theta} - \hat{\boldsymbol{\theta}})^T \mathbf{R}(\boldsymbol{\theta} - \hat{\boldsymbol{\theta}})\right\}.$$

Worked Example 1J: *Sampling from Multivariate Exponential Families*

(a) Let the $q \times 1$ observation vectors $\mathbf{Y}_1, \mathbf{Y}_2, \ldots, \mathbf{Y}_n$ comprise a random sample from the p-dimensional exponential family, with density

$$p(\mathbf{y} \mid \boldsymbol{\theta}) = \exp\left\{A(\boldsymbol{\theta}) + \boldsymbol{\theta}^T \mathbf{t}(\mathbf{y}) + \mathbf{B}(\mathbf{y})\right\} \qquad \left(\boldsymbol{\theta} \in R^p, \mathbf{y} \in R^q\right),$$

where $\boldsymbol{\theta}$ is a $p \times 1$ unknown vector of parameters, $A(\boldsymbol{\theta})$ does not depend upon $\mathbf{y}$, and $\mathbf{t}(\mathbf{y})$ and $B(\mathbf{y})$ do not depend upon $\boldsymbol{\theta}$. Show that the statistic $\mathbf{T} = n^{-1} \sum_{i=1}^n \mathbf{t}(\mathbf{y}_i)$ is sufficient for $\boldsymbol{\theta}$, where $\mathbf{y}_i$ denotes the observed $\mathbf{Y}_i$.

(b) Let $\mathbf{Y}_1, \mathbf{Y}_2, \ldots, \mathbf{Y}_n$ denote a random sample from the p-dimensional multivariate normal distribution, with unknown mean vector $\boldsymbol{\theta}$, unknown covariance matrix $\mathbf{C}$, and density

$$p(\mathbf{y} \mid \boldsymbol{\theta}, \mathbf{C}) = (2\pi)^{-\frac{1}{2}p} |\mathbf{C}|^{-\frac{1}{2}} \exp\left\{ -\tfrac{1}{2}(\mathbf{y} - \boldsymbol{\theta})^T \mathbf{C}^{-1}(\mathbf{y} - \boldsymbol{\theta})\right\}.$$

Find a sufficient statistic for $(\boldsymbol{\theta}, \mathbf{C})$.

Hint: $\sum_{i=1}^n (\mathbf{y}_i - \boldsymbol{\theta})(\mathbf{y}_i - \boldsymbol{\theta})^T = \mathbf{S} + n(\bar{\mathbf{y}} - \boldsymbol{\theta})(\bar{\mathbf{y}} - \boldsymbol{\theta})^T$, where $\mathbf{S} = \sum_{i=1}^n (\mathbf{y}_i - \bar{\mathbf{y}})(\mathbf{y}_i - \bar{\mathbf{y}})^T$. Also, $\mathbf{a}^T \mathbf{C}^{-1} \mathbf{a} = \operatorname{trace}(\mathbf{C}^{-1} \mathbf{a}\mathbf{a}^T)$ for any $p \times 1$ column vector $\mathbf{a}$.

Model Answer 1J:

(a) The likelihood of $\boldsymbol{\theta}$ is

$$\begin{aligned} l(\boldsymbol{\theta} \mid \mathbf{y}) &= \prod_{i=1}^n p(\mathbf{y}_i \mid \boldsymbol{\theta}) \\ &= \exp\left\{ nA(\boldsymbol{\theta}) + \boldsymbol{\theta}^T \mathbf{T}(\mathbf{y}) + \sum_{i=1}^n B(\mathbf{y}_i)\right\}, \end{aligned}$$

where $\mathbf{T}(\mathbf{y}) = n^{-1}\sum_{i=1}^{n}\mathbf{t}(\mathbf{y}_i)$. Therefore,

$$l(\boldsymbol{\theta} \mid \mathbf{y}) = h(\mathbf{y})g\big(\boldsymbol{\theta}, \mathbf{T}(\mathbf{y})\big),$$

where $h(\mathbf{y}) = \exp\left\{\sum_{i=1}^{n} B(\mathbf{y}_i)\right\}$, and

$$g\big(\boldsymbol{\theta}, \mathbf{T}(\mathbf{y})\big) = \exp\left\{nA(\boldsymbol{\theta}) + \boldsymbol{\theta}^T\mathbf{T}(\mathbf{y})\right\}.$$

Hence, by the Neyman–Fisher factorization theorem, $\mathbf{T} = \mathbf{T}(\mathbf{y})$ is sufficient for $\boldsymbol{\theta}$.

(b) The likelihood of $\boldsymbol{\theta}$ and $\mathbf{C}$ is

$$\begin{aligned} l(\boldsymbol{\theta}, \mathbf{C} \mid \mathbf{y}) &= \prod_{i=1}^{n} p(\mathbf{y}_i \mid \boldsymbol{\theta}, \mathbf{C}) \\ &= (2\pi)^{-\frac{1}{2}np}|\mathbf{C}|^{-\frac{1}{2}} \exp\left\{-\frac{1}{2}\sum_{i=1}^{n}(\mathbf{y}_i - \boldsymbol{\theta})^T\mathbf{C}^{-1}(\mathbf{y}_i - \boldsymbol{\theta})\right\}. \end{aligned}$$

Since $(\mathbf{y}_i - \boldsymbol{\theta})^T\mathbf{C}^{-1}(\mathbf{y}_i - \boldsymbol{\theta}) = \text{trace}\left[\mathbf{C}^{-1}(\mathbf{y}_i - \boldsymbol{\theta})(\mathbf{y}_i - \boldsymbol{\theta})^T\right]$,

$$\begin{aligned} \sum_{i=1}^{n}(\mathbf{y}_i - \boldsymbol{\theta})^T\mathbf{C}^{-1}(\mathbf{y}_i - \boldsymbol{\theta}) &= \text{trace}\left[\mathbf{C}^{-1}\sum_{i=1}^{n}(\mathbf{y}_i - \boldsymbol{\theta})(\mathbf{y}_i - \boldsymbol{\theta})^T\right] \\ &= \text{trace}\left[\mathbf{C}^{-1}\left\{\mathbf{S} + n(\bar{\mathbf{y}} - \boldsymbol{\theta})(\bar{\mathbf{y}} - \boldsymbol{\theta})^T\right\}\right]. \end{aligned}$$

Therefore, as a function of $\boldsymbol{\theta}$ and $\mathbf{C}$, $l(\boldsymbol{\theta}, \mathbf{C} \mid \mathbf{y})$ depends only upon data via $\bar{\mathbf{y}}$ and $\mathbf{S}$. Consequently, by the Neyman–Fisher factorization theorem, $(\bar{\mathbf{y}}, \mathbf{S})$ is sufficient for $(\boldsymbol{\theta}, \mathbf{C})$.

Worked Example 1K: *The Likelihood Principle*

Let $Y_1, Y_2, \ldots, Y_n$ comprise a random sample from a normal distribution whose mean θ and variance ϕ are both unknown. You may assume, without proof, that $\bar{Y}$ and $S^2 = \sum_{i=1}^{n}(Y_i - \bar{Y})^2$ are independent, where the sample mean $\bar{Y}$ is normally distributed with mean θ and variance ϕ/n, and S^2/ϕ has a chi-squared distribution with $n - 1$ degrees of freedom.

(a) Describe your likelihood for θ and ϕ when you are told only that $\bar{Y} = \bar{y} = 10.1$ and $S^2 = s^2 = 12.3$, with no further information about the observations, apart from the value $n = 150$ of n.

(b) Show that according to the likelihood principle, you should draw inferences in case (a) as if you had been given the values of all of the observations and then calculated $\bar{y} = 10.1$ and $s^2 = 12.3$. Discuss whether or not you believe that it is reasonable to neglect further information in this way.

Model Answer 1K:

(a) The density of $U = S^2/\phi$ is

$$p(u) = \frac{2^{-\frac{\nu}{2}}}{\Gamma\left(\frac{\nu}{2}\right)} u^{\frac{\nu}{2}-1} \exp\left\{-\frac{u}{2}\right\} \qquad (0 < u < \infty),$$

where $\nu = n - 1$. Consequently the density of S^2 is

$$p\left(s^2 \mid \phi\right) = \frac{2^{-\frac{\nu}{2}}}{\Gamma\left(\frac{\nu}{2}\right)} \phi^{-\frac{\nu}{2}} \left(s^2\right)^{\frac{\nu}{2}-1} \exp\left\{-\frac{s^2}{2\phi}\right\} \qquad \left(0 < s^2 < \infty\right).$$

Since $\bar{Y}$ and S^2 are independent, their joint density is

$$\begin{aligned} p\left(\bar{y}, s^2 \mid \theta, \phi\right) &= (2\pi)^{-\frac{1}{2}} \left(\frac{\phi}{n}\right)^{-\frac{1}{2}} \exp\left\{-\frac{1}{2} n \phi^{-1} (\bar{y} - \theta)^2\right\} p\left(s^2 \mid \phi\right) \\ &= (2\pi)^{-\frac{1}{2}} 2^{-\frac{1}{2}(n-1)} n^{\frac{1}{2}} \phi^{-\frac{1}{2}n} \left(s^2\right)^{-\frac{1}{2}(n+1)} \\ &\quad \times \exp\left[-\frac{1}{2}\phi^{-1}\left\{s^2 + n(\bar{y} - \theta)^2\right\}\right] \\ &\qquad \left(-\infty < \bar{y} < \infty, 0 < s^2 < \infty\right). \end{aligned}$$

For fixed $\bar{y}$ and s^2, this is also the likelihood of θ and ϕ, given $\bar{y}$ and s^2.

(b) Given $y_1, y_2, \ldots, y_n$, the likelihood of θ and ϕ is

$$\begin{aligned} l(\theta, \phi \mid \mathbf{y}) &= (2\pi)^{-\frac{1}{2}n} \phi^{-\frac{1}{2}n} \exp\left[-\tfrac{1}{2}\phi^{-1}\left\{s^2 + n(\bar{y} - \theta)^2\right\}\right] \\ &\qquad (-\infty < \theta < \infty, 0 < \phi < \infty). \end{aligned}$$

As a function of θ and ϕ, this is proportional to the likelihood in (a). Therefore, by the likelihood principle, we should draw the same inference about θ and ϕ in each case.

This is, of course, only valid in the hypothetical situation where we definitely can be sure that the model is true, that is, where the observations are indeed a random sample from a normal distribution. We might otherwise wish to use our individual observations to check the model or to decide whether or not to reject an outlying observation, and this might well affect our inferences for the mean and variance.

(D) *Profile Likelihood*

A $k \times 1$ vector of parameters possesses the likelihood $l(\boldsymbol{\theta} \mid \mathbf{y})$ for $\boldsymbol{\theta} \in \Theta$. Consider a real-valued parameter of interest $\eta = g(\boldsymbol{\theta})$. Then the *profile likelihood* of η is

$$l_p(\eta \mid \mathbf{y}) = \sup l(\boldsymbol{\theta} \mid \mathbf{y}) \qquad (\eta \in \mathcal{A}), \tag{1.5.7}$$

where the supremum should be taken over all $\theta \in \Theta$, satisfying $g(\boldsymbol{\theta}) = \eta$, and $\mathcal{A}$ is the space of all possible values of η. Sprott and Kalbfleisch (1969) and Kalbfleisch and Sprott (1970) recommend the curve (1.5.7) for drawing "marginal inferences" about the parameter of interest η.

In the special case where η denotes the kth element θ_k of $\boldsymbol{\theta}$, the profile likelihood of η is

$$l_p(\eta \mid \mathbf{y}) = \sup_{\theta_1,\ldots,\theta_{k-1}|\eta} l(\boldsymbol{\theta} \mid \mathbf{y}) \qquad (\eta \in \mathcal{A}). \tag{1.5.8}$$

Leonard (1982a), Barndorff-Nielsen (1983), and Butler (1986) recommend replacing (1.5.8) by the *modified profile likelihood*

$$l_M(\eta \mid \mathbf{y}) \propto |\mathbf{R}_\eta|^{-\frac{1}{2}} l_p(\eta \mid \mathbf{y}), \tag{1.5.9}$$

where $\mathbf{R}_\eta$ is the $(k-1) \times (k-1)$ matrix whose (j, l)th element evaluates

$$-\partial^2 \log l(\boldsymbol{\theta} \mid \mathbf{y})/\partial\theta_j \partial\theta_l$$

at conditional maximum of $l(\boldsymbol{\theta} \mid \mathbf{y})$ with respect to $\boldsymbol{\theta}$, given θ_k. Standard releases of the statistical packages *S* and *S*plus yield plots of profile likelihoods, for example, for nonlinear regression problems (Bates and Watts, 1988). Profile likelihoods may be compared with "partial likelihoods" (e.g., Wong, 1986) and appear to retain more information regarding the parameter of interest. Profile likelihoods may also be used in more complicated models where the nuisance parameter is an unknown function, for example, Cox's proportionate hazards model. See Severini and Wong (1992).

(E) *The Expectation Maximization (EM) Algorithm*

Let vectors $\mathbf{y}$ and $\mathbf{z}$ possess joint density $p(\mathbf{y}, \mathbf{z} \mid \boldsymbol{\theta})$, depending upon a $k \times 1$ vector $\boldsymbol{\theta}$ of unknown parameters. Assume, however, that $\mathbf{y}$ is observed, but $\mathbf{z}$ consists of a vector of missing observations. Then it is of interest to maximize, with respect to $\boldsymbol{\theta}$, the likelihood

$$l(\boldsymbol{\theta} \mid \mathbf{y}) = \int p(\mathbf{y}, \mathbf{z} \mid \boldsymbol{\theta})\, d\mathbf{z}, \tag{1.5.10}$$

where the integral should be taken over all possible realizations of $\mathbf{z}$.

Let $\boldsymbol{\theta}^{(r)}$ denote the rth realization of $\boldsymbol{\theta}$ in an iterative process, beginning with some initial vector $\boldsymbol{\theta}^{(0)}$. Then $\boldsymbol{\theta}^{(r+1)}$ may be obtained by maximizing

$$E \log p(\mathbf{y}, \mathbf{z} \mid \boldsymbol{\theta}), \tag{1.5.11}$$

with respect to $\boldsymbol{\theta}$, where the expectation should be taken with respect to the conditional distribution $p(\mathbf{z} \mid \mathbf{y})$ of $\mathbf{z}$, given $\mathbf{y}$, and $\boldsymbol{\theta} = \boldsymbol{\theta}^{(r)}$. This is the EM algorithm, as developed by Dempster, Laird, and Rubin (1977). Wu (1983) describes very general conditions under which the sequence $\boldsymbol{\theta}_1^{(1)}, \boldsymbol{\theta}_2^{(2)}, \ldots$ will converge to at least a local maximum. Lee (1997, p. 252) describes why the method works.

Consider those special cases where the conditional distribution of $\mathbf{y}$, given $\mathbf{z}$, does not depend upon $\boldsymbol{\theta}$, that is, $p(\mathbf{y}, \mathbf{z} \mid \boldsymbol{\theta})$ takes the form $p(\mathbf{y} \mid \mathbf{z})p(\mathbf{z} \mid \boldsymbol{\theta})$. If, furthermore, $\mathbf{z}$ possesses a density of the form

$$p(\mathbf{z} \mid \boldsymbol{\theta}) = \exp\left\{A(\boldsymbol{\theta}) + \mathbf{b}^T(\boldsymbol{\theta})\mathbf{t}(\mathbf{z}) + C(\mathbf{z})\right\}, \tag{1.5.12}$$

belonging to the k-parameter exponential family, where $\mathbf{t}(\mathbf{z})$ is a $k \times 1$ statistic that is sufficient for $\boldsymbol{\theta}$, within this family, then the above representation tells us that the maximum likelihood vector for $\boldsymbol{\theta}$ satisfies the equation

$$E[\mathbf{t}(\mathbf{z}) \mid \boldsymbol{\theta}] = E\left[\mathbf{t}(\mathbf{z}) \mid \boldsymbol{\theta}, \mathbf{y}\right], \tag{1.5.13}$$

where the expectation on the left-hand side should be taken with respect to the distribution of $\mathbf{z}$, given $\boldsymbol{\theta}$, and the expectation on the right-hand side should be regarded as conditional upon both the observation vector $\boldsymbol{\theta}$ and the vector of unknown parameter $\mathbf{y}$.

SELF-STUDY EXERCISES

1.5.a Let $Y_1, \ldots, Y_n$ denote a random sample from the Gamma $G(\gamma, \gamma)$ distribution, with density

$$p(y \mid \gamma) = \frac{1}{\Gamma(\gamma)}\gamma^\gamma y^{\gamma-1}\exp\{-\gamma y\} \qquad (0 < y < \infty, 0 < \gamma < \infty). \tag{1.5.14}$$

Show that this density belongs to the one-parameter exponential family. Find a UMVU estimator for $\phi = \Psi(\gamma) - \log\gamma$, where $\Psi(\gamma) = \partial \log\Gamma(\gamma)/\partial\gamma$, and find its variance. Show that your estimator is maximum likelihood for ϕ.

1.5.b *A Log-Linear Model for Variance Components:* Consider independently distributed sums of squares $S_1^2, \ldots, S_n^2$, with specified degrees of freedom $\nu_1, \ldots, \nu_n$, where, for $i = 1, \ldots, n$, S_i^2/ϕ_i possesses a chi-squared distribution with ν_i degrees of freedom. Let $\alpha_i = \log\phi_i$ and consider the model

$$\alpha_i = \mathbf{x}_i^T\boldsymbol{\beta} \qquad (i = 1, \ldots, n), \tag{1.5.15}$$

where the $\mathbf{x}_i$ are specified $p \times 1$ vectors and $\boldsymbol{\beta}$ is a $p \times 1$ vector of unknown parameters.

(a) Represent the likelihood of $\boldsymbol{\beta}$ in the form

$$l(\boldsymbol{\beta} \mid \mathbf{y}) \propto \exp\left\{\sum_{i=1}^n \nu_i\mathbf{x}_i^T\boldsymbol{\beta} - \frac{1}{2}\sum_{i=1}^n s_i^2 e^{-\mathbf{x}_i^T\boldsymbol{\beta}}\right\}, \tag{1.5.16}$$

where $s_1^2, \ldots, s_n^2$ denote the corresponding observed sum of squares. Find Fisher's information matrix $F_n(\beta)$ and the observed likelihood information matrix $\mathbf{R}$. Show that it is in general untrue that $\mathbf{R} = F_n(\hat{\boldsymbol{\beta}})$, where $\hat{\boldsymbol{\beta}}$ denotes the maximum likelihood vector of $\boldsymbol{\beta}$ (Johnson and Klotz, 1993, note that this result is true within a multivariate exponential family).

(b) Use an exponential series expansion to represent $\log l(\boldsymbol{\beta} \mid \mathbf{y})$ in a Taylor series about an arbitrary $\boldsymbol{\beta} = \boldsymbol{\beta}_0$. Use a second-order approximation to this expansion to give a multivariate normal approximation to the likelihood of $\boldsymbol{\beta}$. Also develop a computational scheme for $\hat{\boldsymbol{\beta}}$.

1.5.c The sufficiency principle (Birnbaum, 1962) tells us that "if $\mathbf{t} = \mathbf{t}(\mathbf{y})$ is sufficient for θ under experiment ϵ, with sampling model $p(\mathbf{y} \mid \boldsymbol{\theta})$, then whenever $\mathbf{t}(\mathbf{y}_1) = \mathbf{t}(\mathbf{y}_2)$,

$$\langle \epsilon, \mathbf{y}_1 \rangle \equiv \langle \epsilon, \mathbf{y}_2 \rangle,$$

in the sense that we should draw the same inference about θ whether $\mathbf{y}_1$ or $\mathbf{y}_2$ is observed. In other words, $\mathbf{y}_1$ and $\mathbf{y}_2$ are equivalent in evidential meaning." Show that the sufficiency principle can equivalently be stated as "for a single experiment ϵ, suppose that there are two realizations y_1 and y_2 of the observation vector $\mathbf{y}$ such that, as a function of $\boldsymbol{\theta}$, $p(\mathbf{y}_1 \mid \boldsymbol{\theta}) \propto p(\mathbf{y}_2 \mid \boldsymbol{\theta})$. Then $\mathbf{y}_1$ and $\mathbf{y}_2$ are equivalent in evidential meaning."

Give an example to illustrate your result. Is this principle reasonable? Note that the sampling model is assumed to be definitely true.

1.5.d In the notation of the previous exercise, the conditionality principle(Cox, 1958; Birnbaum, 1962) states that "if ϵ, and ϵ_2 are two experiments, with sampling models $p_1(\mathbf{y}_1 \mid \boldsymbol{\theta})$ and $p_2(\mathbf{y}_2 \mid \boldsymbol{\theta})$ for the same parameter $\boldsymbol{\theta}$, and $\epsilon = \gamma\epsilon_1 + (1-\gamma)\epsilon_2$ is the mixed experiment, which chooses between ϵ_1 and ϵ_2 with probabilities γ and $(1-\gamma)$, then

$$\langle \epsilon, (\epsilon_i, \mathbf{y}_i) \rangle \equiv \langle \epsilon_i, \mathbf{y}_i \rangle \qquad (i = 1, 2),$$

in the sense that if a randomization device (e.g., the toss of a coin) is used to decide between two straightforward experiments ϵ_1 and ϵ_2, then we should draw inferences about $\boldsymbol{\theta}$, given $\mathbf{y}$, as if our experiment ϵ_i had not been selected by a randomization scheme." Explain this principle in practical terms. Do you find it to be reasonable?

Read Cox (1958) and describe why fixed-size significance testing need not satisfy the conditionality principle. Consider

ϵ_1: Choose $n = 400$ people at random with replacement from the U.S. population and measure their I.Q.s,

and

ϵ_2: Choose $n = 10{,}000$ people at random with replacement from the U.S. population and measure their I.Q.s.

You are not sure which sample size to use, and so flip a coin to decide. Assume that your observations are a random sample upon a normal distribution with unknown mean and known variance $\lambda = 225$. Do you find the conditionality principle reasonable in this case? Fully discuss this in the context of fixed-size tests for $H_0 : \theta = 100$ versus $H_1 : \theta \neq 100$.

1.5.e *Birnbaum's Theorem:* The likelihood principle tells us that for the two experiments ϵ_1 and ϵ_2 in the previous exercise,

$$\langle \epsilon_1, \mathbf{y}_1 \rangle \equiv \langle \epsilon_2, \mathbf{y}_2 \rangle,$$

whenever, as a function of θ, $p_1(\mathbf{y}_1 \mid \theta) \propto p_2(\mathbf{y}_2 \mid \theta)$. Birnbaum's 1962 proof may be summarized as follows:

Proof of Birnbaum's theorem. Consider two simple experiments ϵ_1 and ϵ_2 yielding observations $\mathbf{y}_1$ and $\mathbf{y}_2$, such that as a function of $\boldsymbol{\theta}$,

$$p_1(\mathbf{y}_1 \mid \boldsymbol{\theta}) \propto p_2(\mathbf{y}_2 \mid \boldsymbol{\theta}),$$

where p_1 and p_2 are the specified sampling models associated with ϵ_1 and ϵ_2. Consider the mixed experiment $\epsilon = \frac{1}{2}\epsilon_1 + \frac{1}{2}\epsilon_2$, which selects ϵ_1 or ϵ_2, each with probability $\frac{1}{2}$. Then, under experiment ϵ,

$$l(\boldsymbol{\theta} \mid (\epsilon_i, \mathbf{y}_i)) = \tfrac{1}{2} p_i(\mathbf{y}_i \mid \boldsymbol{\theta}) \qquad (i = 1, 2),$$

so that by our initial assumption,

$$l(\boldsymbol{\theta} \mid (\epsilon_1, \mathbf{y}_1)) \propto l(\boldsymbol{\theta} \mid (\epsilon_2, \mathbf{y}_2)).$$

Consequently, by the sufficiency principle applied to the mixed experiment ϵ,

$$\langle \epsilon, (\epsilon_1, \mathbf{y}_1) \rangle \equiv \langle \epsilon, (\epsilon_2, \mathbf{y}_2) \rangle.$$

Therefore, by the conditionality principle,

$$\langle \epsilon_1, y_1 \rangle \equiv \langle \epsilon_2, y_2 \rangle. \qquad \blacksquare$$

Do you think that this proof provides an overwhelming justification of the likelihood principle? Do you think that the sufficiency principle is simple to accept when applied to a mixed experiment? Fully discuss this in the context of Exercise 1.5.d.

1.5.f Consider a coin with unknown probability of heads θ, and the following three experiments:

$\mathcal{E}_1$: The coin is independently tossed a fixed number $n = 100$ times, and you observe $y = 55$ heads.

$\mathcal{E}_2$: The coin is independently tossed until you observe a fixed number $r = 55$ heads, and you then observe that $n = 100$ tosses have occurred.

$\mathcal{E}_3$: You are independently tossing the coin indefinitely when your spouse comes in with two glasses of Saki. At this point, you notice that you have observed $r = 55$ heads out of $n = 100$ tosses.

By reference to the binomial and negative binomial distributions, show that $\mathcal{E}_1$ and $\mathcal{E}_2$ yield likelihoods for θ that are proportional as functions of θ. Therefore, according to the likelihood principle, you should draw the same inferences about θ in each case. Show, however, that for $\mathcal{E}_1$, the usual unbiased estimate for θ is $55/100$, whereas for $\mathcal{E}_2$ this is $44/99$. Does this concern you?

How would you draw inferences about θ under $\mathcal{E}_3$? Does the stopping rule concern you? If your spouse brought just one glass of Saki, then this would affect your sampling distribution. Should this change your inference about θ?

1.5.g Let θ possess a likelihood proportional to $\theta^y(1-\theta)^{n-y}$. Show that $\lambda = \log\theta - \log(1-\theta)$ possesses a likelihood proportional to $\exp\{\lambda y\}/(1+e^{\lambda})^n$. Show, as a special case of the approximation (1.3.11), that this can be replaced by a normal likelihood with location $\hat{\lambda} = \log(y/(n-y))$ and dispersion $= 1/$ (likelihood information) $= y^{-1} + (n-y)^{-1}$. By expanding the log-likelihood of λ in a Taylor series about $\lambda^* = \log[(y+\frac{1}{2})/(n-y+\frac{1}{2})]$, find an alternative likelihood approximation.

Use m applications of this approximation to find algebraically explicit approximations to the maximum likelihood vector $\hat{\boldsymbol{\beta}}$ and precision matrix $\mathbf{R}$ for the linear logistic model of Exercise 1.3.c.

1.5.h Let $Y_1, Y_2, \ldots, Y_n$ comprise a random sample from the k-parameter exponential family, with density

$$p(y \mid \boldsymbol{\theta}) = \exp\left\{\theta_1 t_1(y) + \cdots + \theta_k t_k(y) - \log D(\boldsymbol{\theta})\right\} \qquad \left(y \in R, \boldsymbol{\theta} \in R^k\right), \tag{1.5.17}$$

where $\boldsymbol{\theta} = (\theta_1, \ldots, \theta_k)^T$ is unknown and $t_1(y), \ldots, t_k(y)$ are specified, for example, they could comprise a basis of B-splines with the linear term in the argument of (1.5.17) denoting a general cardinal spline with p knots (see Atilgan and Bozdogan, 1990), and

$$D(\boldsymbol{\theta}) = \int_{R'} \exp\left\{\theta_1 t_1(y) + \theta_2 t_2(y) + \cdots + \theta_k t_k(y)\right\} dy. \tag{1.5.18}$$

Find a k-dimensional sufficient statistic for $\boldsymbol{\theta}$. Show that when $t_j(y) = y^j$, for $j = 1, \ldots, k$, the first k sample moments comprise a sufficient statistic. Suggest how to choose k.

(The p-parameter exponential family is frequently applied. See Stone, 1991, for some developments. A discrete version, generalizing the binomial distribution, is described by Hsu, Leonard, and Tsui, 1991. It can be justified via Boltzmann's maximum entropy theorem; see Section 3.4. Arnold, 1991, considers multivariate exponential families.)

1.5.i *The Skewed Normal Distribution* (Edgeworth, 1899; Leonard, 1980; and Fernandez and Steel, 1998): Let $Y_1, Y_2, \ldots, Y_n$ denote a random sample from the distribution with density

$$f(y \mid \mu, \sigma_1^2, \sigma_2^2) = \begin{cases} \left\{\frac{2}{\pi(\sigma_1+\sigma_2)^2}\right\}^{\frac{1}{2}} \exp\left\{-\frac{1}{2\sigma_1^2}(y-\mu)^2\right\} & \text{for } -\infty < y \le \mu \\ \left\{\frac{2}{\pi(\sigma_1+\sigma_2)^2}\right\}^{\frac{1}{2}} \exp\left\{-\frac{1}{2\sigma_2^2}(y-\mu)^2\right\} & \text{for } \mu \le y < \infty. \end{cases} \tag{1.5.19}$$

Show that the profile likelihood for the mode μ, when σ_1^2 and σ_2^2 are also unknown, is

$$l_p(\mu \mid \mathbf{y}) \propto \left[S_1(\mu) + S_2(\mu)\right]^{-\frac{3}{2}n} \qquad (-\infty < \mu < \infty), \tag{1.5.20}$$

where

$$S_1(\mu) = \left[\sum_{i:y_i \le \mu} (y_i - \mu)^2\right]^{\frac{1}{3}} \tag{1.5.21}$$

and

$$S_2(\mu) = \left[\sum_{i:y_i > \mu} (y_i - \mu)^2\right]^{\frac{1}{3}}. \tag{1.5.22}$$

Show that the maximum likelihood estimates of the skewness parameter $\rho = \sigma_1/(\sigma_1 + \sigma_2)$ = probability to the left of μ and the scale parameter $\sigma^2 = (\sigma_1 + \sigma_2)^2$ are

$$\hat{\rho} = \frac{S_1(\hat{\mu})}{S_1(\hat{\mu}) + S_2(\hat{\mu})} \tag{1.5.23}$$

and

$$\hat{\sigma}^2 = n^{-1}\{S_1(\hat{\mu}) + S_2(\hat{\mu})\}^2, \tag{1.5.24}$$

where $\hat{\mu}$ maximizes (1.5.20). Note that the moment estimate

$$\hat{\mu}^* = \bar{y} - \sqrt{\frac{2}{\pi}}\hat{\sigma}(1 - 2\hat{\rho}) \tag{1.5.25}$$

provides an initial approximation to $\hat{\mu}$.

1.5.j *A Box–Tiao Robustifying Distribution* (see Box and Tiao, 1973): Let $Y_1, Y_2, \ldots, Y_n$ denote a random sample from the normal mixture, with density

$$p(y \mid \mu, \phi, \sigma_1^2, \sigma_2^2) = \phi\Psi_y(\mu, \sigma_1^2) + (1 - \phi)\Psi_y(\mu, \sigma_2^2), \tag{1.5.26}$$

where $\Psi_y(\mu, \sigma^2)$ denotes a normal density with mean μ and variance σ^2.

This model is probabilistically equivalent to considering missing observations $Z_1, \ldots, Z_n$, which are independent and binary, with common probability ϕ, and such that Y_i is normally distributed with mean μ and variance σ_1^2 if $Z_i = 1$, but possesses variance σ_2^2 if $Z_i = 0$. Find the joint distribution of the pair (Y_i, Z_i) and show that Z_i, given that $Y_i = y_i$, is binary, with probability

$$\phi_i^* = \frac{\phi\Psi_{y_i}(\mu, \sigma_1^2)}{\phi\Psi_{y_i}(\mu, \sigma_1^2) + (1 - \phi)\Psi_{y_i}(\mu, \sigma_2^2)}. \tag{1.5.27}$$

Hence, use the EM algorithm to develop an iterative scheme for the computation of the maximum likelihood estimates of ϕ, μ, σ_1^2, and σ_2^2. Note that Newton–Raphson is quite inefficient for estimating ϕ and that it is very difficult to transform the likelihood function to approximate multivariate normality. (See Titterington, Smith, and Makov, 1985, for a review of mixture methods, and George, 1984, for applications in economics.)

1.5.k Let the current observation vector $\mathbf{y}$ and a future observation vector $\mathbf{z}$ possess joint density $p(\mathbf{y}, \mathbf{z} \mid \boldsymbol{\theta})$. Then the *predictive likelihood* of $\mathbf{z}$ given $\mathbf{y}$ (see Hinkley, 1979; Lejeune and Faulkenberry, 1982) is

$$p^*(\mathbf{z} \mid \mathbf{y}) = \sup_{\theta \mid \mathbf{y}, \mathbf{z}} p(\mathbf{y}, \mathbf{z} \mid \boldsymbol{\theta}). \tag{1.5.28}$$

Moreover, the modified predictive likelihood of $\mathbf{z}$ given $\mathbf{y}$ (see Leonard, 1982a; Davison, 1986; Butler, 1986; Leonard, Tsui, and Hsu, 1990) is

$$\tilde{p}(\mathbf{z} \mid \mathbf{y}) = p^*(\mathbf{z} \mid \mathbf{y})|\mathbf{R}(\mathbf{y}, \mathbf{z})|^{-\frac{1}{2}}, \tag{1.5.29}$$

where

$$\mathbf{R}(\mathbf{y}, \mathbf{z}) = -\frac{\partial^2 \log p(\mathbf{y}, \mathbf{z} \mid \boldsymbol{\theta})}{\partial(\boldsymbol{\theta}\boldsymbol{\theta}^T)} \tag{1.5.30}$$

is evaluated at the realization of $\boldsymbol{\theta}$, which achieves the supremum (1.5.28), given $\mathbf{y}$ and $\mathbf{z}$. The parametrization should be chosen so that $l(\boldsymbol{\theta} \mid \mathbf{y}, \mathbf{z})$ is approximately multivariate normal in $\boldsymbol{\theta}$. Suppose that $\mathbf{y}$ and $\mathbf{z}$ are independent, given θ, and possess binomial distributions with common probability θ and sample sizes n and m. Find the predictive likelihood of $\mathbf{z}$, given $\mathbf{y}$, using the original parametrization. Find the modified predictive likelihood after transforming to $\lambda = \log\theta - \log(1 - \theta)$.

1.5.l *Residual Analysis for the Linear Logistic Model:* Under the assumptions and notation of Exercise 1.3.c, consider cases where none of the y_i or $n_i - y_i$ are less than 5, and let

$$l_i = \log \frac{y_i + \frac{1}{2}}{n_i - y_i + \frac{1}{2}} - \left(y_i + \frac{1}{2}\right)^{-1} + \left(n_i - y_i + \frac{1}{2}\right)^{-1} \tag{1.5.31}$$

and

$$v_i = \left(y_i + \frac{1}{2}\right)^{-1} + \left(n_i - y_i + \frac{1}{2}\right)^{-1}. \tag{1.5.32}$$

Define the *ith residual* by

$$r_i = l_i - \mathbf{x}_i^T \hat{\boldsymbol{\beta}}, \tag{1.5.33}$$

where $\hat{\boldsymbol{\beta}}$ is calculated by Newton–Raphson (see Worked Example 1I), and define the *ith residual variance* by

$$u_i = v_i - \mathbf{x}_i^T \mathbf{R}^{-1} \mathbf{x}_i, \tag{1.5.34}$$

where $\mathbf{R}$ is the exact likelihood information matrix.

(a) *Research level:* Use the developments of Exercise 1.5.g to show that when the model specification including (1.3.17) is true, the distribution of $r_i/u_i^{1/2}$ is approximately standard normal.

(b) Discuss how the result in (a) can be used in practice to investigate the adequacy of a particular model specification of the form (1.3.17). Show, using the approximations of Exercise 1.5.g, that u_i will be positive whenever these approximations are accurate.

(c) Show that as an overall check, AIC would prefer a model of the form (1.3.17) to the model that does not put any constraints on the binomial logits α_i whenever

$$\begin{aligned} \mathbf{t}^T \hat{\boldsymbol{\beta}} &- \sum_{i=1}^{m} n_i \log\left(1 + e^{\mathbf{x}_i^T \hat{\boldsymbol{\beta}}}\right) \\ &- \sum_{i=1}^{m} n_i \left\{p_i \log p_i + (1 - p_i)\log(1 - p_i)\right\} \\ &> m - p, \end{aligned}$$

where $p_i = y_i/n_i$.

Note: Optimal design for the linear logistic model is discussed by Chaloner and Larntz (1989). An interesting diagnostic procedure for related models based upon information theory and entropy is discussed by Soofi (1992). Sequential design measures are proposed by Leonard and Hsu (1996).

1.6 Practical Examples

(A) *Estimating a Gamma Parameter*

For the situation discussed in Section 1.5 (A), we generated $M = 100$ random samples, each of size $n = 5$, from a Gamma $G(\theta, 1)$ density, with $\theta = 5$. The maximum likelihood estimator $\hat{\theta}$ possessed average bias 0.150 and mean squared error 0.932, which should be compared with the larger mean squared error of the unbiased estimator $\theta^* = \bar{y} = 1.023$. This illustrates that although biased for small samples, the maximum likelihood estimator can still perform well when n is finite, with respect to mean squared error.

(B) *Profile Likelihoods for the Physics Data*

The data in Table 1.6.1 describe, with $n = 10$, the regression of the scattering cross section y, against the reciprocal of the total energy x, for the physics data reported by Weisberg (1980, p. 85). The $w_i^{-1/2}$ in the fourth column denote relative weights, suggested by scientific consideration, for the standard errors of the observations. We analyze these data under the modeling assumptions

$$y_i = \beta_1 + \beta_2 x_i + \epsilon_i \qquad (i = 1, 2, \ldots, n), \tag{1.6.1}$$

where the ϵ_i are independent and normally distributed error terms with zero means and error terms $w_i^{-1}\sigma^2$. This model possesses three unknown parameters β_1, β_2, and σ^2, with likelihood

$$l(\beta_1, \beta_2, \sigma^2 \mid \mathbf{y}) \propto (\sigma^2)^{-\frac{1}{2}n} \exp\left\{-\frac{1}{2\sigma^2}\sum_{i=1}^{n} w_i(y_i - \beta_1 - \beta_2 x_i)^2\right\}. \tag{1.6.2}$$

Using the ideas of Section 1.5 (D), we find that the profile likelihoods of β_1, β_2, and σ^2 are, respectively,

$$l_p(\beta_1 \mid \mathbf{y}) \propto \left[\sum_{i=1}^{n} w_i\{y_i - \beta_1 - \hat{\beta}_2(\beta_1)x_i\}^2\right]^{-\frac{1}{2}n}, \tag{1.6.3}$$

$$l_p(\beta_2 \mid \mathbf{y}) \propto \left[\sum_{i=1}^{n} w_i\{y_i - \hat{\beta}_1(\beta_2) - \beta_2 x_i\}^2\right]^{-\frac{1}{2}n}, \tag{1.6.4}$$

and

$$l_p(\sigma^2 \mid \mathbf{y}) \propto (\sigma^2)^{-\frac{1}{2}n} \exp\left\{-\frac{1}{2\sigma^2}\sum_{i=1}^{n} w_i(y_i - \beta_1^* - \beta_2^* x_i)^2\right\}, \tag{1.6.5}$$

where

$$\beta_1^* = \frac{\sum_{i=1}^{n} w_i y_i - \beta_2^* \sum_{i=1}^{n} w_i x_i}{\sum_{i=1}^{n} w_i}, \tag{1.6.6}$$

$$\beta_2^* = \frac{(\sum_{i=1}^{n} w_i)(\sum_{i=1}^{n} w_i x_i y_i) - (\sum_{i=1}^{n} w_i y_i)(\sum w_i x_i)}{(\sum_{i=1}^{n} w_i)(\sum_{i=1}^{n} w_i x_i^2) - (\sum_{i=1}^{n} w_i x_i)^2}, \tag{1.6.7}$$

Table 1.6.1. *Data for physics example.*

i	x_i	y_i	$w_i^{-1/2}$
1	0.345	367	17
2	0.287	311	9
3	0.251	295	9
4	0.225	268	7
5	0.207	253	7
6	0.186	239	6
7	0.161	220	6
8	0.132	213	6
9	0.084	193	5
10	0.060	192	5

$\hat{\beta}_1(\beta_2)$ replaces β_2^* in (1.6.6) by β_2, and

$$\hat{\beta}_2(\beta_1) = \frac{\sum_{i=1}^n w_i x_i y_i - \beta_1 \sum_{i=1}^n w_i x_i}{\sum_{i=1}^n w_i x_i^2}. \tag{1.6.8}$$

Figure 1.6.1 reports the profile likelihood of the slope β_2 and clearly indicates that β_2 is nonzero. Figure 1.6.2 reports the predictive likelihood of y, when $x = 0.060$, using the techniques of Exercise 1.5.k.

(C) *The Regression of Gas Upon Electricity Consumption*

Likelihood methods can also be used to develop nonlinear regression models for data that are too noisy for standard linear model techniques. Consider, for example, the data reported by Tukey (1977, p. 267) and relating to the regression of gas use upon

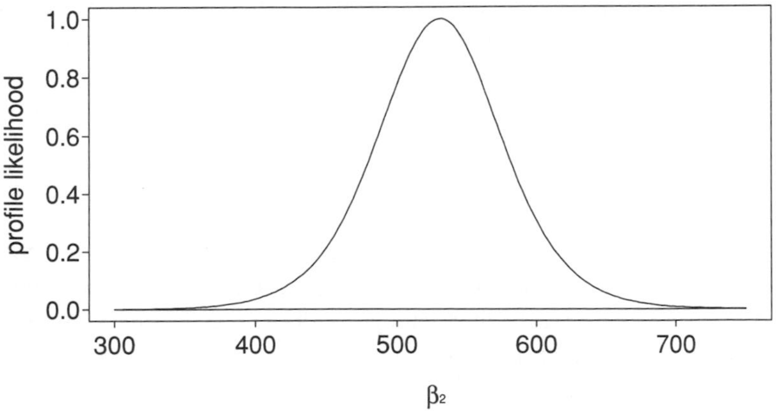

Figure 1.6.1. Profile likelihood of slope parameter.

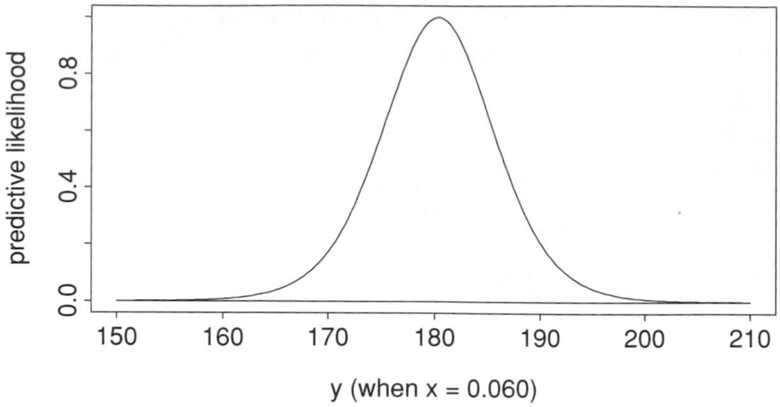

Figure 1.6.2. Predictive likelihood of future observation.

electricity use for $n = 152$ townhouses in New Jersey. Atilgan and Leonard (1987) fitted a mixture of bivariate normal densities of the form

$$f(x, y) = \sum_{i=1}^{m} \phi_i \Psi_{x,y}(\boldsymbol{\mu}_i, \mathbf{C}_i) \tag{1.6.9}$$

to these data, where $\Psi_{x,y}(\boldsymbol{\mu}, \mathbf{C})$ denotes a bivariate normal density with mean vector $\boldsymbol{\mu}$ and covariance matrix $\mathbf{C}$. However, the mean vectors $\boldsymbol{\mu}_1, \ldots, \boldsymbol{\mu}_m$ were specified as $m = p^2$ points on an equally spaced $p \times p$ grid. The x-coordinates of the grid were located at the points $x_{(1)} + (p+1)^{-1} j(x_{(1)} - x_{(n)})j$ for $i = 1, \ldots, p$, and the y-coordinates were located at the points $y_{(1)} + (p+1)j(y_{(n)} - y_{(1)})$ for $j = 1, \ldots, p$, where $x_{(1)}$ and $x_{(n)}$ are the smallest and largest observed electricity consumptions, and $y_{(1)}$ and $y_{(n)}$ are the smallest and largest gas consumptions. The parameters $\phi_1, \ldots, \phi_m$ and $\mathbf{C}_1, \ldots, \mathbf{C}_m$ were then estimated by the EM algorithm (see Section 1.5 (E) and Exercise 1.5.j). The value of $m = p^2$ was chosen by maximizing AIC, defined in (1.1.5), suggesting $m = 25$ and $p = 5$. The conditional expectation with respect to the Y variable, conditional on x, as calculated from this mixture, provides a semiparametric procedure for fitting the regression of y on x. Atilgan and Leonard (1987) compared this curve with Tukey's original data.

1.7 The Midcontinental Rift

Nyquist (1986) and Nyquist and Wang (1988) report ultrasonic measurements of the depth in kilometers of the midcontinental drift over a cross-sectional range of 150 kilometers along the Saint Croix river valley, on the border between Minnesota and Wisconsin. Practical constraints precluded measurements deeper than 7 or 8 kilometers. However, it was of substantial interest to estimate ω, the maximum depth of the rift, together with a flexural parameter α. Geological theory, relating to the bending of beams, suggested that under an "unbroken plate" model, the depth y should be related to the range x by the equation

$$y = \omega\phi(x, \alpha), \tag{1.7.1}$$

where

$$\phi(x, \alpha) = \left(\sin\frac{x}{\alpha} + \cos\frac{x}{\alpha}\right)\exp\left(-\frac{x}{\alpha}\right), \tag{1.7.2}$$

but that for a "broken plate" model, (1.7.2) should be replaced by

$$\phi(x, \alpha) = \left(\cos\frac{x}{\alpha}\right)\exp\left(-\frac{x}{\alpha}\right).$$

The observed depths $y_1, \ldots, y_n$ were taken to be independent and normally distributed with respective means $\mu_1, \ldots, \mu_n$ and common variance σ^2, where

$$\mu_i = \omega\phi(x_i, \alpha) \qquad (i = 1, \ldots, n), \tag{1.7.3}$$

and $x_1, \ldots, x_n$ are the observed ranges. This yields a nonlinear regression model with three unknown parameters α, ω, and σ^2, and likelihood

$$l\big(\alpha, \omega, \sigma^2 \mid \mathbf{y}\big) \propto \left(\sigma^2\right)^{-\frac{1}{2}} \exp\left\{-\frac{1}{2\sigma^2}\sum_{i=1}^{n}\big(y_i - \omega\phi(x_i, \alpha)\big)^2\right\}. \tag{1.7.4}$$

Consequently, the maximum likelihood estimate of ω, given α, is

$$\hat{\omega}(\alpha) = \frac{\sum_{i=1}^{n} y_i\phi(x_i, \alpha)}{\sum_{i=1}^{n} \phi^2(x_i, \alpha)}, \tag{1.7.5}$$

and the maximum likelihood estimate of σ^2, given ω and α, is

$$\hat{\sigma}^2 = \frac{1}{n}\sum_{i=1}^{n}\{y_i - \omega\phi(x_i, \alpha)\}^2. \tag{1.7.6}$$

Therefore, the profile likelihood of α satisfies

$$l_p(\alpha \mid \mathbf{y}) \propto \left[\sum_{i=1}^{n}\{y_i - \hat{\omega}(\alpha)\phi(x_i, \alpha)\}^2\right]^{-\frac{n}{2}} \qquad (0 < \alpha < \infty). \tag{1.7.7}$$

The log of this profile likelihood is reported by Nyquist and Wang for both the broken plate model and the unbroken plate model. The modified profile likelihood (1.5.9) was also considered. This possesses the advantage of yielding distinctly thinner tails. The profile likelihood can be numerically maximized to obtain the maximum likelihood estimate $\hat{\alpha}$ of α. Appropriate substitutions into (1.7.5) and (1.7.6) give the maximum likelihood estimates $\hat{\omega}$ and $\hat{\sigma}^2$ of ω and σ^2. The fitted regression curve $y = \hat{\omega}\phi(x, \hat{\alpha})$ is then used by Nyquist and Wang to estimate the depth of the ridge.

1.8 A Model for Genetic Traits in Dairy Science

Foulley, Gianola, and Im (1990) introduce a model for the genetic evaluation of traits where the observed counts $y_1, \ldots, y_n$ are assumed to be independent and Poisson distributed given their respective means $\mu_1, \ldots, \mu_n$. It is moreover assumed that the $\gamma_i = \log \mu_i$ satisfy

$$\gamma_i = \mathbf{x}_i^T \boldsymbol{\beta} + \mathbf{z}_i^T \boldsymbol{\alpha} \qquad (i = 1, \ldots, n), \tag{1.8.1}$$

where the $\mathbf{x}_i$ and $\mathbf{z}_i$ are specified $p \times 1$ and $q \times 1$ design vectors, and $\boldsymbol{\beta}$ is a $p \times 1$ vector of unknown "fixed effects." Moreover, $\boldsymbol{\alpha}$ is a $q \times 1$ vector of "random effects," which is taken to possess a multivariate normal distribution with zero mean vector and covariance matrix $\sigma^2 \mathbf{A}$. Then σ^2 is the unknown additive genetic variance, and $\mathbf{A}$ is a specified matrix of additive relationships. Tempelman (1993) applied our techniques to this model, using the Laplacian methods of Chapters 5 and 6, and obtained an approximation to the marginal likelihood of σ^2. A variety of Laplacian techniques can be applied to nonlinear models in many disciplines.

In the context of the model in (1.8.1), note the importance of obtaining the approximations in the context of a reduced form of the model, with number of parameters much less than n. Shun and McCullagh (1995) demonstrate difficulties with Laplacian approximations if the number of parameters increases with the number of observations.

1.9 Least Squares Regression with Serially Correlated Errors

Let an $n \times 1$ random vector $\mathbf{y}$ possess a multivariate $N(\mathbf{X}\boldsymbol{\beta}, \mathbf{C})$ distribution, where $\mathbf{X}$ is a specified $n \times p$ matrix, with linearly independent columns, $\boldsymbol{\beta}$ is a $p \times 1$ vector of unknown parameters, and $\mathbf{C} = e^{\gamma_1} \mathbf{V}^{\gamma_2}$, where $\mathbf{V}$ is a specified positive definite matrix, and γ_1 and γ_2 are further unknown parameters. Then, the likelihood of $\boldsymbol{\beta}$, γ_1, and γ_2, given $\mathbf{y}$, is

$$l(\boldsymbol{\beta}, \gamma_1, \gamma_2 \mid \mathbf{y}) \propto \exp\left[-\frac{1}{2} n \gamma_1 - \frac{1}{2} \gamma_2 \sum_{i=1}^{n} \log d_i - \frac{1}{2} e^{-\gamma_1} \sum_{i=1}^{n} e^{-d_i \gamma_2} \left\{ \mathbf{e}_i^T (\mathbf{y} - \mathbf{X}\boldsymbol{\beta}) \right\}^2 \right], \tag{1.9.1}$$

where $\mathbf{V}^{\gamma_2} = \sum_i^n d_i \mathbf{e}_i \mathbf{e}_i^T$, with the d_i representing the eigenvalues of $\mathbf{V}$, and the $\mathbf{e}_i$ denoting the corresponding normalized eigenvectors. Consequently, the maximum likelihood estimator of $\boldsymbol{\beta}$, given γ_2, is the weighted least squares vector

$$\hat{\boldsymbol{\beta}} = \left(\mathbf{X}^T \mathbf{V}^{-\gamma_2} \mathbf{X} \right)^{-1} \mathbf{X}^T \mathbf{V}^{-\gamma_2} \mathbf{y}, \tag{1.9.2}$$

where $\mathbf{V}^{-\gamma_2} = \sum_i^n d_i^{-\gamma_2} \mathbf{e}_i \mathbf{e}_i^T$, and the maximum likelihood estimator $\hat{\gamma}_1$ of γ_1, given γ_2, satisfies

$$e^{\hat{\gamma}_1} = n^{-1} \mathbf{y}^T \mathbf{H}(\gamma_2) \mathbf{y}, \tag{1.9.3}$$

where

$$\mathbf{H}(\gamma_2) = \mathbf{V}^{-\gamma_2} - \mathbf{V}^{-\gamma_2}\mathbf{X}\left(\mathbf{X}^T\mathbf{V}^{-\gamma_2}\mathbf{X}\right)^{-1}\mathbf{X}^T\mathbf{V}^{-\gamma_2}. \tag{1.9.4}$$

The maximum likelihood estimate $\hat{\gamma}_2$ of γ_2 may be obtained by consideration of a numerical plot of the profile likelihood

$$l_p(\gamma_2 \mid \mathbf{y}) \propto \left\{\mathbf{y}^T\mathbf{H}(\gamma_2)\mathbf{y}\right\}^{-\frac{1}{2}n} \exp\left\{-\frac{1}{2}\gamma_2 \sum_{i=1}^{n} \log d_i\right\}. \tag{1.9.5}$$

Leonard and Hsu (1992) introduce the matrix logarithmic and matrix exponential transformation in a Bayesian context, and Chiu, Leonard, and Tsui (1996) use these transformations to develop a generalized linear model for covariance matrices. Previous applications have been confined to mathematical physics (e.g., Bellman, 1971, p. 226). For the current model, $\mathbf{A} = \log \mathbf{C} = \gamma_1 \mathbf{I}_n + \gamma_2 \mathbf{U}$, where $\mathbf{U} = \log \mathbf{V} = \sum_{i=1}^{n} (\log d_i)\mathbf{e}_i\mathbf{e}_i^T$. Note that the logarithmic matrix transformation of a positive definite matrix takes the logs of the eigenvalues and leaves the normalized eigenvectors unchanged.

When the data are nonrandomized, and serially correlated errors may be present, and there is just one set of explanatory variables, that is, $E(y_k) = \mathbf{x}_k^T \boldsymbol{\beta}$ for $k = 1, \ldots, n$, where each $\mathbf{x}_k$ vector depends upon just one scalar x_k, consider taking the (i, j)th element of $\mathbf{U}$ to be $u_{ij} = -(x_i - x_j)^2$. Then $\mathbf{C} = e^{\gamma_1} \exp\{\gamma_2 \mathbf{U}\}$ can be used to model serial correlation; for example, if $\gamma_2 = 0$, then no serial correlation is present. Furthermore, our likelihood procedure does not depend upon the inversion of any high-dimensional matrices once the spectral decomposition of the single matrix $\mathbf{U}$ is obtained.

1.10 Annual World Crude Oil Production (1880–1972)

Moore and McCabe (1989, p. 147) fit the linear regression of $y = \log$ (oil production) on $x =$ years during $n = 21$ particular years between 1880 and 1972. They use the method of least squares and obtain an excellent fit. Their method, however, is open to some conceptual discussion. Under the usual normality assumptions, the method of least squares provides only UMVU estimators if the observations are taken to be independently distributed, with common variance. Whenever the explanatory variable x represents a time variable, the observations may be serially correlated, in which case the method of least squares is open to question.

We applied the methodology of Section 1.9 to these data. The profile likelihood (1.9.5) for γ_2 is reported in Figure 1.10.1. This possesses a maximum at $\hat{\gamma}_2 = -0.1472$ and does not refute the choice $\gamma_2 = 0$, which would justify the Moore–McCabe analysis. The estimated covariance matrix $\hat{C} = e^{\hat{\gamma}_1}\mathbf{V}^{\hat{\gamma}_2}$ of the observations did not yield any correlations greater than 0.15. All other quantities of interest were estimated to be remarkably close to the Moore–McCabe estimates.

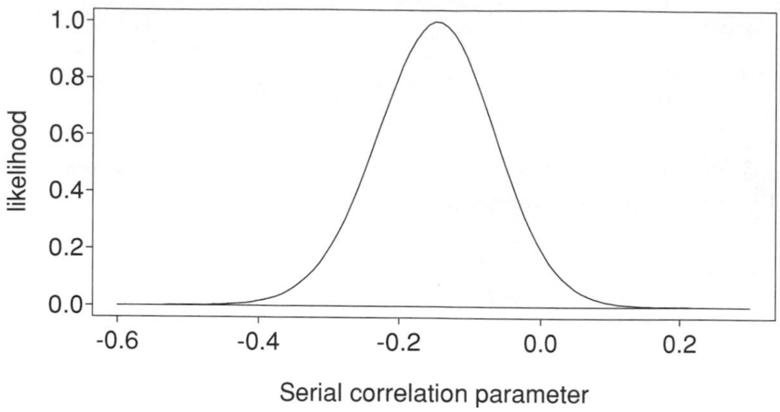

Figure 1.10.1. Likelihood of serial correlation parameter.

Our results suggest that least squares can indeed be appropriate in situations where there are serial correlations in the data but where the data do not contain enough information to identify strong correlations or nonlinear dependencies. The maximum entropy Exercise 3.4.k will also be relevant to this discussion.

Worked Example 1L: *Profile Likelihood in Regression:*

Observations $Y_1, Y_2, \ldots, Y_n$ are independent and normally distributed with common variance ϕ and respective means $\xi_1, \xi_2, \ldots, \xi_n$, satisfying

$$\xi_i = \mathbf{x}_i^T \boldsymbol{\beta} + g_i(\gamma),$$

where the $\mathbf{x}_i$ are specified $p \times 1$ design vectors with $p+1 < n$, $g_i(\gamma)$ are specified functions of an unknown scalar parameter γ, and $\boldsymbol{\beta}$ is a $p \times 1$ vector of further unknown parameters.

(a) Find the profile likelihood of γ.

(b) Describe a procedure based upon (a) for computing the maximum likelihood estimate of $\boldsymbol{\beta}$.

Hint: The value $\hat{\gamma}$ of γ maximizing its profile likelihood is also its maximum likelihood estimate. Let $\hat{\boldsymbol{\beta}}(\gamma)$ denote the maximum likelihood vector of $\boldsymbol{\beta}$ for a given γ. Then $\hat{\boldsymbol{\beta}} = \hat{\boldsymbol{\beta}}(\hat{\gamma})$ is the unconditional maximum likelihood vector of $\boldsymbol{\beta}$. Note further that

$$\sum_{i=1}^{n} (y_i - \mathbf{x}_i^T \boldsymbol{\beta})^2 = S_R^2 + (\boldsymbol{\beta} - \hat{\boldsymbol{\beta}})^T \mathbf{Q} (\boldsymbol{\beta} - \hat{\boldsymbol{\beta}}),$$

where $\mathbf{Q} = \sum_{i=1}^n \mathbf{x}_i \mathbf{x}_i^T$ and $S_R^2 = \sum_{i=1}^n (y_i - \mathbf{x}_i^T \hat{\boldsymbol{\beta}})^2$.

Model Answer 1L:

(a) The joint density of the observations is

$$p(\mathbf{y} \mid \boldsymbol{\beta}, \gamma, \phi) = \prod_{i=1}^{n} (2\pi)^{-\frac{1}{2}} \phi^{-\frac{1}{2}} \exp\left\{-\frac{1}{2\phi}\left(y_i - \mathbf{x}_i^T \boldsymbol{\beta} - g_i(\gamma)\right)^2\right\},$$

$$\propto \phi^{-\frac{1}{2}n} \exp\left\{-\sum_{i=1}^{n} \frac{1}{2\phi}\left(u_i - \mathbf{x}_i^T \boldsymbol{\beta}\right)^2\right\},$$

where $u_i = y_i - g_i(\gamma)$. This likelihood function can be rearranged in the form

$$l(\boldsymbol{\beta}, \gamma, \phi \mid \mathbf{y}) \propto \phi^{-\frac{1}{2}n} \exp\left\{-\frac{1}{2\phi} S(\gamma) - \frac{1}{2\phi}\left(\boldsymbol{\beta} - \hat{\boldsymbol{\beta}}(\gamma)\right)^T \mathbf{Q}\left(\boldsymbol{\beta} - \hat{\boldsymbol{\beta}}(\gamma)\right)\right\},$$

where $\hat{\boldsymbol{\beta}}(\gamma)$ and $S(\gamma)$, respectively, are equal to $\hat{\boldsymbol{\beta}}$ and S_R^2, but with the y_i replaced by the corresponding $u_i = y_i - g_i(\gamma)$. Maximizing this expression with respect to $\boldsymbol{\beta}$, we find that $\hat{\boldsymbol{\beta}}(\gamma)$ is the maximum likelihood vector for $\boldsymbol{\beta}$ and γ and that the profile likelihood for γ and ϕ is

$$l_p(\gamma, \phi \mid \mathbf{y}) \propto \phi^{-\frac{1}{2}n} \exp\left\{-\frac{1}{2\phi} S(\gamma)\right\}.$$

Maximizing with respect to ϕ, we find that $\hat{\phi} = S(\gamma)/n$ is the maximum likelihood estimate for ϕ, given γ, and that

$$l_p(\gamma \mid \mathbf{y}) \propto \{S(\gamma)\}^{-\frac{1}{2}n}.$$

(b) Maximization of the profile likelihood is equivalent to minimization of the residual sum of squares $S(\gamma)$, and this may be completed via a computer search in one dimension, yielding the estimate $\hat{\gamma}$. Then $\hat{\boldsymbol{\beta}}(\hat{\gamma})$ is the maximum likelihood vector for $\boldsymbol{\beta}$.

Worked Example 1M: *The EM Algorithm*

Consider a one-way analysis of covariance (ANCOVA) parallel line model with random effects and unequal replications. Observations Y_{ij} $(i = 1, 2, \ldots, m;\ j = 1, 2, \ldots, n_i)$ are independent, given $\theta_1, \theta_2, \ldots, \theta_m$, and β, and normally distributed with common variance σ_W^2, and means

$$E(Y_{ij} \mid \theta_i, \beta) = \theta_i + \beta(x_{ij} - x_{i\cdot}) \qquad (i = 1, 2, \ldots, m;\ j = 1, 2, \ldots, n_i),$$

with $x_{i\cdot} = n_i^{-1} \sum_{j=1}^{n_i} x_{ij}$, and where the x_{ij} are specified covariates. Moreover, the θ_i comprise a random sample from a normal distribution with mean μ and variance σ_B^2. Note that $\theta_1, \theta_2, \ldots, \theta_m$ are unobservable, and μ, β, σ_B^2, and σ_W^2 are unknown parameters. This model can be used to detect whether an apparently significant difference between the group effects is still significant in the presence of a designated confounding variable. Consequently, the value of σ^2 is of considerable interest.

(a) Find the joint density of the Y_{ij} and the θ_i and express this as a product of a function of σ_W^2 and β, and a function of σ_B^2, σ_W^2, and the θ_i.

(b) By completing the square in θ_i, or referring to Lemma 3.4.1 of Chapter 3, show that given the y_{ij}, the θ_i are conditionally independent, and normally distributed, with respective means

$$\theta_i^* = \left(n_i\sigma_W^{-2} + \sigma_B^{-2}\right)^{-1}\left(n_i\sigma_W^{-2} y_{i\cdot} + \sigma_W^{-2}\mu\right)$$

and variances

$$v_i = \left(n_i\sigma_W^{-2} + \sigma_B^{-2}\right)^{-1}.$$

(c) Find the expectation of the log of your density in (a) with respect to the conditional distribution in (b) but when θ_i^* and v_i are specified.

(d) Hence obtain equations for the values of σ_B^2, σ_W^2, β, and μ maximizing the density of the y_{ij}, unconditional upon the θ_i. Describe how these may be solved iteratively.

Model Answer 1M:

(a)

$$\begin{aligned}
p(\mathbf{y}, \boldsymbol{\theta}) &= p(\mathbf{y} \mid \boldsymbol{\theta}) p(\boldsymbol{\theta}) = \left(\sigma_B^2\right)^{-\frac{1}{2}m}\left(\sigma_W^2\right)^{-\frac{N}{2}} \\
&\quad \times \exp\left\{-\frac{1}{2}\sigma_W^{-2}\sum_{i=1}^{m}\sum_{j=1}^{n_i}\left\{y_{ij} - \theta_i + \beta(x_{ij} - x_{i\cdot})\right\}^2\right. \\
&\qquad\qquad \left. - \frac{1}{2}\sigma_B^{-2}\sum_{i=1}^{m}(\theta_i - \mu)^2\right\} \\
&= \left(\sigma_B^2\right)^{-\frac{1}{2}m}\left(\sigma_W^2\right)^{-\frac{N}{2}} \\
&\quad \times \exp\left\{-\frac{1}{2}\sigma_W^{-2}S(\beta) - \frac{1}{2}\sigma_W^{-2}\sum_{i=1}^{m} n_i(\theta_i - y_{i\cdot})^2\right. \\
&\qquad\qquad \left. - \frac{1}{2}\sigma_B^{-2}\sum_{i=1}^{m}(\theta_i - \mu)^2\right\},
\end{aligned}$$

where $N = \sum_{i=1}^m n_i$, and

$$S(\beta) = \sum_{i=1}^m \sum_{j=1}^{n_i} \left\{ y_{ij} - y_{i\cdot} - \beta(x_{ij} - x_{i\cdot}) \right\}^2 .$$

This density clearly factorizes, as required.

(b) By completing the square in θ_i, as recommended,

$$\begin{aligned} &\sigma_W^{-2} n_i (\theta_i - y_{i\cdot})^2 + \sigma_B^{2} (\theta_i - \mu)^2 \\ &= \left(n_i \sigma_W^{-2} + \sigma_B^{-2}\right)\left(\theta_i - \theta_i^*\right)^2 + \left(n_i^{-1}\sigma_W^2 + \sigma_B^2\right)^{-1}(y_{i\cdot} - \mu)^2 \\ &= v_i^{-1}\left(\theta_i - \theta_i^*\right)^2 + \text{a term not depending upon } \theta_i, \end{aligned}$$

where θ_i^* is defined above. Therefore the result immediately follows.

(c) Since

$$\begin{aligned} \log p(\mathbf{y}, \boldsymbol{\theta}) = & -\frac{m}{2}\log\sigma_B^2 - \frac{N}{2}\log\sigma_W^2 - \frac{1}{2}\sigma_W^{-2} S(\beta) \\ & -\frac{1}{2}\sigma_W^{-2}\sum_{i=1}^m n_i(\theta_i - y_{i\cdot})^2 - \frac{1}{2}\sigma_B^{-2}\sum_{i=1}^m (\theta_i - \mu)^2, \end{aligned}$$

the required expectation is therefore

$$\begin{aligned} E\log p(\mathbf{y}, \boldsymbol{\theta}) = & -\frac{1}{2} m \log\sigma_B^2 - \frac{1}{2} N \log\sigma_W^2 \\ & -\frac{1}{2}\sigma_W^{-2} S(\beta) - \frac{1}{2}\sigma_W^{-2}\sum_{i=1}^m \left(\theta_i^* - y_{i\cdot}\right)^2 \\ & +\frac{1}{2}\sigma_B^{-2}\sum_{i=1}^n \left(\theta_i^* - \mu\right)^2 - \frac{1}{2}\sigma_W^{-2}\sum_{i=1}^m n_i v_i - \frac{1}{2}\sigma_B^{-2}\sum_{i=1}^m v_i . \end{aligned}$$

(d) Keeping the θ_i^* and v_i fixed (for the moment), the expectation in (c) is maximized by the values of β, μ, σ_B^2, and σ_W^2 satisfying

$$\hat{\mu} = \theta_\cdot^*, \qquad \hat{\beta} = \left\{ \sum_{i=1}^m \sum_{j=1}^{n_i} (x_{ij} - x_{i\cdot})^2 \right\}^{-1} \sum_{i=1}^m \sum_{j=1}^{n_i} (x_{ij} - x_{i\cdot})(y_{ij} - y_{i\cdot}),$$

$$\hat{\sigma}_B^2 = \frac{1}{m}\left\{ \sum_{i=1}^m \left(\theta_i^* - \theta_\cdot^*\right)^2 + \sum_{i=1}^m v_i \right\},$$

and

$$\hat{\sigma}_W^2 = \frac{1}{N}\left\{S(\hat{\beta}) + \sum_{i=1}^{m} n_i(\theta_i^* - y_{i\cdot})^2 + \sum_{i=1}^{m} n_i v_i\right\}.$$

The maximum likelihood solution may be obtained by substituting trial values for β, μ, σ_B^2, and σ_W^2 into our expressions for the θ_i^* and v_i, obtaining new values for β, μ, σ_B^2, and σ_W^2 from the four preceding equations, substituting back into the θ_i^* and v_i, and cycling until convergence. It might be necessary to try different initial values in order to search for a global maximum.

2

The Discrete Version of Bayes' Theorem

2.0 Preliminaries and Overview

How can physicians rationalize their thought processes when performing medical diagnosis? How can a prisoner rationalize new information regarding his possible execution? How can you update your probabilities regarding uncertain events when you receive fresh information? A very simple theorem, virtually equivalent to the definition of conditional probability, provides an effective rationale. Generalizations of the procedure described in Section 2.1 define the entire Bayesian paradigm.

Self-Study Exercises 2.1.a and 2.1.b are interesting. The first involves the resolution of the Prisoner's Dilemma via Bayes' theorem. However, solutions of the Let's Make a Deal problem by Bayes' theorem require the most careful interpretation, since they should concentrate attention on the probability of winning. It is easier to maximize the probability of winning, for two different strategies, using the *law of total probability*.

In Exercise 2.1.c and in Section 2.2, Bayes' theorem is applied to the estimation of a discrete-valued parameter, and in Section 2.3 it is applied to model selection. In equation (2.3.2), a method is described for estimating a sampling model as a linear combination of sampling models, and this is applied in Section 2.4 (A) to an analysis of failure time data. The empirical probabilities in (2.3.3) are related to BIC and to the situation where each candidate model contains an unknown vector of parameters. While its justification, at the beginning of Chapter 5, will require a large value of n, this procedure is recommended whenever $n \geq 46$. It can also be justified by finite sample simulations (e.g., Section 2.4 (B)).

Bayes' theorem has been frequently employed in legal cases. See Lindley (1977) for a discussion of the evaluation of forensic evidence, Lindley (1991) for an overview of the role of Bayes' theorem in legal cases, Aitken (1995) for more details, Sections 2.1 and 2.8 for some examples, and Leonard (1999) for further discussions. Bayes' theorem has also been applied to logistic discrimination problems and to related "neural nets," for example, for medical diagnosis. The general methodology is described in Section 2.5, and a useful extension is introduced in Exercise 2.5.c. Sections 2.6 and 2.7 are concerned with Anderson's diagnosis of psychotic patients and with the 1978 Ontario Fetal Metabolic Acidosis study.

For further reading of the Bayesian approach, see O'Hagan (1994) together with his excellent bibliography, Berger (1985) and Smith (1988) for decision-theoretic approaches, Press (1989), Carlin and Louis (1996), and Gelman, Carlin, Stern, and Rubin (1995) for broad-ranging applications, DeGroot (1970) and Bernardo and Smith (1994) for the logical foundations, and Lee (1997) and Lindley (1965) for clear introductions. For some historical background, see Jeffreys (1961).

2.1 Bayes' Theorem

Consider a statistical experiment $\mathcal{E}$, with a sample space S of possible outcomes, and let $\{B_1, B_2 \ldots, B_r\}$ comprise a partition of S. Let $\{p(A);\ A \subseteq S\}$ denote a probability distribution defined on all events in S. Then, for any events A and B in S, with $p(A) > 0$, $p(B \mid A) = p(A \cap B)/p(A)$ denotes the conditional probability that B occurs, given that A is known to occur. Bayes' theorem or, to be precise, a discrete version of a theorem employed by Bayes, in a continuous case for the purpose of inference (Bayes, 1763, Barnard, 1958) simply tells us that

$$p(B_i \mid A) = \frac{p(A \mid B_i)p(B_i)}{p(A)} \qquad (i = 1, 2, \ldots, r), \tag{2.1.1}$$

whenever $p(A) > 0$, where by the law of total probability,

$$p(A) = \sum_{j=1}^{r} p(A \mid B_j)p(B_j). \tag{2.1.2}$$

The Reverend Thomas Bayes (1702–1761), according to an investigation by Dennis V. Lindley of local records, is buried in Bunhill Cemetery, Moorgate, London, and statisticians from all over the world visit his family (Bayes–Cotton) grave. A plaque was added to the top of the grave during the 1960s, and an elderly caretaker, if still alive, will tell you that "Bayes was responsible for getting rockets to the moon," presumably because Bayes' theorem is a component of stochastic control theory (e.g., Aoki, 1975; White, 1975). A picture of Bayes, together with his historic paper, is reproduced by Press (1989). The theorem was originally applied to a problem involving billiard balls. The authorship of Bayes' theorem has been questioned by Stigler (1983), who humorously suggests numerous possible alternative authors, including the little-known English mathematician Saunderson. Dale (1982, 1991) discusses whether Laplace might instead be the original author of Bayes' theorem. It has recently been verified that Bayes matriculated in divinity at the University of Edinburgh in 1719 and was later famous for his defense of the teachings of Sir Isaac Newton. Although he did not graduate, he may well have received instruction from James Gregory, who held the Edinburgh Chair of Mathematics, or from Gregory's deputy and successor Colin MacLaurin.

The theorem provides an effective rationale for medical diagnosis. Consider a disease that is thought to occur in a proportion $\theta = 0.01$ of the population and suppose that a physician observes that out of patients with the disease, 99% possess symptom Z. The physician might well conclude that a patient with symptom Z (e.g., a positive result on a blood test) possesses a high propensity for the disease. However, for a randomly chosen person in the population, let A and B, respectively, denote the events that the person has the symptom and the disease. Then $p(B) = \theta = 0.01$, and $p(A \mid B) = 0.99$. Therefore, Bayes' theorem with $r = 2$, $B_1 = B$, and $B_2 = B^C$ tells us that

$$\begin{aligned} p(\text{disease} \mid \text{symptom}) &= p(B \mid A) \\ &= \frac{p(A \mid B)p(B)}{p(A \mid B)p(B) + p(A \mid B^C)p(B^C)} \\ &= \frac{0.99 \times 0.01}{0.99 \times 0.01 + 0.99 \times p(A \mid B^C)} \\ &= \frac{1}{1 + 100p(A \mid B^C)}. \end{aligned}$$

This diagnostic probability can be surprisingly small, for example, if $p(A \mid B^C) = 1/10$ and $p(B \mid A) = 1/11$, quite contrary to the physician's intuition. Moreover,

$$\begin{aligned} p(\text{disease} \mid \text{no symptom}) &= p(B \mid A^C) \\ &= \frac{p(A^C \mid B)p(B)}{p(A^C \mid B)p(B) + p(A^C \mid B^C)p(B^C)} \\ &= \frac{0.01 \times 0.01}{0.01 \times 0.01 + 0.99 \times \{1 - p(A \mid B^C)\}} \\ &= \frac{1}{9901 - 9900p(A \mid B^C)}. \end{aligned}$$

Therefore, even if the symptom is absent, there is still a small probability that the patient possesses the disease. For example, AIDS/HIV tests yielding false negatives cannot completely safeguard our blood supply. For applications of Bayes' theorem to the diagnosis of jaundice and deep-vein thrombosis, see Knill-Jones (1974) and Emerson (1974). For a survey of application of Bayes' theorem in biostatistics, see Breslow (1990a). Further issues in medicine are discussed by Geisser (1987, 1992) and Geisser and Johnson (1992). In the engineering literature, Kaplan, Garrick, and Bieniarz (1981) and Kaplan (1983) apply the discrete version of Bayes' theorem to nuclear, feed water flow, steam generator, and safety injection problems. For some applications in reliability theory, see Barlow and Singpurwalla (1985) and Barlow and Zhang (1987).

Worked Example 2A: *Bayes' Theorem in Criminal Cases*

Consider a population or sampling frame of N individuals, exactly one of whom is the "Leopard," that is, the unknown perpetrator of a particular crime. Before receiving any evidence, you assign equal probabilities $1/N$ to each of the events $\{i\}$ = {the ith member of the population is the Leopard}, ($i = 1, 2, \ldots, N$).

Represent by Ω a piece of evidence presented at the trial. Let ϕ_i denote your probability that evidence Ω would have occurred, given $\{i\}$. (In cases involving DNA matching, Ω frequently denotes evidence that the defendant gives a perfect match, for example, on five DNA probes, and all probabilities need to be conditional upon the event that the criminal gives a similarly perfect match; see Leonard, 1999.)

(a) Given Ω, show that your probability for $\{i\}$ is

$$\xi_i = \frac{\phi_i}{N\phi.},$$

where $\phi.$ denotes the average of the ϕ_i. ($\phi.$ can be interpreted as the probability of Ω, given that the criminal was a "random man," or an individual selected at random from the population.)

(b) Supposc that $\phi_1 = 1$, $\phi_i = \epsilon$ for $i = 2, 3, \ldots, M$, and $\phi_i = 0$ for $i = M+1, M+2, \ldots, N$. Show that

$$\xi_1 = \frac{1}{1+(M-1)\epsilon}.$$

(c) Comment on the fact that $\xi_1 < \frac{1}{2}$ whenever $\epsilon > \frac{1}{M-1}$.

Note: The probabilities $p\{i\} = 1/N$ can be referred to as *prior* probabilities and the $\xi_i = p(\{i\} \mid \Omega)$ as *posterior* probabilities.

Model Answer 2A:

(a) By Bayes' theorem,

$$\xi_i = p(\{i\} \mid \Omega) = \frac{p(\Omega \mid \{i\})p(\{i\})}{p(\Omega)} = \frac{N^{-1}\phi_i}{p(\Omega)},$$

where, by the law of total probability,

$$p(\Omega) = \sum_{i=1}^{N} p(\Omega \mid \{i\})p(\{i\}) = N^{-1}\sum_{i=1}^{N}\phi_i = \phi..$$

Consequently,

$$\xi_i = \frac{\phi_i}{N\phi.}.$$

(b) In this case, $N\phi. = \sum_{i=1}^{N}\phi_i = 1+(M-1)\epsilon$, and the result follows, as $\phi_1 = 1$.

(c) You would need $\xi_1 > \frac{1}{2}$ to convict based upon the balance of odds, even though $\phi_1 = 1$ gives a perfect match. However, this will not occur if your common probability ϵ relating to each of the further $M-1$ nonexcluded members of the population exceeds $1/(M-1)$. (In cases where an inappropriate version of Bayes' theorem is used, a value of $\phi_1 = 1$ can give an alleged $\xi_1 > 99.99$.)

SELF-STUDY EXERCISES

2.1.a *The Prisoner's Dilemma* (e.g., Poundstone, 1992): Consider three prisoners A, B, and C, exactly one of whom will be pardoned; the other two will be executed. Let A, B, and C, respectively, denote the events that prisoner A, B, or C will be pardoned. In the absence of any other evidence to the contrary, it is reasonable to assume the "prior probabilities" $p(A) = p(B) = p(C) = \frac{1}{3}$.

The warden now enters prisoner A's cell and tells him that B will be executed, that is, the event $W(A, B)$ occurs. Prisoner A thinks that he now possesses an increased probability $\frac{1}{2}$ of being pardoned. However, use Bayes' theorem to show that

$$p\big(A \mid W(A, B)\big) = \tfrac{1}{3},$$

under the assumptions

$$p\big(W(A, B) \mid A\big) = p\big(W(A, C) \mid A\big) = \tfrac{1}{2},$$

$$p\big(W(A, C) \mid B\big) = 1,$$

and

$$p\big(W(A, B) \mid C\big) = 1.$$

Are these assumptions reasonable? Note that the information provided by the warden does not influence prisoner A's information at all.

Prisoner C is listening through a hole in the wall. What is his updated probability of execution?

2.1.b *Let's Make a Deal:* Consider the Monty Hall television game, with three doors A, B, and C. There are goats behind two doors, and a car behind the third door. Let A, B, and C, respectively, denote the events that the car is behind door A, B, or C. You may initially assume that $p(A) = p(B) = p(C) = \frac{1}{3}$.

A contestant wins upon correctly guessing which door conceals the car, and initially points to one of the three doors. Assume that the host then always points to one of the other doors and correctly informs the contestant that this conceals a goat. The contestant may then either *stick*, that is, keep to his original choice, or *switch*, that is, indicate that the car is behind the third door, not already pointed to. Let W denote the event that the contestant wins.

Use the law of total probability to show that under the *stick* strategy, $p(W) = \frac{1}{3}$, whereas under the *switch* strategy, $p(W) = \frac{2}{3}$, and that these probabilities do not depend upon how the host chooses between the other two doors.

(This solution was kindly provided to us by Professor Bob Barmisch of the University of Wisconsin, Department of Electrical Engineering. It is tempting to apply Bayes' theorem along the lines indicated in the Prisoner's Dilemma. The most careful interpretation of the conditioning events, however, is then required. See also Morgan et al., 1991.)

2.1.c Observations $y_1, \ldots, y_n$ constitute a random sample from a standard t-distribution with ν degrees of freedom, where ν is unknown. Before viewing $\mathbf{y} = (y_1, \ldots, y_n)^T$, you possess "prior probability" $(e - 1)e^{-j}$ that ν is equal to j, for $j = 1, 2, \ldots$. Show that after viewing $\mathbf{y}$, you possess "posterior probability" taking the form

$$\frac{e^{-j} l_j(\mathbf{y})}{\sum_{g=1}^{\infty} e^{-g} l_g(\mathbf{y})}$$

that ν is equal to j, for $j = 1, 2, \ldots$, and describe $l_j(\mathbf{y})$. Describe the mean of your posterior distribution for ν. Does this provide a reasonable way of estimating ν?

2.1.d A robbery at Harrods in London is observed to have been committed by a Royal Marine.

(a) You are informed that a population of $N = 20{,}000$ Royal Marines had access to Harrods at the time of robbery. Without any further information, what is your only fair prior probability ϕ that any particular marine is the robber?

(b) Evidence is then produced that the robber had a Union Jack tattooed on his forehead (this feature will be referred as the I.D. evidence). You are informed that exactly $M = 20$ out of the $N = 20{,}000$ marines have this emblem. In the absence of further information, what is your posterior probability ϕ^* that any marine out of these 20 is the robber?

(c) The computer records of the N marines are then randomly sampled until a marine is discovered with the I.D. evidence. He is immediately arrested and charged with the robbery. No further evidence is submitted. The likelihood ratio R^* is reported, where

$$R^* = \frac{p(\text{ I.D. evidence} \mid \text{defendant})}{p(\text{ I.D. evidence} \mid \text{random marine})} = \frac{1}{M/N} = 1{,}000.$$

The prosecuting counsel judges that the odds are 1,000 to 1 against the defendant when compared with a random man from the population. Is this fair? Does this involve the prosecutor's fallacy? Does it involve the defender's fallacy? See Section 2.8, part (a).

(d) A restatement of Bayes' theorem, known as the Essen-Möller formula (Essen-Möller, 1938), tells us that the suspect's posterior probability of guilt is

$$\lambda^* = \frac{R\lambda}{R\lambda + 1 - \lambda},$$

where R is a separately defined likelihood ratio, and λ is the prior probability of guilt. Equivalently,

$$\frac{\lambda^*}{1-\lambda^*} = \frac{R\lambda}{1-\lambda}.$$

Show that in the Royal Marine case, we must have $R = (N-1)/(R-1)$. Interpret this result.

2.1.e Consider the Essen-Möller formula of Exercise 2.1.d for general λ, when

$$R = \frac{1}{p(\text{ I.D. evidence} \mid \text{random marine})},$$

but where the random marine is chosen from the $N - 1$ members of the population, excluding the defendant. Hence R^{-1} averages the $N - 1$ probabilities

$$\phi_i = p(\text{ I.D. evidence} \mid i\text{th member}) \qquad (i = 2, 3, \ldots N),$$

for the $N - 1$ individuals. Show that the Essen-Möller formula is then equivalent to a version of Bayes' theorem (2.1.1), which assigns prior probability λ to the defendant and prior probabilities $(1 - \lambda)/(N - 1)$ to each of the remaining $N - 1$ members of the population. Would you regard the choice $\lambda = \frac{1}{2}$ as neutral? Would a choice of λ exceeding N^{-1} indicate the presence of prior evidence against the defendant?

2.2 Estimating a Discrete-Valued Parameter

A fair coin is tossed until the rth head is recorded and the number m of coins tossed is observed and reported. If r is not reported, then how do you draw inferences about r from the data? Based upon negative binomial assumptions, the likelihood of r, given m, is

$$l(r \mid m) \propto {}^{m-1}C_{r-1} = \frac{(m-1)!}{(r-1)!(m-r)!} \qquad (r = 1, \ldots, m). \tag{2.2.1}$$

You may assign a countable sequence $\phi_1, \phi_2, \ldots$ of probabilities, summing to unity, such that ϕ_i denotes your prior probability that $r = i$, $(i = 1, 2, \ldots)$. This specification may be based upon your information or beliefs about r, before observing m. Then, by Bayes' theorem, the corresponding conditional probabilities, given that you observe a specified value of m, are

$$p(r \mid m) = \frac{\phi_r / \big[(r-1)!(m-r)!\big]}{\sum_{i=1}^{m} \big\{\phi_i / \big[(i-1)!(m-i)!\big]\big\}} \qquad (r = 1, \ldots, m). \tag{2.2.2}$$

The values in (2.2.2) are your posterior probabilities on r. For example, under the particular specification

$$\phi_r \propto r^{-1} \qquad (r = 1, \ldots, m^*), \tag{2.2.3}$$

we have

$$p(r \mid m) \propto \frac{1}{r!(m-r)!} \propto {}^{m}C_r = \frac{m!}{r!(m-r)!} \qquad \big(r = 1, \ldots, \min\{m, m^*\}\big). \tag{2.2.4}$$

It follows that, whenever $m \leq m^*$, the "posterior distribution" of r is binomial, with probability $\frac{1}{2}$ and sample size m. The mean of this distribution is $E(r \mid m) = \frac{m}{2}$, and the latter is a reasonable point estimate of r. However, the posterior standard deviation is $\frac{1}{2}\sqrt{m}$, so that based upon just a single replication, we would not expect $\frac{1}{2}m$ to provide a particularly accurate estimate of r.

There are many applications of Bayesian methodologies that place prior probabilities on the number of unknown fish in a lake, for example, using hypergeometric models (see Roberts, 1967, and Smith and Sedransk, 1982). It is possible to estimate any unknown population size by using a random sampling/tagging/resampling procedure.

Worked Example 2B: *A Coin Lands on Its Edge*

A coin has probability θ of landing on its edge. Professors Cook, Bolton, and Smyth make the separate assumptions that $\theta = 10^{-10}$, $\theta = 10^{-11}$, and $\theta = 10^{-12}$. You assign prior probabilities of $\frac{3}{8}$ to each of Professors Cook's and Bolton's hypotheses, and probability $\frac{1}{4}$ to Professor Smyth's. The coin is then tossed and lands on its edge. Find your posterior probabilities for the three professors' hypotheses.

Model Answer 2B:

By the law of total probability, the unconditional probability that the coin lands on its edge is

$$\begin{aligned} p(\text{edge}) &= 10^{-10}\left(1 \times \tfrac{3}{8} + 10^{-1} \times \tfrac{3}{8} + 10^{-2} \times \tfrac{1}{4}\right) \\ &= 4.15 \times 10^{-9}. \end{aligned}$$

Consequently, the probability that Professor Cook's hypothesis is correct is

$$\begin{aligned} p(\text{Cook} \mid \text{edge}) &= \frac{p(\text{edge} \mid \text{Cook})}{p(\text{edge})} \\ &= \frac{10^{-10} \times \frac{2}{8}}{4.15 \times 10^{-9}} = \frac{0.375}{0.415} = 0.904. \end{aligned}$$

Similarly,

$$p(\text{Bolton} \mid \text{edge}) = \frac{10^{-1} \times \frac{3}{8}}{0.415} = 0.090,$$

and

$$p(\text{Smyth} \mid \text{edge}) = \frac{10^{-2} \times \frac{1}{4}}{0.415} = 0.006.$$

Historical note: In 1972, during his second lecture in statistics at the University of Warwick, Professor Jeffrey Harrison tossed a coin, which landed on its edge. This event, together with subsequent correspondence by Arthur Koestler and Tom Leonard in the *Sunday Times* of London, is fully discussed by Vaughan (1979).

2.3 Applications to Model Selection

You wish to choose between r fully specified models, for example, with densities $p_1(\mathbf{y}), \ldots, p_r(\mathbf{y})$, for describing the variability underlying your observation vector $\mathbf{y} = (y_1, \ldots, y_n)^T$, for $\mathbf{y} \in S$. Before viewing the data, you possess subjective probabilities $\phi_1, \ldots, \phi_r$, summing to unity, where for $j = 1, \ldots, r$, ϕ_j denotes your probability that the jth model is the most appropriate. For example, in the absence of further information to distinguish between the models, the choices $\phi_j = r^{-1}$, for $j = 1, \ldots, r$, are reasonable. Alternatively, if the models are nested, $\phi_j \propto e^{-\alpha j}$ may be reasonable for some subjective choice of α. The ϕ_j may be referred to as your "prior probabilities." Then, your "posterior probability" that the jth model is the most appropriate is denoted by

$$\phi_j^* = \frac{\phi_j p_j(\mathbf{y})}{\sum_{g=1}^{r} \phi_g p_g(\mathbf{y})} \qquad (j = 1, \ldots, r). \tag{2.3.1}$$

In a time-series context, Harrison and Stevens (1976) demonstrate that it is unnecessary to constrain attention to a particular model, but that you can keep updating, in time, your posterior probabilities for different models. Zellner, Hong, and Min (1991) apply similar ideas to econometric data. Note that the "posterior mean value function" of your sampling density is

$$p^*(\mathbf{u}) = \frac{\sum_{j=1}^{r} \phi_j(\mathbf{u}) p_j(\mathbf{y})}{\sum_{j=1}^{r} \phi_j p_j(\mathbf{y})} \qquad (\mathbf{u} \in S). \tag{2.3.2}$$

With a broad enough range of choices of your trial models $p_1(\mathbf{y}), \ldots, p_r(\mathbf{y})$, (2.3.2) can provide a "nonparametric" estimate of the true sampling density and hence can suggest a model not included in the original r choices. Suppose next that $p_1(\mathbf{y}) = f_1(\mathbf{y} \mid \boldsymbol{\theta}_1), \ldots, p_r(yv) = f_r(yv \mid \boldsymbol{\theta}_r)$ depend upon r vectors $\boldsymbol{\theta}_1, \ldots, \boldsymbol{\theta}_r$ of unknown parameters, where $\boldsymbol{\theta}_j$ is a $k_j \times 1$ vector, for $j = 1, \ldots, r$. Then, following Schwarz (1978), and under certain regularity conditions, the posterior probability that the jth model is most appropriate can be approximated for large enough n and whatever prior information is available for the unknown parameters by

$$\phi_j^* = \frac{\left(\frac{n}{2\pi}\right)^{-\frac{1}{2}k_j} \phi_j \hat{p}_j(\mathbf{y})}{\sum_{g=1}^{r} \left(\frac{n}{2\pi}\right)^{-\frac{1}{2}k_g} \phi_g \hat{p}_g(\mathbf{y})} \qquad (j = 1, \ldots, r), \tag{2.3.3}$$

where $\hat{p}_j(\mathbf{y}) = f_j(\mathbf{y} \mid \hat{\boldsymbol{\theta}}_j)$, and $\hat{\boldsymbol{\theta}}_j$ denotes the maximum likelihood vector for $\boldsymbol{\theta}_j$. The approximation in (2.3.3) is valid, as $n \to \infty$, whenever the result in (1.3.6) is true for each $\boldsymbol{\theta}_j$, as $n \to \infty$, with k_j fixed (see Section 6.1). For finite values of n, the adjustments in (2.3.3), which relate to the Bayesian information criterion (1.1.6), provide sensible pragmatic modifications to (2.3.1) when there are different numbers of unknown parameters in the r initial models. Owing to the sensitivity problems, with respect to the extra assumptions needed for a more precise analysis (e.g., Exercise 3.10.f), (2.3.3) comprises the best formal suggestion we are able to make, when comparing different models, for finite n. This approach contrasts with Hill (1986), who introduces more subjective considerations when selecting a model, and with the Bayes factor approach (e.g., McCulloch and Rossi, 1992).

Worked Example 2C: *Model Choice for Point Processes*

Arrivals during the time interval $(0, a)$ occur in a Poisson process with intensity function $\lambda(t)$ for $t \in (0, a)$. If $N(t_1, t_2)$ denotes the number of arrivals in interval (t_1, t_2),

(1) for any $(t_1, t_2) \subseteq (0, a)$, $N(t_1, t_2)$ has a Poisson distribution, with mean

$$\int_{t_1}^{t_2} \lambda(t)\, dt,$$

and

(2) for any nonoverlapping intervals (t_1, t_2) and (t_1^*, t_2^*) contained in $(0, a)$, $N(t_1, t_2)$ and $N(t_1^*, t_2^*)$ are independent.

If n arrivals are observed in $(0, a)$, at times $x_1, x_2, \ldots, x_n$, it is known (e.g., Leonard, 1978) that the likelihood of λ, that is, the joint density/p.m.f. of x_1, $x_2, \ldots, x_n$, and n, given λ, is

$$l(\lambda \mid \mathbf{x}, n) = \left\{ \prod_{i=1}^{n} \lambda(x_i) \right\} \exp\left\{ - \int_0^a \lambda(t)\, dt \right\}.$$

Following Steinijans (1976), consider special models of the form

$$g(t) = \log \lambda(t) = \beta_0 + \beta_1 t + \beta_2 t^2 + \cdots + \beta_q t^q.$$

You possess equal prior probabilities $\phi_0 = p(q = 0) = \frac{1}{2}$ and $\phi_1 = p(q = 1) = \frac{1}{2}$ on the two models where $q = 0$ and $q = 1$.

(a) Find the maximum likelihood estimate and maximized likelihood of β_0 when $q = 0$.

(b) Find the likelihood of β_0 and β_1 when $q = 1$. Show that the maximum likelihood estimate $\hat{\beta}_1$ of β_1 maximizes the profile likelihood

$$l_p(\beta_1 \mid \mathbf{x}, n) \propto \beta_1^n \left(e^{\beta_1 a} - 1\right)^{-n} \exp\{n\bar{x}\beta_1\}$$

and describe the maximum likelihood estimate $\hat{\beta}_0 = \hat{\beta}_0(\beta_1)$ of β_0, given β_1.

(c) Describe an approximation to your posterior probability that $q = 0$, that is, that the Poisson process is homogeneous. Your approximation should be valid for large n, whatever prior information is available for β_0 and β_1, and should be algebraically explicit in terms of $\hat{\beta}_1$.

Model Answer 2C:

(a) When $q = 0$, $g(t) = \beta_0$, so that $\lambda(t) = e^{\beta_0}$. Substituting into the likelihood for λ, we find that the likelihood of β_0 is given by

$$l(\beta_0 \mid \mathbf{x}, n) = \exp\left\{ n\beta_0 - \int_0^a e^{\beta_0} dt \right\} = \exp\left\{ n\beta_0 - a e^{\beta_0} \right\}.$$

This is maximized when $\beta_0 = \log(n/a)$, and the maximized likelihood is

$$l\left(\hat{\beta}_0 \mid \mathbf{x}, n\right) = \left(\frac{n}{a}\right)^n e^{-n}.$$

(b) When $q = 1$, $g(t) = \beta_0 + \beta_1 t$, so that $\lambda(t) = e^{\beta_0 + \beta_1 t}$. Therefore, the likelihood of β_0 and β_1 is

$$\begin{aligned} l(\beta_0, \beta_1 \mid \mathbf{x}, n) &= \exp\left\{ n\beta_0 + n\bar{x}\beta_1 - e^{\beta_0} \int_0^a e^{\beta_1 t} dt \right\} \\ &= \exp\left\{ n\beta_0 + n\bar{x}\beta_1 - \beta_1^{-1} e^{\beta_0} \left(e^{\beta_1 a} - 1\right) \right\}. \end{aligned}$$

This is maximized, for fixed β_1, when

$$\beta_0 = \hat{\beta}_0(\beta_1) = \log\left(\frac{n\beta_1}{e^{\beta_1 a} - 1}\right).$$

Consequently,

$$l_p(\beta_1 \mid \mathbf{x}, n) = e^{-n} n^n \beta_1^n \left(e^{\beta_1 a} - 1\right)^{-n} \exp\{n\bar{x}\beta_1\}.$$

Then $\hat{\beta}_0 = \hat{\beta}_0(\hat{\beta}_1)$, where $\hat{\beta}_1$ maximizes the profile likelihood.

(c) Adaptation of (2.3.3) shows that your approximate posterior probability is

$$\begin{aligned}
\phi_0^* &= \frac{\frac{1}{2}\left(\frac{n}{2\pi}\right)^{-\frac{1}{2}} l\left(\hat{\beta}_0 \mid \mathbf{x}, n\right)}{\frac{1}{2}\left(\frac{n}{2\pi}\right)^{-\frac{1}{2}} l\left(\hat{\beta}_0 \mid \mathbf{x}, n\right) + \frac{1}{2}\left(\frac{n}{2\pi}\right)^{-1} l\left(\hat{\beta}_0, \hat{\beta}_1 \mid \mathbf{x}, n\right)} \\
&= \frac{\left(\frac{n}{2\pi}\right)^{\frac{1}{2}} l\left(\hat{\beta}_0 \mid \mathbf{x}, n\right)}{\left(\frac{n}{2\pi}\right)^{\frac{1}{2}} l\left(\hat{\beta}_0 \mid \mathbf{x}, n\right) + l\left(\hat{\beta}_0, \hat{\beta}_1 \mid \mathbf{x}, n\right)} \\
&= \frac{\left(\frac{n}{2\pi}\right)^{\frac{1}{2}} \left(\frac{n}{a}\right)^n e^{-n}}{\left(\frac{n}{2\pi}\right)^{\frac{1}{2}} \left(\frac{n}{a}\right)^n + l\left(\hat{\beta}_0, \hat{\beta}_1 \mid \mathbf{x}, n\right)} \\
&= \frac{\left(\frac{n}{2\pi}\right)^{\frac{1}{2}}}{\left(\frac{n}{2\pi}\right)^{\frac{1}{2}} + R},
\end{aligned}$$

where

$$R = a^n \hat{\beta}_1^n \left(e^{a\hat{\beta}_1} - 1\right)^{-n} \exp\left\{n\bar{x}\hat{\beta}_1\right\}.$$

SELF-STUDY EXERCISES

2.3.a *Investigating Overdispersion:* Let $y_1, \ldots, y_m$ denote independent frequencies, which under model M_1 possess a binomial distribution, with probability 0.5 and sample size n. Under model M_2, each independent frequency y_i possesses the beta binomial distribution obtained by collapsing the following two stages:

Stage I: y_i possesses a binomial distribution with probability θ_i.

Stage II: θ_i possesses a beta distribution with probability 0.5 and specified variance λ.

Then λ relates to the overdispersion of $y_1, \ldots, y_m$ when compared with frequencies from a binomial distribution. Model M_2 provides an overdispersion model (e.g., Paul and Plackett, 1978). You have prior probabilities $1 - \epsilon$ and M_1 and ϵ on M_2. Find your posterior probability on M_1 and your posterior mean for the common probability mass function of the y_i. Indicate how you would extend your results to the situation where λ is unknown. Discuss experimental designs under which overdispersion is likely, or unlikely, to occur.

By reference to an approximation to the distribution of a chi-squared goodness-of-fit statistic, described by Paul and Plackett and by Leonard (1977a), under model M_2, find an explicit approximation to the maximum likelihood estimate of λ. Can the chi-squared statistic indicate lack of randomization at the design stage?

Note: Other overdispersion models and related quasi-likelihood methods are discussed by Albert and Pepple (1989).

2.3.b Models $M_1, M_2, \ldots, M_r$ are nested and depend upon respective parameter vectors $\boldsymbol{\theta}_1, \ldots, \boldsymbol{\theta}_r$, which possess respective dimensionality $q_1 = 1, q_2 = 2, \ldots, q_r = r$, and are unknown. The prior probability ϕ_j that model M_j is true is proportional to e^{-j} for $j = 1, \ldots, r$. Show that the model maximizing the corresponding approximate posterior probability (2.3.3) also maximizes the general information criterion (1.1.4), for some α, and find α.

2.3.c Observations $y_1, \ldots, y_n$ are independent and normally distributed with respective means $\mu_1, \ldots, \mu_n$ and common unknown variance σ^2. You assign equal prior probabilities to the linear regression model

$$M_1:\ \mu_i = \beta_0 + \beta_1 x_i \qquad (i = 1, \ldots, n)$$

and to the quadratic model

$$M_2:\ \mu_i = \beta_0 + \beta_1 x_i + \beta_2 x_i^2 \qquad (i = 1, \ldots, n),$$

where β_0, β_1, and β_2 are unknown and $x_1, \ldots, x_n$ are specified explanatory variables. Let S_j^2 denote the residual sum of squares for model M_j for $j = 1, 2$. Show from (2.3.3) that your empirical posterior probability that M_1 is appropriate is denoted by

$$\phi_1^* = \left\{1 + \left(\frac{n}{2\pi}\right)^{-\frac{1}{2}} \left(\frac{S_2^2}{S_1^2}\right)^{-\frac{n}{2}}\right\}^{-1}, \tag{2.3.4}$$

and hence show that

$$\phi_1^* < \left\{1 + \left(\frac{n}{2\pi}\right)^{-\frac{1}{2}}\right\}^{-1}. \tag{2.3.5}$$

2.4 Practical Examples

(A) *Selecting among Probability Distributions for Failure Time Data*

The data in Table 2.4.1 were reported by Quesenberry and Kent (1982) and describe $n = 100$ cycles-to-failure times for samples of yarn. Quesenberry and Kent assume that the observations constitute a random sample from a two-parameter distribution with density $f(y)$ for $0 < y < \infty$. They consider three possible choices:

(a) *Log-normal:* $f(y) = \frac{1}{(2\pi\sigma^2)^{\frac{1}{2}}|y|} \exp\left\{-\frac{1}{2\sigma^2}(\log y - \mu)^2\right\}$.

(b) *Gamma:* $f(y) = \frac{\beta^\alpha}{\Gamma(\alpha)} y^{\alpha-1} \exp\{-\beta y\}$.

(c) *Weibull:* $f(y) = \alpha\beta y^{\beta-1} \exp\{-\alpha y^\beta\}$.

Table 2.4.1. *Cycles-to-failure for yarn.*

86	146	251	653	98	249	400	292	131	169	175	176	76	264	15
364	195	262	88	264	157	220	42	321	180	198	38	20	61	121
282	224	149	180	325	250	196	90	229	166	38	337	65	151	341
40	40	135	597	246	211	180	93	315	353	571	124	279	81	186
497	182	423	185	229	400	338	290	398	71	246	185	188	568	55
55	61	244	20	284	393	396	203	829	239	236	286	194	277	143
198	264	105	203	124	137	135	350	193	188					

It is straightforward to fit each of these models by maximum likelihood. The suprema of the log-likelihoods were (a) -631.76, (b) -625.24, and (c) -625.20 in these three cases, suggesting that the Weibull model is slightly superior to the Gamma, with the log-normal running a poor third. Assume prior probabilities $\phi_1 = \phi_2 = \phi_3 = \frac{1}{3}$ on each of the three candidates. Then the corresponding empirical posterior probabilities in (2.3.3) become $\phi_1^* = 0.00072$, $\phi_2^* = 0.48832$, and $\phi_3 = 0.51095$.

Curves (a), (b), and (c) in Figure 2.4.1 describe the fitted sampling densities under assumptions (a), (b), and (c). Curve (d) describes the compromise (2.3.2), which weights curves (a), (b), and (c) according to the empirical posterior probabilities. Curve (d) can be interpreted as a "semiparametric" estimate of the sampling density. It should be interpreted with some caution, since it depends upon eight distinct parameters (the six parameters appearing in (a), (b), and (c), plus two distinct mixing probabilities).

(B) *Computer Simulations for Regression Models*

Computer simulations were performed to investigate the empirical posterior probabilities (2.3.3) for model M_1 of Exercise 2.3.c, with $n = 100$, and $x_1, \ldots, x_{100}$ equally

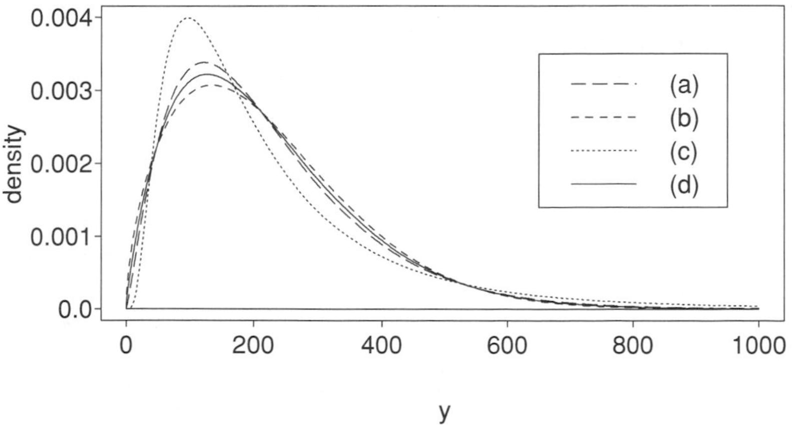

Figure 2.4.1. Estimated failure time densities: (a) Gamma; (b) Weibull; (c) log-normal; (d) semiparametric compromise.

spaced on the interval [0, 1]. Ten thousand samples of size $n = 100$ were first generated from model M_2, with $\beta_0 = \beta_1 = \beta_2 = \sigma = 1$. The average value of $\phi_2^* = 1 - \phi_1^*$ was 0.9456. Therefore, the criterion (2.3.3) strongly suggests retention of a quadratic model when compared with a simpler linear model. However, increasing σ makes the data more noisy, and a linear model, with one less parameter, becomes more appropriate. For example, with $\beta_0 = \beta_1 = \beta_2$ and $\sigma = 4$, the average value of ϕ_2^* on 10,000 simulations was just 0.3906. Also BIC correctly chose a quadratic model only 24.55% of the time, when compared with 97.84% on the first set of simulations.

The simulations were repeated, but regarding M_1 as true, with $\beta_0 = \beta_1 = \sigma = 1$. The average value of $100\phi_1^*$ on 10,000 simulations was 69.8%, and BIC correctly chose the linear model 90.27% of the time. These percentages were not affected much by increasing σ. When $\beta_0 = \beta_1 = 1$ and $\sigma = 10$, they were respectively replaced by 69.65% and 89.92%.

2.5 Logistic Discrimination and the Construction of Neural Nets

Consider the data $\{(y_i, \mathbf{x}_i);\ \text{for } i = 1, \ldots, n\}$ for n individuals where y_i is a binary response and $\mathbf{x}_i$ is a specified $p \times 1$ design vector. For example, let y_i equal unity if the ith patient possesses a particular disease, and zero otherwise. Let the $\mathbf{x}_i$ comprise measurements, for the ith patient, relating to p different symptoms. It is possible to analyze data of this type by using the linear logistic model described in Exercise 1.3.c, but with $n_1 = n_2 = \cdots = n_m = 1$. However, for noisy data sets, it may be difficult to model the choice of regression function (1.3.17). Moreover, when the observations are binary rather than binomial, with larger sample sizes, it is more difficult to check any particular model against the data. For example, standard tests for fit like the chi-squared goodness-of-fit test do not possess finite sample properties approximating the asymptotic properties, and the theoretical justifications for using AIC and BIC in (1.1.5) and (1.1.6) do not hold.

Computer scientists alternatively analyze this type of data by using linear programming and the theory of "neural nets" (e.g., Mangasarian, 1968 and 1992, and Mangasarian, Setiono, and Wolberg, 1989). The procedures developed in this section are also "neural nets" with input vector $\mathbf{x}_i$ and output $p(y_i = 1 \mid \mathbf{x}_i)$, a number measured on a unit scale. Bishop (1995) summarizes Bayesian theory in the context of neural nets and artificial intelligence.

It is possible to use the $\mathbf{x}_i$'s for which $y_i = 1$ to model some joint density $f_1(\mathbf{x})$, for example, for the symptom vectors of patients with the disease, and to use the $\mathbf{x}_i$'s for which $y_i = 0$ to model another joint density $f_0(\mathbf{x})$, for example, for the symptom vectors of those patients without the disease. These joint densities can often be readily checked against the data.

Consider a randomly chosen individual from among the n and suppose that the physician knows the numbers n_1 and n_0 of patients with and without the disease, such that $n_1 + n_0 = n$. Then the physician's prior probability that the randomly chosen patient has the disease is $\phi = n_1/n$. His corresponding posterior probability that the patient has the disease, given that the patient has symptom vector $\mathbf{x}$, is

$$\phi^* = \phi^*(\mathbf{x}) = \frac{n_1 f_1(\mathbf{x})}{n_1 f_1(\mathbf{x}) + n_0 f_0(\mathbf{x})}. \tag{2.5.1}$$

Consequently, the logit $\alpha^* = \log \phi^* - \log(1 - \phi^*)$ satisfies

$$\alpha^* = \alpha^*(\mathbf{x}) = \log \frac{n_1}{n_0} + \log \frac{f_1(\mathbf{x})}{f_0(\mathbf{x})}. \tag{2.5.2}$$

The right-hand side of (2.5.2) is the logistic discrimination function and replaces the linear logistic regression function (1.3.17), which takes the special form $\alpha^* = \mathbf{x}^T \boldsymbol{\beta}$. For example, following Anderson (1975), assume that for $i = 0, 1$, $f_i(\mathbf{x})$ is a multivariate normal $N(\boldsymbol{\mu}_i, \mathbf{C}_i)$ density, with mean vector $\boldsymbol{\mu}_i$ and covariance matrix $\mathbf{C}_i$. Then the $\boldsymbol{\mu}_i$ and $\mathbf{C}_i$ can be estimated by sample quantities calculated from the $\mathbf{x}$ vectors for the two subpopulations of patients, and

$$\begin{aligned} \alpha^* = \alpha^*(\mathbf{x}) &= \log \frac{n_1}{n_0} - \frac{1}{2} \log |\mathbf{C}_1| + \tfrac{1}{2} \log |\mathbf{C}_0| \\ &\quad - \frac{1}{2}(\mathbf{x} - \boldsymbol{\mu}_1)^T \mathbf{C}_1^{-1} (\mathbf{x} - \boldsymbol{\mu}_1) \\ &\quad + \frac{1}{2}(\mathbf{x} - \boldsymbol{\mu}_0)^T \mathbf{C}_1^{-1} (\mathbf{x} - \boldsymbol{\mu}_0) && (2.5.3) \\ &= \beta_0 + \mathbf{x}^T \boldsymbol{\beta} + \mathbf{x}^T \mathbf{B} \mathbf{x}, && (2.5.4) \end{aligned}$$

where

$$\begin{aligned} \beta_0 &= \log \frac{n_1}{n_0} - \frac{1}{2} \log |\mathbf{C}_1| + \frac{1}{2} \log |\mathbf{C}_0| \\ &\quad - \frac{1}{2} \boldsymbol{\mu}_1^T \mathbf{C}^{-1} \boldsymbol{\mu}_1 + \frac{1}{2} \boldsymbol{\mu}_0^T \mathbf{C}^{-1} \boldsymbol{\mu}_0, && (2.5.5) \\ \boldsymbol{\beta} &= \mathbf{C}_1^{-1} \boldsymbol{\mu}_1 - \mathbf{C}_0^{-1} \boldsymbol{\mu}_0, && (2.5.6) \end{aligned}$$

and

$$\mathbf{B} = -\frac{1}{2} \mathbf{C}_1^{-1} + \frac{1}{2} \mathbf{C}_0^{-1}. \tag{2.5.7}$$

This provides a direct way of modeling a multiple quadratic regression function, with explicit estimators for the coefficients. For example, the nondiagonal elements of $\mathbf{B}$ in (2.5.7) are the coefficients of the multiplicative interaction terms. Standard procedures for investigating equality of two multivariate normal covariance matrices (e.g., Anderson, 1974, Ch. 10) can be used to investigate whether (2.5.4) can be reduced to the linear model $\alpha^* = \beta_0 + \mathbf{x}^T \boldsymbol{\beta}$. If indeed $\mathbf{C}_1 = \mathbf{C}_0 = \mathbf{C}$, then $\boldsymbol{\beta} = \mathbf{C}^{-1}(\boldsymbol{\mu}_1 - \boldsymbol{\mu}_0)$, so that the linear coefficients can be investigated via standard procedures for comparing two multivariate normal mean vectors.

If $f_1(\mathbf{x})$ and $f_0(\mathbf{x})$ are not multivariate normal, then (2.5.2) can be used to model some complicated regression functions and nonlinear interaction structures. For example, if $p = 1$ and both $f_1(x)$ and $f_0(x)$ denote skewed normal densities (see Exercise 1.5.i), with different parameters, then the logistic regression function (2.5.2) becomes piecewise quadratic (see Low et al., 1981, and Section 2.7).

Worked Example 2D: *Logistic Discrimination Analysis for Survival Times*

Consider n patients with disease D, of whom n_0 possess complication Q and $n_1 = n - n_0$ do not. The recovery times for patients with Q constitute a random sample from a distribution with c.d.f. $F_0(x) = 1 - \exp\{-\beta_0 x^\alpha\}$. Similarly, the common c.d.f. for patients without Q is $F_1(x) = 1 - \exp\{-\beta_1 x^\alpha\}$.

(a) A randomly chosen patient with the disease has recovery time x. Find the probability ϕ^* that the patient has complication Q.
(b) Show that $\gamma^* = \log\phi^* - \log(1 - \phi^*)$ takes the form $A + Bx^\alpha$ and find A and B.
(c) Show that $A = 0$ corresponds to $n_0\beta_0 = n_1\beta_1$ and that $B = 0$ corresponds to $\beta_0 = \beta_1$.

Model Answer 2D:

(a) For patients with complication C, recovery time density is

$$f_0 = \frac{\partial F_0(x)}{\partial x} = \alpha\beta_0 x^{\alpha-1} \exp\big\{-\beta_0 x^\alpha\big\}.$$

Similarly, for patients without complication C,

$$f_1 = \alpha\beta_1 x^{\alpha-1} \exp\big\{-\beta_1 x^\alpha\big\}.$$

Therefore,

$$\phi^* = \frac{n_0 f_0(x)}{n_0 f_0(x) + n_1 f_1(x)} = \frac{n_0\beta_0 \exp\big\{-\beta_0 x^\alpha\big\}}{\sum_{i=0}^{1} \beta_i \exp\big\{-\beta_i x^\alpha\big\}}.$$

(b) Therefore, $\gamma^* = \log(n_0\beta_0/n_1\beta_1) - (\beta_0 - \beta_1)x^\alpha$. So $A = \log(n_0\beta_0/n_1\beta_1)$ and $B = \beta_1 - \beta_0$.
(c) Follows immediately.

SELF-STUDY EXERCISES

2.5.a Under the formulation of Section 2.5, with $p = 1$, suppose that for patients with a disease, x is $N(\mu_1, \sigma^2)$ distributed, and for patients without the disease, x is $N(\mu_2, \sigma^2)$ distributed. Show that the logistic discrimination function in (2.5.2) is linear in x. Show how a two-sample t-statistic can be used to investigate whether or not the linear term is zero. Show how a two-sample F-statistic can be used to check whether a quadratic term should be added to the model.

2.5.b Suppose now that for general p, each $\mathbf{x}_i$ vector consists of p binary responses $x_{i1}, \ldots, x_{ip}$ denoting which of p symptoms are present or absent, for the ith patient. For patients with the disease, suppose that $x_{ij} = 1$ with probability θ_{j1} and $x_{ij} = 0$ with probability $1-\theta_{j1}$. For patients without the disease, suppose that $x_{ij} = 1$ with probability θ_{j2} and $x_{ij} = 0$ with probability $1-\theta_{j2}$. Show that the logistic discrimination function in (2.5.2) is linear in $x_1, x_2, \ldots, x_p$. Show that it is possible to investigate whether any particular linear coefficient is zero, using standard procedures for comparing two binomial proportions.

Do you think that independence of the binary responses is a reasonable assumption in practice? Do you think that a linear discrimination function is reasonable?

2.5.c *A Polychoric Model for Nonindependent Binary Responses* (e.g., Lord and Novick, 1986, pp. 337–349; Albert, 1992): Consider a vector $\mathbf{x} = (x_1, \ldots, x_p)^T$ of possibly nonindependent binary responses, and suppose that the distribution of $\mathbf{x}$ may be described in the following two stages:

Stage I: For $j = 1, 2, \ldots, p$, the distribution of x_j, given θ_j, is binary with probability θ_j. Given $\theta_1, \theta_2, \ldots, \theta_p$, the binary responses $x_1, \ldots, x_p$ are independent.

Stage II: Consider probits $\alpha_1, \ldots, \alpha_p$ satisfying $\theta_j = \Phi(\alpha_j)$ for $j = 1, 2, \ldots, p$, where $\Phi(\cdot)$ denotes the standard normal distribution function. Assume that $\boldsymbol{\alpha} = (\alpha_1, \ldots, \alpha_p)^T$ possesses a multivariate normal $N(\boldsymbol{\mu}, \mathbf{C})$ distribution.

Express the unconditional probability mass function of $\mathbf{x}$ in terms of the cumulative distribution function of the multivariate normal distribution. Show how the EM algorithm described in Section 1.5 (E) can be used to compute the maximum likelihood estimates of $\boldsymbol{\mu}$ and $\mathbf{C}$, unconditional on $\boldsymbol{\alpha}$, given a random sample $\mathbf{x}_1, \ldots, \mathbf{x}_n$ of observation vectors, from this distribution.

Show how this formulation may be applied to a logistic discrimination analysis in order to model nonlinear interaction terms when the explanatory variables are binary. By regarding some of the α's as observed, show how this formulation can be used to model nonindependent binary responses together with some further explanatory variables, which are measurements.

2.6 Anderson's Prediction of Psychotic Patients

Consider data $\{(y_i, \mathbf{x}_i);\ \text{for } i = 1, \ldots, n\}$ for n individuals where each y_i is a binary response. Anderson (1975) considered measurements $x_1 = $ size and $x_2 = $ shape for a psychological test on a group of 25 normal ($y = 0$) and 25 ($y = 1$) psychotic patients. He used the method outlined in Section 2.5, with $p = 2$, and obtained a quadratic discrimination function (2.5.4) by fitting bivariate normal distributions to the observations on (x_1, x_2) for normal and for psychotic patients. However, he employed an iterative procedure to estimate the mean vectors and covariance matrices, and this is perhaps more efficient than using sample quantities.

Anderson provides a logistic-quadratic discriminant curve classifying those values of x_1 and x_2 for which his fitted $\alpha^*(\mathbf{x}) = \alpha^*(x_1, x_2)$ is equal to zero, that is, such that his fitted probability $\phi^*(\mathbf{x})$ in (2.5.1) is equal to one half. Therefore, patients with (x_1, x_2) values to the left of this curve are judged to possess a higher-than-even chance of being psychotic, while those with (x_1, x_2) values to the right of the curve are judged to have a higher than even chance of being normal. His curve is well validated by the current data, since only 4 out of the 25 patients are misclassified. His logistic linear discriminant curve (obtained by assuming equality of the two bivariate normal covariance matrices) misclassifies 3 patients, but appears to be too conservative for diagnosing psychotic patients.

Atilgan and Leonard (1987, and our Section 1.6 (C)) estimated the distribution for (x_1, x_2) by mixtures of m bivariate densities in each of the two subpopulations. The choice $m = 5$ optimizes AIC. The new "AIC discriminant curve" misclassifies just 2 out of the 50 patients.

2.7 The Ontario Fetal Metabolic Acidosis Study

In 1978, the Ontario Fetal Metabolic Acidosis data set, relating to medical records for about $n = 2{,}000$ mothers and their babies who had attended the intensive care unit of Kingston General Hospital, was analyzed. The dependent variable for this quite noisy, nonrandomized, data set was

$y =$ a buffer base reading of the level of acidosis in a blood sample taken from the umbilical cord during labor.

A value $y \leq 36.1$ units was at that time taken as a definition of the presence of the condition "fetal metabolic acidosis" or "fetal asphyxia," and babies with $y > 38.5$ units were thought to be at negligible risk from the types of postnatal symptoms caused by low acidosis. For the whole data set, $y \leq 36.1$ was observed in about 20% of the cases in intensive care. It was of interest to be able to predict high-risk cases, at the instant prior to labor.

The three main predictors for this condition were at the time thought to be (1) prematurity, (2) overdue, overweight babies, and (3) presence of meconium-stained fluid ((3) had previously been taken as a definition of the fetal metabolic acidosis). Consideration of the data by multiple regression and contingency table analysis showed negligible correlation/association between gestational age and y, and between birthweight and y, and the inclusion of a multiplicative interaction term scarcely reduced the residual sum of squares below the total sum of squares. For cases where meconium-stained fluid was present, the percentage of babies with $y \leq 36.1$ increased only from 20% to under 25%. None of the other twenty or so explanatory variables available yielded an obvious association with y.

The data set appeared, at first sight, to contain negligible information content. This was certainly true in the sense of yielding accurate point predictions for the y variable. However, the data set turned out to be very rich in information content in terms of making probability statements about y, given the estimated gestational age (length of pregnancy) and the birthweight (as predicted from an ultrasonic measurement of the child's skull). Our analysis demonstrates how Bayes' theorem can be used to extract interesting conclusions from noisy data sets.

The $n = 85$ babies in the sample with gestational age ≤ 33 weeks were initially considered. Scatterplots were prepared describing these data at three different levels of

Table 2.7.1. *Estimation of parameters of birthweight distributions.*

Gestational age	Buffer base	$\hat{\mu}$	$\hat{\sigma}_1^2$	$\hat{\sigma}_2^2$	Sample size
	low	1.086	0.094	0.249	22
$\leq$ 33 weeks	median	1.280	0.069	0.934	14
	high	1.456	0.082	0.745	49
	low	1.621	0.059	0.232	25
34 – 35 weeks	median	2.453	0.123	0.228	13
	high	2.216	0.121	0.221	51
	low	1.832	0.017	0.122	40
36 – 37 weeks	median	2.815	0.195	0.319	52
	high	2.518	0.175	0.583	117
	low	2.514	0.294	0.517	38
38 weeks	median	2.941	0.289	0.335	47
	high	3.237	0.362	0.133	140
	low	3.182	0.347	0.503	62
39 weeks	median	3.938	0.396	0.404	65
	high	3.289	0.236	0.209	167
	low	3.339	0.326	0.337	131
40 – 41 weeks	median	3.513	0.361	0.223	181
	high	3.475	0.227	0.260	459
	low	3.649	0.197	0.140	59
42 $\leq$ weeks	median	3.359	0.118	0.457	66
	high	3.385	0.131	0.380	159

the buffer base reading y. For example, the 22 babies with $y \leq 36.1$ were therefore defined as suffering from fetal metabolic acidosis. The scatterplots were observed to shift slightly to the right, as y increases, suggesting that lighter babies are at risk among these premature babies.

The skewed normal distribution of Exercise 1.5.i was used to fit the observed birthweights at the three different levels of the buffer base reading. The parameters of the skewed normal distribution were estimated by iterative solution of equations (1.5.23), (1.5.24), and (1.5.25). The consequent estimates for μ, σ_1^2, and σ_2^2 are described in the first three rows of Table 2.7.1. We then estimated the probability that randomly chosen out of the $n = 85$, a baby with birthweight x would possess fetal metabolic acidosis, from Bayes' theorem,

$$\phi^*(x) = \frac{n_1 f_1(x)}{n_1 f_1(x) + n_2 f_2(x) + n_3 f_3(x)}, \tag{2.7.1}$$

where $n_1 = 22, n_2 = 14$, and $n_3 = 49$, and the $f_i(x)$ denote our three fitted skewed normal distributions. This produced curve (a) in Figure 2.7.1, which plots our probabilities ϕ^* of fetal metabolic acidosis as a function of the birthweight x. A later, smoothed and updated, version of this curve is reported by Low et al. (1981).

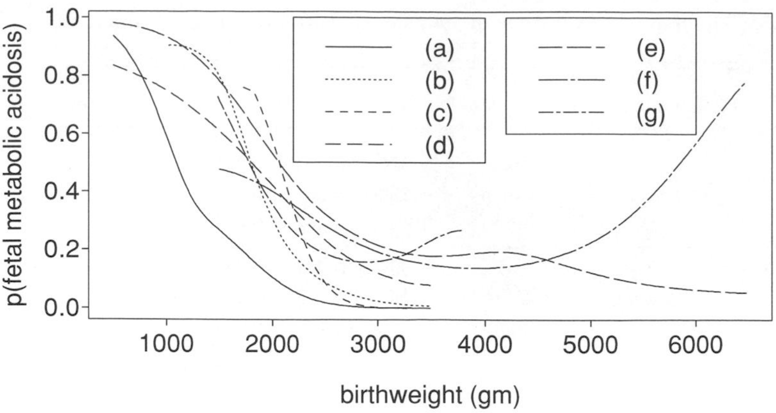

Figure 2.7.1. The probability of fetal metabolic acidosis, as a function of birthweight, for gestation ages (a) ≤ 33 weeks; (b) 34–35 weeks; (c) 36–37 weeks; (d) 38 weeks; (e) 39 weeks; (f) 40–41 weeks; (g) ≥ 42 weeks.

Our estimated probability that $y \leq 36.1$ is high for low birthweight x and low for high birthweight x. Curve (a) describes a "cross-over point" just above 1,700 gm. Premature babies with projected birthweights below the cross-over point are thought to possess a higher risk of fetal metabolic acidosis when compared with the risk of 20% for a typical baby in intensive care.

Curves (b), (c), (d), and (e) in Figure 2.7.1 simply plot the probabilities of fetal metabolic acidosis against birthweight, but for increasing gestational ages up to 39 weeks. The corresponding cross-over points were 2,165 gm, 2,335 gm, 2,612 gm, and 3,129 gm. Curves (f) and (g), for gestational ages exceeding 39 weeks, demonstrated considerably less relationship between the probability that $y < 36.1$ and birthweight. Overall, we had discovered a complex nonlinear interaction between birthweight and gestational age in its effect upon the probability of fetal metabolic acidosis. It would have been difficult to model this interaction using other techniques.

At this point in the analysis, the physicians discovered that our "cross-over points" virtually identically matched the already published tenth percentiles of the birthweight distributions, for different gestational ages, when taken across the whole population of Ontario. We therefore immediately consolidated our discovery that the babies at risk are those who are "light in the womb," in other words, the interuterine growth-retarded babies. The population birthweight distributions are approximately normal. The skewness of the data in our samples is created by the admissions policy of the intensive care unit.

Our conclusions were further checked out against the data sets by "cross-validating" the cross-over points x^* at each gestational age. For babies with gestational age ≤ 34 weeks, Low et al. (1981) reported the slightly larger data set

	$y \leq 36.1$	$y > 36.1$	
$x < x^*$	32	26	58
$x \geq x^*$	28	124	152

The 58 babies with $x < x^*$ have on average an estimated 55.17% chance of possessing fetal metabolic acidosis and should therefore be screened for postnatal complications. The 52 babies with $x \geq x^*$ have only a chance of around 18.42% of possessing fetal metabolic acidosis and should therefore be treated like typical intensive care patients.

As with all nonrandomized studies, our conclusions could be made spurious by the presence of "lurking variables." (See Section 1.2 (H), together with the famous Oldenberg Stork and impurity, frothing, and yield examples reported by Box, Hunter, and Hunter, 1978, p. 8 and p. 493.) However, none of the other 20 or so variables recorded in our data set was found to substantively affect our probabilities. The physicians interpreted this as indicating that fetal metabolic acidosis is not a condition that suddenly occurs just prior to or during labor, but that it is rather related to an unknown condition or "lurking variable" that is present in the womb throughout pregnancy (the effect of the mother smoking or the mother's chemical dependency provide just two possibilities).

SELF-STUDY EXERCISES

2.7.a Nine months to the day after a blackout in New York City on November 9th, 1965, *The New York Times* reported a sudden upsurge in the number of births. What would you infer regarding a possible increased rate of conception during the blackout? Gestational ages (lengths of pregnancy) are subject to random variation, and their standard deviation should be measured in weeks rather than days (see Section 2.7 and Izenman and Zabell, 1976).

2.7.b There are usually more births on Mondays and Tuesdays because of delays in Caesarean sections over the weekend. Carefully discuss whether induced births are possible "lurking variables" in the context of the New York blackout situation. The data in Table 2.7.2 are relevant.

2.7.c In a parentage case, the alleged father drove through a small rural town one evening and gave the mother a ride home. The following evidence was presented by the prosecution:

(a) Based upon a genetic blood test and a 50% prior probability, there was a posterior probability well exceeding 99% that the defendant was the father.

(b) Based upon ultrasound readings relating to projected weight in the womb taken several months afterwards, the baby must have been conceived on the night of the ride.

Table 2.7.2. *Numbers of births in New York, August 1–25, 1966.*

	Day						
Week	Monday	Tuesday	Wednesday	Thursday	Friday	Saturday	Sunday
1	451	468	429	448	466	377	344
2	488	438	455	468	462	405	377
3	451	497	458	429	434	410	351
4	467	508	432	426			

The defense introduced the New York blackout example, detailed discussions of Bayes' theorem, and difficulties in evaluating unknown genotype frequencies in the likelihood ratio. What do you think?

2.8 Practical Guidelines

The following practical guidelines relate to applications of Bayes' theorem for the evaluation of evidence and employ the notation of Worked Example 2A. See also Exercises 2.1.d and 2.1.e. The guidelines contrast with current standard practice in legal cases, for example, for genetic testing and DNA profiling, which gives rise for social concern at an international level.

(a) The ξ_i and ϕ_i may also be taken to be conditional probabilities, given any piece of evidence or background information Ω_0, which would not affect your prior assumption that each member of S is equally likely to be the Leopard.

(b) If more relevant evidence Ω^* is collected, in addition to Ω, then the current posterior probabilities ξ_i can be updated to new posterior probabilities by employing them as prior probabilities on a further application of Bayes' theorem (2.1.1). This should refer to new choices of the ϕ_i, where ϕ_i is now the probability $p(\Omega^* \mid \{i\}, \Omega, \Omega_0)$ and replaces $p(A \mid B_i)$ in (2.1.1).

(c) The two pieces of evidence Ω and Ω^* should not be regarded as independent, given $\{i\}$ and Ω_0, unless this assumption is definitely known to be true.

(d) An incorrect assumption of independence of Ω and Ω^*, given $\{i\}$ and Ω_0, would involve the oversimplification $p(\Omega^* \mid \{i\}, \Omega, \Omega_0) = p(\Omega^* \mid \{i\}, \Omega_0)$. This could substantially inflate the apparent overall evidence if repeated for a sequence of evidential items (e.g., the Collins case, see Freedman, Pisani, and Purves, 1991, pp. 233–234).

(e) If Ω and Ω^* are indeed independent, then simply reapply Bayes' theorem, since apparently correct shortcuts can give different answers. (For example, independence of events for each individual in a population does not necessarily imply independence of these events for a random man. Consequently, likelihood ratios relating to separate pieces of evidence can only be multiplied together after checking the correctness of the related probabilistic formulas.)

(f) It is absolutely essential to apply Bayes' theorem to probabilities relating to some sample space S, for example, the population of possible suspects, and to keep checking that any sets of probabilities, on the elements of S, add to unity. Shortcuts involving a random or "typical man" can be misleading.

(g) When studying a probability or conditional probability for one suspect, always list the corresponding probabilities for all other members of S (many of those probabilities may be equal for sets of suspects).

(h) Study the prosecutor's fallacy (Aitken, 1995, p. 37) and the so-named defender's fallacy (Aitken, 1995, pp. 38–39). The prosecutor's fallacy tells us not to confuse $p(\text{evidence} \mid \text{guilt})$ with $p(\text{guilt} \mid \text{evidence})$. The defender's fallacy tells us that a low value for $p(\text{guilt} \mid \text{evidence})$ may still be associated with a high "weight of evidence" or logarithm of the "likelihood ratio." In the context of Worked

Example 2A, the likelihood ratio when $\phi_1 = 1$ (suggesting a perfect match for the defendant who is identified with the first individual in the population) is

$$R = \frac{N - 1}{N\phi. - 1}. \tag{2.8.1}$$

This is the correct choice of R for the version of Bayes' theorem in part (d) of Exercise 2.1.d, but with the prior probability λ set equal to N^{-1}. In the context of the "defender's fallacy," note the unique role of our prior probability N^{-1} in situations where there is no previous evidence. The corresponding posterior probability of guilt

$$\xi_1 = \frac{N^{-1}R}{N^{-1}R + 1 - N^{-1}} = \frac{R}{R + N - 1} = \frac{1}{N\phi.} \tag{2.8.2}$$

appears to be more directly interpretable than R itself, thus resolving a suggestion by Aitken that the defender's fallacy is not really a fallacy. If this formula is applied to a population of N suspects, then the N in the prior probability N^{-1} should always equate with the number N of terms averaged when calculating $\phi.$ For example, when $N = 2$, $\xi_1 = 1/(1 + \phi_2)$, where ϕ_2 should be identified with the second suspect rather than a random member of a large population. Your choice of R itself should result from the laws of conditional probability, and should hence be interwoven with both the prior information and your choice of sample space. Note also that Bayes' theorem provides a valid procedure not only for the combination of more than one piece of evidence but also if there is no prior evidence apart from the population size of suspects.

(i) Work these ideas through simple numerical examples, for example, Exercise 2.1.d.

(j) These methods should be used only in situations where, based upon the available evidence, all probabilities and conditional probabilities can be specified by experts according to the laws of probability. Some types of evidence may preclude this. It typically would be unrealistic to ask a judge or jurors to become directly involved in this subtle assessment exercise.

(k) In situations where the evaluation of the ϕ_i depends upon unknown population parameters and refers to data sampled from the population, only random sampling should be performed (to justify a simple sampling model and to provide a representative sample), and a statistical technique should be employed that adequately compensates for the remaining uncertainty in the unknown parameters.

(l) Techniques adequately reflecting the heterogeneity of the ϕ_i across the population are encouraged, since an assumption of homogeneity can lead to quite misleading results (Leonard, 1999).

3

Models with a Single Unknown Parameter

3.0 Preliminaries and Overview

The most convincing advantages of the Bayesian approach will not become completely evident until we consider models with several parameters in Chapters 5 and 6. However, for a single parameter θ, contained in a specified sampling model, it is possible

(a) to use Bayesian intervals to draw finite sample inferences about θ in nonlinear situations (see Section 3.2, Exercises 3.2.b and 3.2.d, Sections 3.3 (B), 3.4, and 3.6), where other philosophies would either involve large sample methods or remain incomplete;
(b) to obtain point estimates with good mean squared error properties or with good frequency properties under a variety of loss functions, which would otherwise yield intractable non-Bayesian analyses (see Section 3.8);
(c) to formally incorporate prior information about θ, if this is available (see Sections 3.1–3.5), and to draw inferences, with good frequency properties, for example, if the prior information is vague (see Section 3.6);
(d) to handle parameter constraints, for example, $\theta > 0$ in a logical manner (see Section 3.1 and Exercises 3.4.m and 3.4.n);
(e) to make conditional probability statements about θ, given the observations actually observed, and which concur with the likelihood principle (Sections 3.2–3.6);
(f) to investigate hypotheses for θ in an extremely simple manner that permits the discernment of both "statistical" and "practical" significance (see Sections 3.2, 3.3 (B), and 3.8);
(g) to construct test statistics with convincing finite sample frequency power properties (Exercises 3.6.h and 3.10.g);
(h) to make predictive probability statements about future observations, for example, in nonlinear situations (Sections 3.2 and 3.3 (B));
(i) to still proceed if there is only partial prior information about θ, via Boltzmann's maximum entropy theorem (3.4.1); and
(j) to make sensible real-life decisions in the face of uncertainty about θ (Section 3.8). See also Ferguson (1967).

All results will depend upon the specification of a prior distribution for θ, and this may be quite difficult in practice. (There is no such entity as an "ignorance prior"; even improper prior distributions represent particular prior information.) Is this really a practical disadvantage? Perhaps the specification of a prior distribution is synonymous with the level of complexity of assumption (*input*) needed to draw precise conclusions (*output*), and perhaps without this complexity it is not possible to use a sampling model to draw completely convincing finite sample conclusions from a finite set of data points. In other words, the prior is needed to catalyze a sensible interaction between the data and the sampling model. Without a prior, the level of inductive thought needed to draw such conclusions may well be beyond our capacity. For example, the likelihood function for θ, on its own, is not enough to yield many of the practical advantages described above.

There are analogies with quantum theory, where it is necessary to assume that relative to the observer, the path of a particle must be defined probabilistically, and with fitting a curve to a finite set of points. Without making some assumptions about the curve, it is simply impossible to decide between a complexity of alternative curves. Similarly, without defining a prior probability distribution for θ, it might be impossible to develop a theory that decides between different types of posterior conclusion about θ, so that ad hoc or inductive conclusions would be needed.

Some of the prior distributions discussed in this chapter are "conjugate," that is, the posterior distribution belongs to the same simple family as the prior distribution. However, we can find no overwhelming reason for using a conjugate prior, and a variety of prior distributions without this special property are also considered. It is, for example, possible to seek a transformation of the form $\lambda = g(\theta)$ such that it is sensible to take λ to possess a normal prior distribution, or some mixture of normal distributions. This key idea will parallel the normal likelihood approximation (3.4.4). The latter is the scalar version of (1.3.11), which motivated transformations to normality in Section 1.3. This ongoing theme will be resumed in Chapters 5 and 6.

3.1 The Bayesian Paradigm

(A) *The Inductive Synthesis*

Leonard (1980, 1983) was influenced by the data analysis described in Section 2.7, and by later conversations with George E. P. Box, to investigate the "inductive modeling process," where inductive thought and data analysis are needed to develop and check plausible models (for example, the diagnostic model in (2.7.1)), in relation to the "scientific background"; mathematics and deductive thought are then used to analyze these models (e.g., the analysis in Exercise 1.5.i of the sampling model in equation (1.5.19)). Box (1983) confirms that applied statisticians should iterate between modeling and a parametric analysis. Too much concentration on deduction can reduce insight, and too much concentration on induction can reduce focus. However, in the current section, we can largely concentrate on deduction and just consider a parametric analysis, with the exception that your intuition may be needed to help construct a meaningful prior distribution. The iterative inductive/deductive modeling process proposed by Leonard (1980) and Box (1980) should be regarded as an essential ingredient of Aitken's inductive synthesis. Although a prior distribution assists the deductive reasoning necessary for parametric inference, the modeling process quite frequently needs to refer to inductive and scientific reasoning.

Let the observation vector $\mathbf{y} = (y_1, \ldots, y_n)^T$ denote the numerical realization of a random vector $\mathbf{Y} = (Y_1, \ldots, Y_n)$, which is taken to possess density or probability mass function $p(\mathbf{y} \mid \theta)$, for $\mathbf{y} \in S$ and $\theta \in \Theta$. However, assume that Θ is some subset of the real line R^1. Also, assume that some iteration in the modeling process has already occurred and that for the moment at least, $p(\mathbf{y} \mid \theta)$ is taken to be completely specified, except that θ is unknown. In this case, $l(\theta \mid \mathbf{y})$ may be taken to summarize the sampling information, given the assumed truth of the sampling model.

(B) *Prior Information*

Before observing $\mathbf{y}$, most scientists will possess some prior information about θ, such as previous data sets, knowledge, or subjective scientific advicc, which might suggest plausible values for θ. For example, when considering the unknown proportion θ of the at-risk population of Wisconsin that possesses the AIDS/HIV virus, you might, upon reflection, say that "my best guess is 0.02, with a standard error of 0.015." An (informative) Bayesian is a scientist who believes that he can represent his prior information on θ by a probability distribution defined on all events (measurable, of course) in Θ. However, we shall see that Bayesians can also handle vague situations where relatively little prior information is utilized during the inferential process. An informative Bayesian can calibrate his prior probabilities via the auxiliary experiment introduced in Exercise 1.1.k. If he can calibrate his probabilities rationally, then they should, at least in principle, satisfy the laws of probability, in particular, Kolmogorov's axiom or the countable additivity property.

(C) *Assessing a Prior Distribution*

If θ possesses a beta distribution with parameters α and β, and density

$$\pi(\theta) = \frac{\Gamma(\alpha+\beta)}{\Gamma(\alpha)\Gamma(\beta)}\theta^{\alpha-1}(1-\theta)^{\beta-1} \qquad (0 < \theta < 1, 0 < \alpha, \beta < \infty), \tag{3.1.1}$$

then the prior mean, or expectation, of θ is

$$\xi = E(\theta) = \frac{\alpha}{\alpha+\beta}, \tag{3.1.2}$$

and the prior variance is

$$\phi = \operatorname{var}(\theta) = \frac{\xi(1-\xi)}{\gamma+1}, \tag{3.1.3}$$

where $\gamma = \alpha + \beta$ is known as the prior sample size. Note that $\gamma = \phi^{-1}\xi(1-\xi) - 1$, $\alpha = \gamma\xi$, and $\beta = \gamma(1-\xi)$. Therefore, choices of 0.02 for ξ and 0.015 for $\phi^{\frac{1}{2}}$ would, in the absence of further restrictions and subject to the beta assumption, suggest $\gamma = 86.11$, $\alpha = 1.72$, and $\beta = 84.39$. Therefore, this prior information is roughly equivalent to the information that if you observed a random sample of about 86 individuals out of the population of at-risk individuals in Wisconsin, you would believe that out of the 86, between 1 and 2 (your prior frequency is 1.72) possessed the AIDS/HIV virus, and the remainder (with prior frequency 84.39) didn't. Summing of the two "prior frequencies" α and β gives the prior sample size $\gamma = \alpha + \beta = 86.11$. The "prior sample size" therefore measures the strength of the prior information.

Under the assumption (3.1.1), the prior probability that $\theta \in (a, b)$ is

$$p(a < \theta < b) = \frac{B_b(\alpha, \beta) - B_a(\alpha, \beta)}{B_1(\alpha, \beta)}, \tag{3.1.4}$$

where

$$B_a(\alpha, \beta) = \int_0^a \theta^{\alpha-1}(1-\theta)^{\beta-1} d\theta \tag{3.1.5}$$

denotes the unnormalized incomplete beta function. For (3.1.1) to provide a reasonable prior density, the probabilities in (3.1.4) would then need to roughly match your prior opinions for any choice of a and b. If you also incorporate an opinion that, say, $\theta > 0.01$, then the above beta distribution can be truncated to the range $0.01 < \theta < 1$, and α and β can be adjusted to ensure that $\xi = 0.02$ and $\phi^{\frac{1}{2}} = 0.015$ match the mean and standard deviation of your truncated beta distribution.

In a situation where you instead possess the information that out of a random sample of $n = 1{,}000$ at-risk inhabitants of Britain, $y = 44$ possessed the AIDS/HIV virus, yielding a point estimate of $p = 0.044$ for the proportion of people in Britain with the AIDS/HIV virus, you might decide to set $\xi = 0.044$ because you have no reason to suppose that the rate in Wisconsin is different from the rate in Britain. However, the prior sample size could be reduced to, say, $\gamma = 250$ because you are unsure how relevant this information is in Wisconsin. This gives $\alpha = 11$, $\beta = 239$, and $\phi^{\frac{1}{2}} = 0.013$.

The assessment of a beta prior distribution is discussed further by Chaloner and Duncan (1983). However, any probability distribution on $\Theta = \{\theta : 0 < \theta < 1\}$ might provide a reasonable prior distribution for θ, depending upon the prior information available. In practice, it is stretching credulity to anticipate that a scientist will be able to fully and uniquely specify this probability distribution without imbedding just a few specific pieces of information into a convenient parametric family. It sometimes suffices to specify the prior distribution without overwhelming precision, particularly when the parameter space Θ is bounded, or when the likelihood function contains substantial information about θ. Note that when considering a population proportion, another possibility is to take $\lambda = \text{logit}(\theta) = \log\theta - \log(1-\theta)$ to possess a normal $N(\mu, \sigma^2)$ prior distribution. Alternatively, the probit $\lambda = \Phi^{-1}(\theta)$ and/or mixtures of normal distributions can be considered.

(D) *Other Prior Information*

As a further example, let $\theta =$ the proportion of the population of California that likes cheese. Given, say, a prior mean of $\xi = 0.3$ and a prior sample size of $\gamma = 20$,

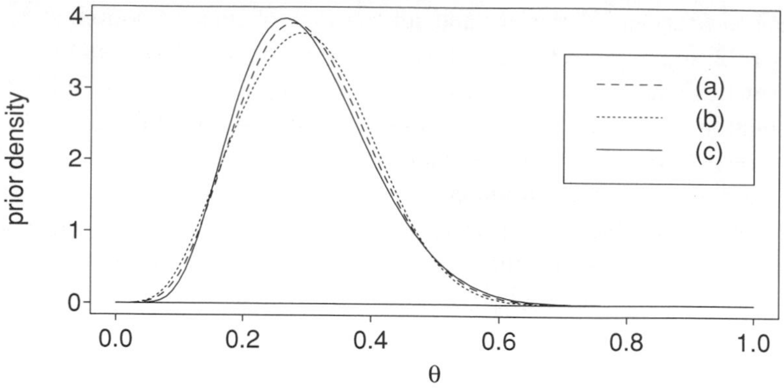

Figure 3.1.1. Prior densities for a population proportion: (a) beta density ($\alpha = 6$, $\beta = 14$); (b) mixture of two beta densities; (c) logistic normal density.

it is possible to use a beta prior distribution with parameters $\alpha = 6$ and $\beta = 14$, yielding a prior standard deviation of 0.1. The corresponding prior density is described by curve (a) of Figure 3.1.1. There are, however, many prior densities with mean 0.3 and standard deviation 0.1. For example, curve (b) of Figure 3.1.1 is a 50% mixture of two beta curves, the first with parameters $\alpha = 6$ and $\beta = 18$, and the second with parameters $\alpha = 10.27$ and $\beta = 19.07$. Curve (c) also matches the first two moments by taking $\lambda = \log\theta - \log(1-\theta)$ to possess a $N(-0.9, 0.25)$ distribution. Note that curve (c) yields prior median $e^{-0.9}/(1+e^{-0.9}) = 0.313$ for θ. When deciding between the choices (a), (b), and (c), it would be useful to specify at least one prior probability; prior probabilities are easy to incorporate via the logistic normal curve (see Exercise 3.1.f).

It has to be recognized that some prior information may be so diverse or eclectic that it may be impossible to represent this information by a probability distribution on Θ. Also, in some studies, the main purpose of prior scientific or medical information may be for it to be refuted by the current data (e.g., the prior information discussed in Section 2.7 for the fetal metabolic acidosis study). Refutation of prior information can lead to scientific discovery. In formal terms, the Bayesian paradigm can, however, incorporate only prior information of a type that can be represented by probabilities, for example, via a process like calibration with a spinning pointer (Exercise 1.1.k), that do indeed satisfy the laws of probability. Note that Press (1989, Ch. 4) considers the subjective assessment of prior distributions in greater detail, with an excellent bibliography. He believes, for example, that it is possible to assess accurately the probability of nuclear war. Tversky and Kahneman (1974), Hogarth (1980), and Kahneman, Slovic, and Tversky (1982) take this one stage further by considering psychological factors relating to subjective probability assessment. Smith (1988, Ch. 4) discusses a relative frequency approach for assessing subjective probabilities.

According to Dickey (1976), "Savage's principle of precise measurement" implies that it is only necessary to specify a prior distribution imprecisely. This principle needs to be checked numerically in any special case, in particular when the sample size is small to moderate. Dickey and Chen (1985) propose further techniques for subjective probability modeling. West (1984) proposes Bayesian procedures that handle outliers and in West (1988) discusses the modeling of expert opinion. More detailed bibliographies on subjective elicitation are provided by Kadane and Wolfson (1998), O'Hagan (1998), and Craig et al. (1998), and in the discussions of these papers.

Worked Example 3A: *Eliciting a Weibull Prior Distribution*

Let θ denote the average recovery time for patients with disease D. Your prior distribution for θ is such that with $\alpha > 0$, $\phi = \theta^\alpha$ possesses an exponential distribution, with c.d.f. $F(\phi) = 1 - \exp\{-\beta\phi\}$, for $0 < \phi < \infty$, so that θ possesses a Weibull distribution, with positive parameters α and β, and c.d.f. $\Lambda(\theta) = 1 - \exp\{-\beta\theta^\alpha\}$ for $0 < \theta < \infty$. Based upon your prior information, your first quartile for θ is $Q_1 = 8$ days and your third quartile is $Q_3 = 15$ days. Find the prior median Q_2 of θ.

Model Answer 3A:

As $Q_1 = 8$, Q_2, and $Q_3 = 15$ are the three quartiles

$$\exp\left\{-\beta 8^\alpha\right\} = 0.75, \tag{3.1.6}$$

$$\exp\left\{-\beta Q_2^\alpha\right\} = 0.50, \tag{3.1.7}$$

$$\exp\left\{-\beta 15^\alpha\right\} = 0.25. \tag{3.1.8}$$

Transforming both sides of (3.1.6) and (3.1.8) yields

$$\log\beta + 2.079\alpha = -1.246$$

and

$$\log\beta + 2.708\alpha = 0.327.$$

Consequently, $\alpha = 2.501$ and $\log\beta = -6.446$. Since

$$\log\beta + \alpha\log Q_2 = -0.367,$$

we therefore have $\log Q_2 = 2.431$ and $Q_2 = 11.370$.

Worked Example 3B: *Eliciting a Log-Normal Prior Distribution*

Let θ denote the number of purchases at a grocery store on Halloween. Your prior distribution for $\gamma = \log\theta$ is normal with mean μ and variance σ^2, so that

$$E(\theta) = \xi = \exp\left\{\mu + \tfrac{1}{2}\sigma^2\right\}$$

and

$$\text{var}(\theta) = \phi = \xi^2\left(e^{\sigma^2} - 1\right).$$

Your prior information suggests that the mean and standard deviation of θ are $\xi = 250$ and $\phi^{\frac{1}{2}} = 25$, respectively. Find μ and σ^2, together with your prior probability that θ exceeds 250.

Model Answer 3B:

As

$$\phi = \xi^2\left(e^{\sigma^2} - 1\right),$$

we have

$$\sigma^2 = \log\left(1 + \frac{\phi}{\xi^2}\right).$$

Since,

$$\xi = \exp\left\{\mu + \tfrac{1}{2}\sigma^2\right\},$$

it follows that

$$\mu = \log\xi - \frac{1}{2}\log\left(1 + \frac{\phi}{\xi^2}\right).$$

Consequently,

$$\sigma^2 = \log\left(1 + \frac{25^2}{250^2}\right) = 0.00995$$

and

$$\mu = \log 250 - \tfrac{1}{2}\sigma^2 = 5.51649.$$

Also

$$\begin{aligned} p(\theta > 250) &= p(\gamma > \log 250) \\ &= 1 - \Phi\left(\frac{\log 250 - \mu}{\sigma}\right) \\ &= 1 - \Phi\left(\frac{\sigma}{2}\right) = 1 - \Phi(0.0449) = 0.4801. \end{aligned}$$

SELF-STUDY EXERCISES

3.1.a Let θ = proportion of the world's population in 2050 A.D. that is Chinese. Based upon your current information, assess your probability distribution for θ.

3.1.b Let θ = average beer consumption in pints per day of Englishmen during the year 3001 A.D. Assuming that your prior distribution for θ is normal, assess your prior standard deviation for θ and your prior probability that θ exceeds 2 pints.

3.1.c Let ξ = average current survival time, in years, for patients with the AIDS/HIV virus. Assuming that $\theta = 1/\xi$ possesses a Gamma distribution, assess your current probability distribution for ξ.

3.1.d Read DeGroot (1974) and Lindley, Tversky, and Brown (1979) and discuss how it might be possible to combine the prior distributions of a number p of experts or to reach a consensus among a group of experts. You should distinguish between the two situations where (a) the p prior distributions relate to the same piece of information I; (b) the p prior distributions relate to separate pieces of information $I_1, \ldots, I_p$. Is it reasonable to average the p prior distributions in case (a) and to take products of p related prior likelihoods (see Section 3.2) in case (b)?

3.1.e Let $\Theta = \{1, 2, \ldots, k\}$ and let q_{ij} denote the ith expert's prior probability that $\theta = j$, $(i = 1, \ldots, p;\ j = 1, \ldots, k)$. Each expert has a common information base I. Then the vectors $\mathbf{q}_i = (q_{i1}, \ldots, q_{ik})^T$ can be regarded as statistical data. Assume that $\mathbf{q}_1, \mathbf{q}_2, \ldots, \mathbf{q}_p$ compose a random sample and that each $\mathbf{q}_i$ follows the Dirichlet distribution with parameters $\alpha\xi_1, \ldots, \alpha\xi_p$, and density $p(\mathbf{q})$, where

$$p(\mathbf{q}) = p(q_1, \ldots, q_k) = \frac{\Gamma(\alpha)}{\prod_{j=1}^{k} \Gamma(\alpha\xi_j)} \prod_{j=1}^{k} q_j^{\alpha\xi_j}$$

$$(\mathbf{q} \in S_U, \boldsymbol{\xi} \in S_U, 0 < \alpha < \infty),$$

confined to the unit simplex S_U. Then $\mathbf{q}_i$ has mean vector $\boldsymbol{\xi} = (\xi_1, \ldots, \xi_k)^T$, and its jth component has variance $\mathrm{var}(q_j) = \xi_j(1 - \xi_j)/(\alpha + 1)$, for $j = 1, \ldots, k$. Show that the average vector $\mathbf{q}_. = p^{-1} \sum_{i=1}^{p} \mathbf{q}_i$ is an unbiased estimate of $\boldsymbol{\xi}$. Do the elements of $\mathbf{q}_.$ therefore provide a reasonable common prior distribution for θ? Would it be more sensible to report just a separate set of posterior probabilities for each expert and see what differences are created by the differences in the prior assumptions?

(Other methods along these lines are described by Gilardoni, 1989.)

3.1.f When eliciting a prior distribution from a scientist for a binomial proportion θ, the scientist states that $p(\theta \leq \theta_1) = \xi_1$ and $p(\theta \leq \theta_2) = \xi_2$, where $\xi_1 < \xi_2$.

Show how this information may be represented by the logistic normal prior, which takes $\gamma = \mathrm{logit}(\theta) = \log(\theta) - \log(1 - \theta) \sim N(\mu, \sigma^2)$, and express μ and σ^2 in terms of $\xi_1, \xi_2, \alpha_1 = \log(\theta_1)$, and $\alpha_2 = \log(\theta_2)$.

If instead the prior information consists of five prior probabilities, investigate conditions under which the prior distribution can be represented by a mixture of two logistic normal distributions.

3.2 Posterior and Predictive Inferences

(A) *Bayes' Theorem*

Under the assumptions of Section 3.1, the posterior density of θ is

$$\pi(\theta \mid \mathbf{y}) \propto \pi(\theta) l(\theta \mid \mathbf{y}) \qquad (\theta \in \Theta), \tag{3.2.1}$$

where the proportionality sign "$\propto$" denotes "proportional, as a function of θ" and $\pi(\theta)$ denotes the prior density, that is, the density of the prior distribution of θ. The constant of proportionality can depend upon $\mathbf{y}$, but not upon θ. It should be chosen to ensure that $\int_\Theta \pi(\theta \mid \mathbf{y}) = 1$. Then (3.2.1) provides a mathematical representation of the statement

$$\text{POSTERIOR INFORMATION} = \text{PRIOR INFORMATION} + \text{SAMPLING INFORMATION}, \tag{3.2.2}$$

which may be spelled out in words as

$$\text{POSTERIOR DENSITY} \propto \text{PRIOR DENSITY} \times \text{LIKELIHOOD}. \tag{3.2.3}$$

Therefore the posterior density summarizes the total information, after viewing the data, and provides a basis for posterior inferences regarding θ. Edwards (1972) proposes a *prior likelihood* $l(\theta)$ rather than a prior density, for example, based upon hypothetical prior observations, and he describes $\log l(\theta)$ as the "prior support." This has some appeal, because the posterior likelihood $l^*(\theta \mid \mathbf{y})$ and posterior support $\log l^*(\theta \mid \mathbf{y})$ satisfy

$$\log l^*(\theta \mid \mathbf{y}) = \log l(\theta) + \log l(\theta \mid \mathbf{y}) \tag{3.2.4}$$

and provide a precise representation of (3.2.2). Edwards's approach does not involve probability distributions of θ and hence is free from the formal measure-theoretic requirement (e.g., Ash, 1972, Ch. 2) of the specification of a dominating measure.

Equation (3.2.1) is a consequence of Bayes' theorem that can, more pedantically, be written as

$$\pi(\theta \mid \mathbf{y}) = \frac{\pi(\theta) p(\mathbf{y} \mid \theta)}{p(\mathbf{y})} \qquad (\theta \in \Theta), \tag{3.2.5}$$

where

$$p(\mathbf{y}) = \int_\Theta p(\mathbf{y}, \theta)\, d\theta \tag{3.2.6}$$

$$= \int_\Theta \pi(\theta) p(\mathbf{y} \mid \theta)\, d\theta \tag{3.2.7}$$

denotes the "prior predictive distribution for $\mathbf{y}$," that is, it is your subjective distribution for $\mathbf{y}$, before viewing the data, when θ is unknown. Note that Zellner (1988) provides an interesting description of Bayes' theorem in terms of maximum entropy, and optimal information processing. These ideas from information theory are related to Lindley's information measure I_L (Lindley, 1956). Note that I_L is the expected entropy distance between the prior and the posterior, so that

$$I_L = \int_S p(\mathbf{y}) \int_\Theta \pi(\theta \mid \mathbf{y}) \log\left\{\frac{\pi(\theta \mid \mathbf{y})}{\pi(\theta)}\right\} d\theta d\mathbf{y}.$$

There are many further applications of this measure of the expected amount of information in the statistical experiment. For example, Bernardo (1979) and Bernardo and

Ramon (1998) summarize the literature of "reference priors" that refer to this criterion. Note the alternative representation

$$I_L = \int_{\Theta} \pi(\theta) \int_S p(\mathbf{y} \mid \theta) \log \frac{p(\mathbf{y} \mid \theta)}{p(\mathbf{y})} \, d\mathbf{y} \, d\theta.$$

Consequently, the concept of entropy is imbedded into the Bayesian as well as the likelihood literature.

Bayesians regard the parameter θ as a random variable; its probability distribution represents your uncertainty in θ. Is this any less reasonable than the assumption, in Section 1.1, that the observed vector $\mathbf{y}$ is some realization of a random vector $\mathbf{Y}$? Bayesians are able to apply all properties of probability when considering the joint distribution of $\mathbf{Y}$ and θ, and any related conditional or marginal distribution. Note that the Neyman–Fisher factorization theorem of Section 1.5 (C) implies that if a sufficient statistic $t(\mathbf{y})$ exists for θ, then the posterior density (3.2.1) or (3.2.5) will depend upon the data only via this sufficient statistic. Note, furthermore, that if the prior density $\pi(\theta)$ remains the same for two experiments ϵ_1 and ϵ_2 for the same parameter θ, yielding respective likelihoods $l_1(\theta \mid \mathbf{y}_1) \propto l_2(\theta \mid \mathbf{y}_2)$, then the posterior densities $\pi_1(\theta \mid \mathbf{y}_1) \equiv \pi_2(\theta \mid \mathbf{y}_2)$ based on the two experiments are equivalent in evidential meaning, so that inferences based just on the posterior density will satisfy the likelihood principle (see Exercises 1.5.e and 1.5.f). Dickey (1973) discusses the scientific reporting of Bayesian results and suggests that the posterior distribution should be reported for a variety of choices of prior distribution.

(B) *Box's Model-Checking Criterion*

Box (1980) recommends using Bayes' theorem (3.2.1) for conditional inferences, given the truth of the assumed model, and then considers the prior predictive probability

$$\alpha_p = \text{prob}\left[\log p(\mathbf{Y}) \leq \log p(\mathbf{y})\right] \tag{3.2.8}$$

for checking the model, where $\mathbf{y}$ denotes the numerical realization of the random observation vector $\mathbf{Y}$ and $p(\mathbf{y})$ is itself the prior predictive density or p.m.f. in (3.2.7). He recommends refutation of the current model if α_p is judged to be too small, that is, if the observed $\log p(\mathbf{y})$ is too small in the sense that it is too far to the left in relation to the left tail of its prior predictive distribution. The criterion (3.2.8) depends upon the choice of prior $\pi(\theta)$ and, in fact, investigates the reasonability of the combination of the sampling model and prior, in relation to the data. You are also left with the task of judging how small α_p should be, in relation to sample size, before deciding whether or not to refute the model. (See also, Exercise 6.1.a.) Box also recommends the consideration of similar prior predictive probabilities when $\log p(\mathbf{y})$ is replaced by other diagnostic statistics. Rubin (1984) suggests alternative versions of (3.2.8) based upon the posterior predictive distribution of the next paragraph. However, we are left with the open question of how you should check your model against a completely general alternative. Both lateral thinking and a careful comparison with the scientific background are important components.

(C) *Predictive Distributions*

Suppose now that $\mathbf{z}$ consists of an $m \times 1$ vector $\mathbf{z}$ of future observations, such that $\mathbf{y}$ and $\mathbf{z}$ are assumed to possess joint density or probability mass function $p(\mathbf{y}, \mathbf{z} \mid \theta)$ for $\mathbf{y} \in S, \mathbf{z} \in S^*$, and $\theta \in \Theta$, where S and S^* consist of all possible realizations of $\mathbf{y}$ and $\mathbf{z}$, and Θ denotes the parameter space. In many examples, $\mathbf{y}$ and $\mathbf{z}$ will be independent, given the same parameter θ, in which case $p(\mathbf{y}, \mathbf{z} \mid \theta)$ will take the form $p(\mathbf{y} \mid \theta)p(\mathbf{z} \mid \theta)$. Then the (posterior) "predictive density" or predictive probability mass function of $\mathbf{z}$, given $\mathbf{y}$, is

$$p(\mathbf{z} \mid \mathbf{y}) = \int_\Theta p(\mathbf{z} \mid \theta, \mathbf{y})\pi(\theta \mid \mathbf{y})\, d\theta \qquad (\mathbf{y} \in S, \mathbf{z} \in S^*), \tag{3.2.9}$$

where $p(\mathbf{z} \mid \theta, \mathbf{y})$ denotes the conditional sampling density of $\mathbf{z}$, given $\mathbf{y}$, and reduces to $p(\mathbf{z} \mid \theta)$ when $\mathbf{y}$ and $\mathbf{z}$ are independent, given θ.

(D) *Inferences for a Binomial Probability*

Assume that a cell frequency y is taken to possess a binomial distribution with probability θ and sample size n and that the prior distribution of θ is taken to be beta with parameters α and β. As functions of θ, $l(\theta \mid y) \propto \theta^y(1-\theta)^{n-y}$ and $\pi(\theta) \propto \theta^{\alpha-1}(1-\theta)^{\beta-1}$. Therefore, from (3.2.1), the posterior density of θ, given y, is

$$\pi(\theta \mid y) \propto \theta^{\alpha+y-1}(1-\theta)^{\beta+n-y-1} \qquad (0 < \theta < 1,\ 0 < \alpha, \beta < \infty). \tag{3.2.10}$$

This is a beta density, with updated parameters $\alpha^* = \alpha + y$ and $\beta^* = \beta + n - y$. Consequently, from standard properties of the beta distribution,

$$\pi(\theta \mid y) = \frac{1}{B_1(\alpha + y, \beta + n - y)}\theta^{\alpha+y-1}(1-\theta)^{\beta+n-y-1}, \tag{3.2.11}$$

where $B_1(\alpha, \beta) = \Gamma(\alpha+\beta)/\Gamma(\alpha)\Gamma(\beta)$. Since the prior mean of θ is $\xi = \alpha/(\alpha+\beta)$, the posterior mean is

$$E(\theta \mid y) = \xi^* = \frac{\alpha^*}{\alpha^* + \beta^*} = \frac{\alpha + y}{\alpha + \beta + n} = \frac{np + \gamma\xi}{n + \gamma}, \tag{3.2.12}$$

where $p = y/n$ and $\gamma = \alpha + \beta$. The estimator for θ in (3.2.12) combines the prior and posterior information of θ. It takes the form of a weighted average of the sample proportion p and the prior mean ξ. The weights are the actual sample size n and the prior sample size γ. This weighted average form is usually available only when the prior distribution is "conjugate," that is, when the prior and posterior densities assume similar functional forms, but with different parameters.

As the prior variance of θ is $\xi(1-\xi)/(\gamma+1)$, the posterior variance is

$$\operatorname{var}(\theta \mid y) = \frac{\xi^*(1-\xi^*)}{\gamma^* + 1}, \tag{3.2.13}$$

where ξ^* is defined in (3.2.12), and $\gamma^* = \gamma + n$ denotes the "posterior sample size."

It is not always true that $\mathrm{var}(\theta \mid y) < \mathrm{var}(\theta)$. Indeed, the posterior variance can be larger when p is very large in relation to ξ. In such situations, the prior estimate will be effectively refuted by the sampling information, causing the posterior distribution to provide less precise information than the prior distribution.

Updating the prior probability in (3.1.4), the posterior probability that $\theta \in (a, b)$ can be computed from

$$p(a < \theta < b \mid \mathbf{y}) = I_b(\alpha^*, \beta^*) - I_a(\alpha^*, \beta^*), \tag{3.2.14}$$

where

$$I_a(\alpha, \beta) = \frac{B_a(\alpha, \beta)}{B_1(\alpha, \beta)} \tag{3.2.15}$$

denotes the normalized incomplete beta function, with $B_a(\alpha, \beta)$ satisfying (3.1.5). Equation (3.2.14) provides the only analytic procedure we know for making a valid probabilistic conditional inference for a binomial probability θ, given $\mathbf{y}$ and for finite n, which does not refer to some approximation. For example, the fiducial approach (Fisher, 1935; Edwards, 1976; Wilkinson, 1977) and structural approach (Fraser, 1966) do not yield suggestions in the binomial situation that specifically address finite sample size.

(E) *Bayesian Intervals*

Since the probability (3.2.14) can be calculated for any interval (a, b), and the posterior density $\pi(\theta \mid y)$ can be sketched, as a function of θ, there really is no need to accord with frequency convention by constraining attention to 95% and 99% intervals.However, the $100(1 - \epsilon)\%$ equal-tailed Bayesian interval for θ is $(g(\frac{1}{2}\epsilon), g(1 - \frac{1}{2}\epsilon))$, where $g(\epsilon)$ satisfies

$$\epsilon = I_{g(\mathcal{E})}(\alpha^*, \beta^*), \tag{3.2.16}$$

and $I_a(\alpha, \beta)$ satisfies (3.2.15). Highest posterior density (HPD) regions (e.g., Berger, 1985, p. 140) are not in general equal tailed, and do not possess an easy invariance property, when considering nonlinear transformations of the parameters. Numerical examples of HPD regions are described in Sections 3.3 (B) and 3.5 (D).

(F) *Investigating Hypotheses*

Suppose, next, that you wish to investigate a null hypothesis $H_0 : \theta = \theta_0$ against the data y and prior information. Then you may do this simply by drawing the posterior density $\pi(\theta \mid y)$ and by distinguishing the position of $\theta = \theta_0$ on your θ axis. You can then make an applied judgment, for example, from the shape of the prior density, its variability, and the position of $\theta = \theta_0$ in relation to the tails of the density, to decide whether or not H_0 is reasonable. One feature of the posterior distribution is the "Bayesian significance probability"

$$\epsilon_0 = p(\theta \leq \theta_0 \mid y) = I_{\theta_0}(\alpha^*, \beta^*), \tag{3.2.17}$$

but all other features of the posterior density should be considered. Consideration of the magnitude of ϵ_0 parallels the frequency notion of judging "statistical significance." Further consideration of $\pi(\theta \mid y)$ and θ_0 in relation to the scientific background helps us to judge "practical significance." For example, if $\gamma^* = (n + \gamma)$ is very large, $\pi(\theta \mid y)$ is closely concentrated about ξ^*, and θ_0 is very close to ξ^*, a small value of ϵ_0 or $1 - \epsilon_0$ may be meaningless in terms of practical differences. Alternatively, if γ^* is small, the posterior density may be very flat, and we may not wish to refute H_0 based upon such a small amount of information, even if either ϵ_0 or $1 - \epsilon_0$ is small. This approach to hypothesis testing avoids Lindley's paradox (Lindley, 1957), since it does not place a positive prior probability on H_0. We find ourselves currently unable to recommend placing positive probabilities on values of parameters with otherwise continuous prior densities, since the properties of the posterior distribution can then become surprisingly counterintuitive (see Worked Example 3D) in the manner it is influenced by the prior assumptions. This is an area that requires further investigation and subtle interpretation.

If the results of your investigation of H_0 are to be interpreted objectively and, say, used by society to assume the truth of a fresh piece of knowledge, then (i) your posterior conclusions should be reconsidered under a broad range of prior assumptions, including the vague prior assumptions of Section 3.6; (ii) the sampling distribution should be objectively justified; for example, independence assumptions typically need to be justified by randomization at the design stage (see Box and Guttman, 1966); (iii) the frequency properties of the statistical procedures should be considered (see, for example, the last two paragraphs of Section 3.6); (iv) selective reporting of results should be avoided (see Dawid and Dickey, 1977); (v) the susceptability of the conclusions to the effects of lurking variables should be considered; and (vi) the investigation should be replicated as often as possible.

(G) *Other Posterior Estimates*

The posterior mean ξ^* in (3.2.12) is not the only summary of the posterior distribution (3.2.10), which might reasonably be taken to estimate θ. Another popular choice is the posterior mode or generalized maximum likelihood estimate $\tilde{\theta}$. In our example, $\tilde{\theta} = (\alpha + y - 1)/(\beta + n - y - 1)$ is the value of θ maximizing the posterior density (3.2.10) as a function of θ. Furthermore, the posterior median θ_M satisfies

$$0.5 = p(\theta \leq \theta_M \mid y) = I_{\theta_M}(\alpha^*, \beta^*) \qquad (3.2.18)$$

and can therefore be computed by inverting an incomplete beta function.

(H) *Prediction of a Binomial Frequency*

Suppose now that given y, we wish to predict a future frequency z, which is independent of y, given θ, and possesses a binomial distribution, with probability θ and sample size m. Then the predictive probability mass function of z, given y, is

$$\begin{aligned} p(z \mid y) &= \int_0^1 p(z \mid \theta)\pi(\theta \mid y)\, d\theta \\ &= \frac{{}^mC_z}{B_1(\alpha^*, \beta^*)} \int_0^1 \theta^{z+\alpha^*-1}(1-\theta)^{m-z+\beta^*-1} d\theta \\ &= \frac{{}^mC_z B_1(z+\alpha^*, m-z+\beta^*)}{B_1(\alpha^*, \beta^*)} \qquad (z = 0, 1, \ldots, m), \end{aligned} \tag{3.2.19}$$

where ${}^mC_z = m!/\{z!(m-z)!\}$. Again, the Bayesian paradigm seems to provide the only convincing rationale for finite sample predictive inferences of this type. The predictive mean of z, given $\mathbf{y}$, is

$$\begin{aligned} E(z \mid \mathbf{y}) &= \underset{\theta|y}{E}\left[E(z \mid \theta)\right] \\ &= \underset{\theta|y}{E}(m\theta) = m\xi^* \\ &= \frac{m(np+\gamma\xi)}{n+\gamma}, \end{aligned} \tag{3.2.20}$$

and the predictive variance of z, given y, is

$$\begin{aligned} \operatorname{var}(z \mid y) &= \underset{\theta|y}{E}\left[\operatorname{var}(z|\theta)\right] + \underset{\theta|y}{\operatorname{var}}\left[E(z \mid \theta)\right] \\ &= \underset{\theta|y}{E}\left[m\theta(1-\theta)\right] + \underset{\theta|y}{\operatorname{var}}[m\theta] \\ &= m\xi^*(1-\xi^*) - m \operatorname{var}(\theta \mid y) + m^2 \operatorname{var}(\theta \mid y) \\ &= m\frac{\gamma^*+m}{\gamma^*+1}\xi^*(1-\xi^*). \end{aligned} \tag{3.2.21}$$

Predictive distributions for a variety of problems in forensic science are described by Aitken et al. (1996) and Izenman et al. (1998).

Worked Example 3C: *Sharp Null Hypotheses*

Let Y be binomial with probability θ and sample size n. Given $Y = y$, you wish to investigate $H_0: \theta = \theta_0$ versus $H_1: \theta \neq \theta_0$, where θ_0 is specified in advance. You could investigate H_0 via the Bayesian significance probability (3.2.17), based upon continuous beta prior and posterior distributions. In Section 3.6, this will be shown, in a special case, to match the classical frequency approach. However, suppose that you intend to employ a more complicated prior distribution as follows:

(1) $p(H_0) = \phi$, $p(H_1) = 1 - \phi$.
(2) Given H_1, θ is beta distributed with parameters α and β.

Based upon this prior assessment,

(a) express the posterior probability $\phi^* = p(H_0 \mid y)$ in the form

$$\phi^* = \frac{\phi R}{\phi R + 1 - \phi}$$

and fully describe R (note that R is the Bayes factor, or posterior odds ratio);

(b) fully describe the posterior distribution of θ.

Model Answer 3C:

(a) From the discrete version of Bayes' theorem,

$$\phi^* = p(H_0 \mid y) = \frac{p(H_0)p(y \mid H_0)}{p(H_0)p(y \mid H_0) + p(H_1)p(y \mid H_1)} = \frac{\phi R}{\phi R + 1 - \phi},$$

where

$$R = \frac{p(y \mid H_0)}{p(y \mid H_1)}.$$

Now,

$$p(y \mid H_0) = {}^nC_y \theta_2^y (1 - \theta_0)^{n-y}$$

and

$$p(y \mid H_1) = \underset{\theta \mid y}{E}\left[{}^nC_y \theta^y (1 - \theta)^{n-y}\right],$$

where the expectation is with respect to a Gamma distribution for θ, with parameters α and β. With $B(\alpha, \beta) = B_1(\alpha, \beta) = \Gamma(\alpha)\Gamma(\beta)/\Gamma(\alpha + \beta)$, this yields the analytic expression

$$\begin{aligned} p(y \mid H_1) &= {}^nC_y \int_0^1 \theta^y (1 - \theta)^{n-y} \frac{\theta^{\alpha-1}(1 - \theta)^{\beta-1}}{B(\alpha, \beta)}\, d\theta \\ &= \frac{{}^nC_y}{B(\alpha, \beta)} \int_0^1 \theta^{\alpha+y-1}(1 - \theta)^{\beta+n-y-1}\, d\theta \\ &= \frac{{}^nC_y\, B(\alpha + y, \beta + n - y)}{B(\alpha, \beta)}, \end{aligned}$$

giving

$$R = \frac{B(\alpha, \beta)}{B(\alpha + y, \beta + n - y)} \theta_0^y (1 - \theta_2)^{n-y}.$$

(b) Conditional on H_1, the conjugate analysis leading to (3.2.10) tells us that the distribution of θ, in the posterior assessment, is beta with parameters $\alpha + y$ and $\beta + n - y$. Since $p(H_1 \mid y) = 1 - \phi^*$, the posterior distribution of θ is now fully defined.

Worked Example 3D: *Lindley's Paradox*

By reference to Worked Example 3C,

(a) show that R can be arranged in the form

$$R = \frac{\pi_1(\theta_0 \mid y)}{\pi_1(\theta_0)},$$

where $\pi_1(\theta)$ and $\pi_1(\theta \mid y)$, respectively, denote the prior and posterior densities of θ, given H_1. (Note that $\phi^* > \phi$ whenever $R > 1$, so that $R = 1$ can be regarded as a neutral value for the Bayes factor. This can, however, occur when θ_0 lies in the tails of the conditional posterior density $\pi_1(\theta \mid y)$. In this case, H_0 is refutable, since $\pi_1(\theta \mid y)$ is the unconditional posterior density under the continuous prior density $\pi_1(\theta)$. The logarithm of the Bayes factor is nevertheless regarded by Good, 1991, as the only valid weight of evidence. See Aitken, 1995, p. 109, for discussions in a forensic science context.)

(b) When n is large and $p = y/n$ is not close to either zero or unity, Johnson (1967) shows that $\pi_1(\theta \mid y)$ may be approximated by a normal density for θ with mean p and variance $v = p(1-p)/n$. Use this result to find a large sample approximation to R. Comment upon the strong dependence of your result upon the prior assumptions.

(c) Show that for large n, $R > 1$ whenever the test statistic $z_0 = (p - \theta_0)/v^{\frac{1}{2}}$ satisfies $|z| < \max(\sqrt{k}, 0)$ for some k, and express k in terms of n, p, and $\pi_1(\theta_0)$.

(d) If $\pi_1(\theta_0)$ is a uniform density on the interval (0, 1), show how this result can create a paradox when compared with fixed-size tests.

Model Answer 3D:

(a) By Bayes' theorem,

$$\pi_1(\theta \mid y) = \frac{p(y \mid \theta)\pi_1(\theta)}{p_1(y)},$$

where $p_1(y) = p(y \mid H_1)$. In particular,

$$\pi_1(\theta_0 \mid y) = \frac{p(y \mid \theta_0)\pi_1(\theta_0)}{p_1(y)} = R\pi_1(\theta_0),$$

where $R = p(y \mid \theta_0)/p_1(y)$. Therefore, R can be arranged in the form

$$R = \frac{\pi_1(\theta_0 \mid y)}{\pi_1(\theta_0)}.$$

From the preceding worked example, $\pi_1(\theta)$ is beta with parameters α and β, and $\pi_1(\theta \mid y)$ is beta with parameters $\alpha + y$ and $\beta + n - y$.

(b) Using the stated normal approximation (large n),

$$R \sim \frac{(2\pi v)^{-\frac{1}{2}} \exp\{-\frac{1}{2}v^{-1}(\theta_0 - p)^2\}}{\pi_1(\theta_0)}.$$

The numerator does not depend upon the prior, and the denominator is extremely sensitive to the choice of prior assumptions. Hence R is particularly dependent upon the prior assumptions when n is large.

(c) $R > 1$ whenever $-\log R < 0$, that is,

$$\begin{aligned} v^{-1}(\theta_0 - p)^2 &< -\log(2\pi v) - 2\log \pi_1(\theta_0) \\ &= k^* - 2\log \pi_1(\theta_0), \end{aligned}$$

where $k^* = -\log(2\pi v) = -\log\{2\pi p(1-p)/n\}$. Hence, the result follows in required form, with $k = k^* - 2\log \pi_1(\theta_0)$.

(d) As $\pi_1(\theta_0) = 1$, $R > 1$ whenever $|z| < \sqrt{k^*}$. When, say, $\sqrt{(k^*)} = 3$, this corresponds to a classical test with approximate significance level 0.0026, so that $|z| = 3$ would refute H_0. However, $|z| = 3$ corresponds to $R = 1$ so that the Bayes factor equally favors H_0 and H_1. The value $\sqrt{k^*} = 3$ occurs when $n = 2\pi p(1-p)e^9 \simeq 16{,}206\,\pi p(1-p)$.

Worked Example 3E: *Bayesian Inference and Prediction for an Unknown Range*

Suppose that $y_1, y_2, \ldots, y_{n+1}$ are a random sample from a uniform distribution on the range $(0, \theta)$, but that y_{n+1} is unobserved. The prior distribution of $\gamma = \log\theta$ is normal with mean μ and variance σ^2.

(a) Find the posterior distribution of γ, given $\mathbf{y}_n = (y_1, y_2, \ldots, y_n)$, and show that this distribution is truncated normal.

(b) Find the predictive distribution of y_{n+1}, given $\mathbf{y}_n$.
Hint: $n\gamma + (\gamma - \mu)^2/2\sigma^2 = (\gamma - \mu + n\sigma^2)^2/2\sigma^2 + n\mu - n^2\sigma^2/2$.

Model Answer 3E:

(a) The likelihood of θ, given $\mathbf{y}_n$, is

$$l(\theta \mid \mathbf{y}_n) = \theta^{-n} I[z \le \theta],$$

where $z = \max(y_1, y_2, \ldots, y_n)$. Consequently the likelihood of γ is

$$\begin{aligned} \pi(\gamma \mid \mathbf{y}_n) &\propto l(\gamma \mid \mathbf{y}_n) \exp\big\{-(\gamma-\mu)^2/2\sigma^2\big\} \\ &\propto \exp\big\{-n\gamma - (\gamma-\mu)^2/2\sigma^2\big\} I[\gamma \ge \log z]. \end{aligned}$$

The quantity $n\gamma + (\gamma - \mu)^2/2\sigma^2$ is minimized when $\gamma = \mu - n\sigma^2$. Consequently, or by using the hint, the exponential contribution to the posterior density is proportional to a normal density with mean $\gamma^* = \mu - n\sigma^2$ and variance σ^2, that is,

$$\pi(\gamma \mid \mathbf{y}_n) = K_z^{-1} \exp\left\{-\frac{1}{2\sigma^2}(\gamma - \gamma^*)^2\right\} \qquad (\log z < \gamma < \infty),$$

where,

$$K_z^{-1} = (2\pi\sigma^2)^{\frac{1}{2}} \left\{1 - \Phi\left(\frac{\log z - \gamma^*}{\sigma}\right)\right\}.$$

(b) Setting $\theta = e^\gamma$ in the expression

$$p(y_{n+1} \mid \theta) = \theta^{-1} I[y_{n+1} \leq \theta]$$

gives

$$p(y_{n+1} \mid \gamma) = e^{-\gamma} I[\gamma \geq \log y_{n+1}].$$

Consequently,

$$\begin{aligned} p(y_{n+1} \mid \mathbf{y}_n) &= \int_{-\infty}^{\infty} p(y_{n+1} \mid \gamma)\pi(\gamma \mid \mathbf{y}_n)\, d\gamma \\ &= K_z \int_{\log z^*}^{\infty} \exp\left\{-\gamma - \frac{1}{2\sigma^2}(\gamma - \gamma^*)^2\right\} d\gamma, \end{aligned}$$

where $z^* = \max[z, y_{n+1}]$. However, by the hint,

$$\gamma + \frac{1}{2\sigma^2}(\gamma - \gamma^*)^2 = \frac{1}{2\sigma^2}(\gamma - \tilde{\gamma})^2 + \gamma^* - \frac{\sigma^2}{2}.$$

Therefore, with $K_z^* = K_z \exp\{-\frac{1}{2}\gamma^* + \frac{1}{4}\sigma^2\}$, we have

$$\begin{aligned} p(y_{n+1} \mid \mathbf{y}_n) &= K_z^* \int_{\log z^*}^{\infty} \exp\left\{-\frac{1}{2\sigma^2}(\gamma - \tilde{\gamma})^2\right\} d\gamma \\ &= (2\pi\sigma^2)^{\frac{1}{2}} K_z^* \left\{1 - \Phi\left(\frac{\log z^* - \gamma^*}{\sigma}\right)\right\} \\ &= \begin{cases} (2\pi\sigma^2)^{\frac{1}{2}} K_z^* \left\{1 - \Phi\left(\frac{\log z - \gamma^*}{\sigma}\right)\right\} & (0 < y_{n+1} < z) \\ (2\pi\sigma^2)^{\frac{1}{2}} K_z^* \left\{1 - \Phi\left(\frac{\log y_{n+1} - \gamma^*}{\sigma}\right)\right\} & (z < y_{n+1} < \infty). \end{cases} \end{aligned}$$

SELF-STUDY EXERCISES

3.2.a You random sample $n = 1{,}000$ people, without replacement, from the at-risk Wisconsin population and are able to observe that exactly $y = 41$ out of the 1,000 possess the AIDS/HIV virus. Based upon a binomial approximation to the hypergeometric distribution, take y to be the realization of a binomial random variable, with probability θ, and sample size $n = 1{,}000$. Assume a beta prior distribution for the population proportion θ, with parameters $\alpha = 1.72$ and $\beta = 84.39$. Find and sketch the posterior density of θ and calculate the posterior mean and mode of θ. Would the posterior information appear to refute the null hypothesis $H_0 : \theta = 0.02$?

A further $m = 3$ people are about to be chosen at random from the at-risk Wisconsin population. Find the predictive probabilities that z, the number out of the three people who have the AIDS/HIV virus, is equal to 0, 1, 2, or 3, and find the predictive mean and variance of z.

3.2.b For the data in Exercise 3.2.a, compute (a) the Bayesian p-value, that is, the posterior probability that $\theta \le 0.02$, (b) 95% and 99% equal-tailed Bayesian intervals for θ, and (c) 95% and 99% HPD regions for θ (these are the intervals with shortest width for the stated probability coverage).

3.2.c Let observations $y_1, \ldots, y_n$ constitute a random sample from an exponential distribution with parameter θ, mean $\xi = \theta^{-1}$, and variance ξ^2. Suppose that the prior distribution of θ is Gamma with parameters α and β, mean $\gamma = \alpha/\beta$, and variance α/β^2. Show that the posterior distribution of θ is Gamma with parameters $\alpha^* = \alpha + n$ and $\beta^* = \beta + n\bar{y}$, where $\bar{y}$ denotes the sample mean. Show that the posterior mean and variance of θ are

$$E(\theta \mid \mathbf{y}) = \frac{\alpha + n}{\beta + n\bar{y}} \tag{3.2.22}$$

and

$$\operatorname{var}(\theta \mid y) = \frac{\alpha + n}{(\beta + n\bar{y})^2}. \tag{3.2.23}$$

When will the posterior variance of θ be less than the prior variance? Find the posterior mode of θ and the posterior mode, mean, and variance of ξ.

Note: The Gamma prior density for θ integrates to unity, yielding the important integral result

$$\int_0^\infty \theta^{\alpha-1} \exp\{-\beta\theta\}\, d\theta = \beta^{-\alpha}\Gamma(\alpha), \tag{3.2.24}$$

for any positive values of α and β, where $\Gamma(\alpha)$ is defined in Exercise 1.1.c. You can use this result to find the moments, when they exist, of $\xi = \theta^{-1}$.

3.2.d In Exercise 3.2.c, show that in the prior assessment,

$$p(\theta \le a) = G_\alpha(\beta a), \tag{3.2.25}$$

where

$$G_\alpha(a) = \int_0^a \frac{\theta^{\alpha-1}}{\Gamma(\alpha)} \exp\{-\theta\}\, d\theta \tag{3.2.26}$$

denotes the normalized incomplete Gamma function. Describe $p(\theta \le a \mid \mathbf{y})$ in similar fashion.

Consider the parameter $\lambda = \log\theta = -\log\xi$. Show how to obtain 95% and 99% equal-tailed Bayesian intervals for λ, and 95% and 99% HPD intervals for λ.

3.2.e In Exercise 3.2.c, suppose instead that $\lambda = \log \theta$ has a prior distribution that is normal with mean μ and variance σ^2. Find an equation for the posterior mode of λ. Use (1.3.11) to find a normal approximation to the likelihood of λ. Based upon this approximation, find an explicit approximation to the posterior mode of λ.

3.3 Practical Examples

(A) *The Posterior Density of a Genotype Frequency*

As a continuation of the example in Section 1.2 (A), let θ denote the proportion of people in England with genotype Z. Assume that based upon a random sample of size $n = 2{,}500$, a sample proportion, or sample genotype frequency, $p = 50/2{,}500 = 0.02$ is observed (this yields an estimated standard error $\sqrt{0.02 \times 0.98/2{,}500} = 0.0028$). Suppose however, that the prior mean is $\xi = 0.06$.

Curve (a) of Figure 3.3.1 describes the (normalized) likelihood function of θ. Curves (b), (c), and (d) denote beta prior densities with mean 0.06 and respective "prior sample sizes" $\gamma = 400$, $\gamma = 1{,}000$, and $\gamma = 2{,}000$, that is, prior standard deviations 0.0118, 0.0075, and 0.0053. Curves (e), (f), and (g) denote the posterior densities corresponding to the respective priors (b), (c), and (d). They incorporate the subjective information from the prior, with the objective information in the likelihood, in a logical manner. The posterior densities (e), (f), and (g) possesses respective means 0.0294, 0.0324, 0.0378 and standard deviations 0.0031, 0.0030, 0.0028. The corresponding "posterior sample sizes" are $\gamma^* = 2{,}900$, $\gamma^* = 3{,}500$, and $\gamma^* = 4{,}500$, respectively.

Note, however, that the posterior standard deviation need not be less than either the prior standard deviation or the estimated standard error (0.0028) of $\hat{\theta}$ if ξ is very different from $\hat{\theta}$ (see Exercise 3.2.c for a similar result when evaluating an exponential

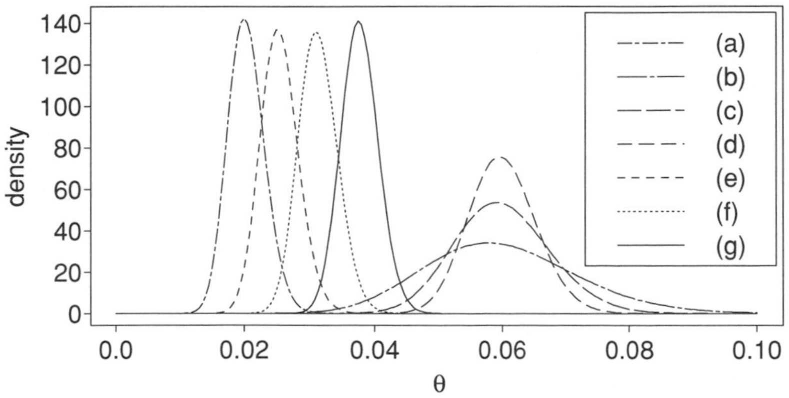

Figure 3.3.1. Posterior inferences for a genotype frequency: (a) normalized likelihood function; (b) beta prior density (mean = 0.06, $\gamma = 400$); (c) beta prior density (mean = 0.06, $\gamma = 1{,}000$); (d) beta prior density (mean = 0.06, $\gamma = 2{,}000$); (e) posterior density ($\gamma^* = 2{,}900$); (f) posterior density ($\gamma^* = 3{,}500$); (g) posterior density ($\gamma^* = 4{,}500$).

parameter). For example, if $\alpha = 1{,}950$ and $\beta = 50$, then $\xi = 0.975$, with prior standard deviation 0.0035. However, the posterior mean is then 0.4444, with standard deviation 0.0074.

(B) *Inferences for Poisson Data, and Quality Control*

Devore (1991, p. 243) reports data on the number of scratches per item for $n = 150$ newly manufactured items. A total of 317 scratches were observed, yielding a sample mean of $\bar{y} = 2.113$, with estimated standard error $\bar{y}/\sqrt{n} = 0.173$. If the observations are regarded as a random sample from a Poisson distribution, with mean θ, the likelihood function of θ is therefore

$$l(\theta \mid \mathbf{y}) \propto \theta^{n\bar{y}} \exp\{-n\theta\} \tag{3.3.1}$$

$$\propto \theta^{317} \exp\{-150\theta\} \qquad (0 < \theta < \infty). \tag{3.3.2}$$

The (normalized) likelihood is described by curve (a) in Figure 3.3.2. Using the notation of Exercise 3.2.c, note that if the prior distribution of θ is Gamma with positive parameters α and β, then the posterior distribution of θ is Gamma with parameters $\alpha + n\bar{y} = \alpha + 317$ and $\beta + n = \beta + 150$. Curve (c) in Figure 3.3.2 describes the posterior density when the prior distribution is Gamma with parameters $\alpha = 40$ and $\beta = 10$, yielding prior mean $\xi = \alpha/\beta = 4.0$ and standard deviation $\alpha^{\frac{1}{2}}/\beta = 0.632$. The posterior distribution is then Gamma with parameters 357 and 160, that is, mean 2.231 and standard deviation 0.118.

Curve (b) in Figure 3.3.2 describes the limiting posterior density when $\alpha = 0.5$, but as $\beta \to 0$. As will be discussed in greater detail in Section 3.5, this "Jeffreys' prior" can be appropriate in situations where there is only "vague" prior information about θ. Note that the posterior density for θ is now Gamma with parameters 317.5 and 150 and, using the representation in Exercise 3.2.d, that the posterior probability, that $\theta \le 2.0$, is $\epsilon_0 = p(\theta \le 2.0 \mid \mathbf{y}) = 0.162$.

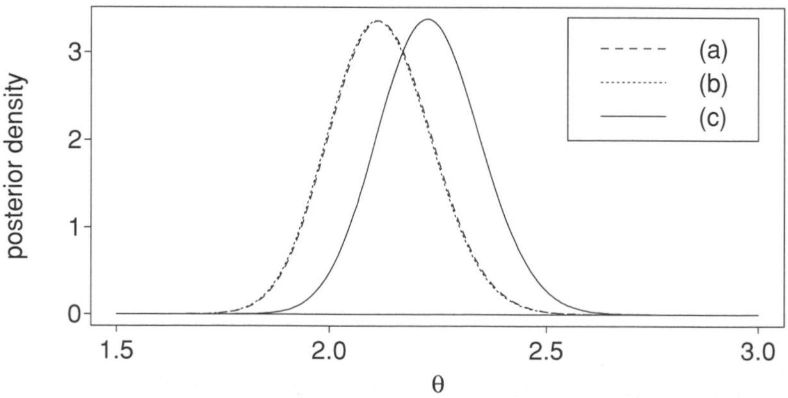

Figure 3.3.2. Posterior inference for a Poisson mean: (a) normalized likelihood; (b) posterior density (Jeffreys' prior); (c) posterior density (Gamma prior, $\alpha = 40$, $\beta = 10$).

Table 3.3.1. *Predictive probabilities for a Poisson frequency.*

y	Case A	Case B
0	0.1213	0.0933
1	0.2550	0.2205
2	0.2689	0.2615
3	0.1897	0.2072
4	0.1006	0.1235
5	0.0429	0.0590
6	0.0153	0.0236
7	0.0047	0.0081
8	0.0013	0.0024
9	0.0003	0.0007
10	0.0001	0.0002

Paralleling (3.2.17), ϵ_0 can be regarded as a "Bayesian significance probability" for investigating the null hypothesis $H_0 : \theta = \theta_0$, where $\theta_0 = 2.0$. For example, in a quality-control context, a manufacturing process can be regarded as "out of control" if it is thought that $\theta > \theta_0$. In the current situation, an appraisal of ϵ_0 together with curve (b) of Figure 3.3.2 tells us that whereas there is some evidence to suggest that the process is out of control, this evidence is by no means conclusive. The interval (1.890, 2.236) provides an equal-tailed 95% Bayesian interval for θ. This may be compared with (1.886, 2.351), the 95% HPD region for θ. The corresponding 99% regions are (1.823, 2.435) and (1.819, 2.431).

Under the posterior density (b), the predictive density for a further frequency y_{151} averages a Poisson distribution with mean θ, with respect to a Gamma distribution with parameters 317.5 and 150. The probabilities for the corresponding Pascal distribution are described, under Case B, in the third column of Table 3.3.1. In the second column, we describe, under Case A, the probabilities for the predictive distribution of y_{151} when, instead, $\alpha = 40$ and $\beta = 10$. The third column predicts that 57.5% of all future observations will not exceed the value 2.

Note, from (3.2.20) and (3.2.21), that under the Gamma posterior, with parameters 317.5 and 150, the predictive mean of y_{151} is $317.5/150 = 2.117$, with predictive variance $E(\theta \mid \mathbf{y}) + \mathrm{var}(\theta \mid \mathbf{y}) = 2.117 + 0.014 = 2.131$, yielding a predictive standard deviation equal to 1.460.

(C) *Approximating the Posterior Distribution of a Binomial Logit*

Let θ possess the posterior distribution described in (3.2.10), for a binomial probability, based upon a beta prior with parameters α and β. Then Lindley (1965) tells us that the posterior density of $\lambda = \mathrm{logit}(\theta)$ is approximately $N[\log(\alpha^*/\beta^*), (\alpha^*)^{-1} + (\beta^*)^{-1}]$, with $\alpha^* = \alpha + y$ and $\beta^* = \beta + n - y$. This approximation can be quite accurate if

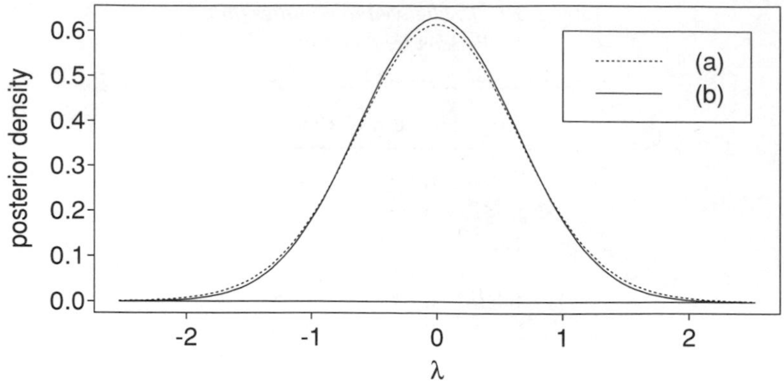

Figure 3.3.3. The posterior density of a binomial logit (under beta prior assumptions): (a) exact posterior density ($\alpha^* = 5$, $\beta^* = 5$); (b) Lindley's normal approximation.

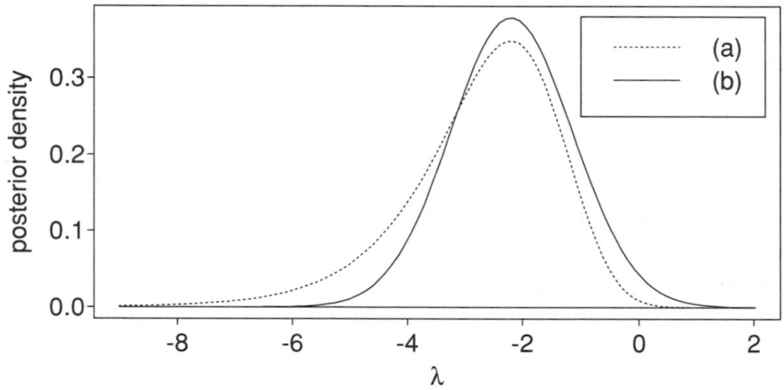

Figure 3.3.4. The posterior density of a binomial logit (under beta prior assumptions): (a) exact posterior density ($\alpha^* = 1$, $\beta^* = 9$); (b) Lindley's normal approximation.

$\alpha^* \geq 5$ and $\beta^* \geq 5$. For example, curve (a) in Figure 3.3.3 describes the exact posterior density when $\alpha^* = \beta^* = 5$, while curve (b) describes Lindley's approximate density.

The approximation can be less accurate when either α^* or β^* is less than 5. For example, Figure 3.3.4 describes a similar comparison when $\alpha^* = 1$ and $\beta^* = 9$.

3.4 Inferences for a Normal Mean with Known Variance

Take $y_1, \ldots, y_n$ to denote a random sample from a normal distribution with unknown mean θ and known variance τ^2. Then the likelihood of θ, given $\mathbf{y} = (y_1, \ldots, y_n)^T$, is, as a function of θ,

$$l(\theta \mid \mathbf{y}) \propto \exp\left\{-\frac{1}{2\tau^2}\sum_{i=1}^{n}(y_i - \theta)^2\right\} \qquad (-\infty < \theta < \infty). \tag{3.4.1}$$

This likelihood may be expressed more simply by noting that

$$\sum_{i=1}^{n}(y_i - \theta)^2 = s^2 + n(\bar{y} - \theta)^2, \tag{3.4.2}$$

where $\bar{y}$ is the sample mean and $s^2 = \sum_{i=1}^{n}(y_i - \bar{y})^2$. Consequently, as a function of θ,

$$l(\theta \mid \mathbf{y}) \propto \exp\left\{-\frac{n}{2\tau^2}(\theta - \bar{y})^2\right\} \qquad (-\infty < \theta < \infty). \tag{3.4.3}$$

The τ^2 known situation will seldom arise in practice. However, in many other single-parameter models, a preliminary normalizing transformation can produce a parameter λ with exact likelihood $l(\lambda \mid \mathbf{y})$ and approximate likelihood

$$l^*(\lambda \mid \mathbf{y}) \propto \exp\left\{-\frac{n}{2\vartheta}(\lambda - \hat{\lambda})^2\right\} \qquad (-\infty < \lambda < \infty), \tag{3.4.4}$$

where $\hat{\lambda}$ denotes the maximum likelihood estimate of λ and

$$n\vartheta^{-1} = \frac{-\partial^2 \log l(\lambda \mid \mathbf{y})}{\partial \lambda^2}\Big|_{\lambda = \hat{\lambda}} \tag{3.4.5}$$

denotes the likelihood information. Note that (3.4.4) is a special case of the multivariate normal approximation (1.3.11). For example, if $\lambda = \log\theta - \log(1-\theta)$, where θ possesses a binary, binomial, or negative binomial likelihood proportional to $\theta^y(1-\theta)^{n-y}$, then $\hat{\lambda} = \log(y/(n-y))$ and $n^{-1}\vartheta = y^{-1} + (n-y)^{-1}$. Some refinements to this approximation are discussed in Exercise 1.5.g. Alternatively, if U is an estimator of a variance component ϕ and $\nu U/\phi$ possesses a sampling distribution that is chi-squared with ν degrees of freedom, where ν is specified, then $\log U$ is approximately normally distributed with mean $\lambda = \log\phi$ and variance $2\nu^{-1}$. Refinements to this approximation are described by Bartlett and Kendall (1946). An approximate normal likelihood for the logarithm of an exponential parameter is considered in Exercise 3.2.e.

Therefore, our suggestions of prior distributions and posterior inferences for θ in the normal situation with τ^2 specified will also be appropriate for a suitably transformed parameter λ in many other single-parameter models. Just replace $\bar{y}$ by $\hat{\lambda}$ and τ^2 by ϑ and realize that the posterior inferences will be approximate. However, exact posterior inferences for λ can alternatively be computed by using one-dimensional numerical integrations.

For a normal mean θ, first consider the conjugate prior distribution, which is normal with mean μ and variance σ^2. This may be justified by Boltzmann's maximum entropy theorem (Cercignani, 1988; Rosenkrantz, 1989). Suppose that you can specify both μ and σ^2 but nothing else about your prior distribution. Then you might choose the prior density $\pi(\theta)$ that maximizes the entropy

$$\varphi(\pi) = -\int_0 \pi(\theta)\log\pi(\theta)\,d\theta, \tag{3.4.6}$$

but subject to the mean and variance of $\pi(\theta)$ being equal to your specified values for μ and σ^2. A straightforward application of the calculus of variations (e.g., Bellman,

1971, Ch. 4) tells you that your optimal prior distribution is normal $N(\mu, \sigma^2)$. This is a special case of Boltzmann's theorem:

Theorem 3.4.1: *The density $\pi(\theta)$ that maximizes $\varphi(\pi)$, subject to the constraints*

$$E[t_j(\theta)] = t_j \qquad (j = 1, \ldots, p),$$

takes the p-parameter exponential family form

$$\pi(\theta) \propto \exp\left\{\lambda_1 t_1(\theta) + \lambda_2 t_2(\theta) + \cdots + \lambda_p t_p(\theta)\right\} \qquad (\theta \in \Theta),$$

where $\lambda_1, \lambda_2, \ldots, \lambda_p$ can be determined, via the p-constraints, in terms of $t_1, \ldots, t_p$.

Proof of Theorem 3.4.1. We will obtain a generalized derivativc of

$$\varphi^*(\pi) = \varphi(\pi) + \lambda \int_\Theta \pi(\theta)\, d\theta + \sum_{j=1}^{p} \lambda_j \int_\Theta t_j(\theta)\pi(\theta)\, d\theta.$$

Setting this generalized derivative equal to the zero function and choosing the Lagrange multipliers λ and $\lambda_1, \ldots, \lambda_p$ to ensure that $\int_\Theta \pi(\theta) = 1$ and our p further constraints are satisfied will provide an equation for our required solution. Note that for any scalar ϵ and a nonzero function $U(\theta)$,

$$\left.\frac{\partial\varphi^*(\pi + \epsilon u)}{\partial\epsilon}\right|_{\epsilon=0} = -\int_\Theta u(\theta) \log \pi(\theta)\, d\theta + (\lambda - 1)\int_\Theta u(\theta)\, d\theta + \sum_{j=1}^{p} \lambda_j \int_\Theta t_j(\theta) u(\theta)\, d\theta.$$

This derivative is equal to zero, for all nonzero functions $U(\theta)$, whenever

$$\log \pi(\theta) = \lambda - 1 + \sum_{j=1}^{p} \lambda_j t_j(\theta).$$

Taking exponentials and absorbing $\exp(\lambda - 1)$ into the constant of proportionality give the required result, as (3.4.6) is a convex function of π. ■

Under the maximum entropy $N(\mu, \sigma^2)$ prior distribution, μ can be specified as a prior estimate of θ, and σ denotes the prior standard deviation. With $\sigma^2 = \tau^2/\kappa$, $\kappa = \tau^2/\sigma^2$ is the "prior sample size," and $\kappa^{-1} = \sigma^2/\tau^2$ is the "signal-to-noise" ratio. Then the posterior density of θ is

$$\pi(\theta \mid \mathbf{y}) \propto \exp\left\{-\frac{1}{2\sigma^2}(\theta - \mu)^2 - \frac{n}{2\tau^2}(\theta - \bar{y})^2\right\} \qquad (-\infty < \theta < \infty). \tag{3.4.7}$$

Lemma 3.4.1 (Completing the square): *For any constants A, B, a, and b,*

$$A(\theta - a)^2 + B(\theta - b)^2 = (A + B)(\theta - \theta^*)^2 + \left(A^{-1} + B^{-1}\right)^{-1}(a - b)^2, \tag{3.4.8}$$

where

$$\theta^* = (A + B)^{-1}(Aa + Bb). \tag{3.4.9}$$

Exercise: Prove this!

Lemma 3.4.1 tells you that as a function of θ,

$$\pi(\theta \mid \mathbf{y}) \propto \exp\left\{ -\tfrac{1}{2}v^{-1}(\theta - \theta^*)^2\right\} \qquad (-\infty < \theta < \infty), \tag{3.4.10}$$

where

$$\theta^* = \frac{n\tau^{-2}\bar{y} + \sigma^{-2}\mu}{n\tau^{-2} + \sigma^{-2}} = \frac{n\bar{y} + \kappa}{n + \kappa} \tag{3.4.11}$$

and

$$v^{-1} = n\tau^{-2} + \sigma^{-2} = \tau^{-2}(n + k). \tag{3.4.12}$$

In other words, θ is a posteriori normally $N(\theta^*, v)$ distributed. This is the maximum entropy distribution, given the posterior mean ξ^* and variance v. The expressions in (3.4.11) define the posterior mean, mode, and median of θ, since these are identical for the normal distribution. They all equal the weighted average

$$\theta^* = \rho\bar{y} + (1 - \rho)\mu,$$

where $\rho = n\tau^{-2}/(n\tau^{-2} + \sigma^{-2})$ describes the "reliability" of $\bar{y}$ as an estimator of θ. In (3.4.11), $\bar{y}$ and μ are weighted according to the relative values of the likelihood precision or information $n\tau^{-2}$ and the prior precision σ^{-2}. By a slight extension of Exercise 3.8.g, a similar weighted average form will be appropriate whenever the posterior mean is a linear function of a one-dimensional sufficient statistic.

Equation (3.4.12) tells us that the posterior precision v^{-1} equals the sum of the sampling precision and the prior precision. Therefore the posterior variance is, in this special case, less than both the prior variance σ^2 and the sampling variance $n^{-1}\tau^2$ of $\bar{y}$. Note that the posterior standard deviation of θ is $\tau/\sqrt{n + \kappa}$, which is less than both $\sigma = \tau/\sqrt{\kappa}$ and the sampling standard deviation $\tau/\sqrt{n}$ of $\bar{y}$. Consideration of equation (3.2.23) and Exercise 3.2.c tells you that this is not generally the case.

Posterior probabilities can be calculated from

$$p(a < \theta < b \mid \mathbf{y}) = \Phi\left(\frac{b - \theta^*}{\sqrt{v}}\right) - \Phi\left(\frac{a - \theta^*}{\sqrt{v}}\right), \tag{3.4.13}$$

where Φ denotes the standard normal distribution function. Also 95% and 99% equal-tailed Bayesian intervals are $(\theta^* - 1.96v^{\frac{1}{2}}, \theta^* + 1.96v^{\frac{1}{2}})$ and $(\theta^* - 2.576v^{\frac{1}{2}}, \theta^* + 2.576v^{\frac{1}{2}})$, and these are also highest posterior density intervals. A Bayesian significance probability for investigating $H_0: \theta = \theta_0$ is $\epsilon_0 = p(\theta \leq \theta_0 \mid \mathbf{y}) = \Phi[(\theta_0 - \theta^*)/v^{\frac{1}{2}}]$, but remember to consider also θ_0 in relation to a sketch of the posterior density. A further sample mean $\bar{z}$, which is independent of $\bar{y}$, given θ, and normally $N(\theta, m^{-1}\tau^2)$ distributed, possesses a predictive distribution, given $\mathbf{y}$, which is normal $N(\theta^*, v + m^{-1}\tau^2)$.

Under a general prior density $\pi(\theta)$ for our normal mean θ, the posterior density of θ is

$$\pi(\theta \mid \mathbf{y}) \propto \pi(\theta) \exp\left\{-\frac{n}{2\tau^2}(\theta - \bar{y})^2\right\} \qquad (-\infty < \theta < \infty). \tag{3.4.14}$$

Note that the prior predictive density of the sample mean $\bar{y}$ is

$$\begin{aligned} p(\bar{y}) &= \int_{-\infty}^{\infty} p(\bar{y} \mid \theta)\pi(\theta)\, d\theta \\ &= \left(2\pi\tau^2\right)^{-\frac{1}{2}} \int_{-\infty}^{\infty} \exp\left\{-\frac{n}{2\tau^2}(\theta - \bar{y})^2\right\} \pi(\theta)\, d\theta. \end{aligned} \tag{3.4.15}$$

Then the first two derivatives of $\log p(\bar{y})$, with respect to $\bar{y}$, satisfy

$$\frac{\partial \log p(\bar{y})}{\partial \bar{y}} = -n\tau^{-2}\bar{y} + n\tau^{-2}E(\theta \mid \mathbf{y}) \tag{3.4.16}$$

and

$$\frac{-\partial^2 \log p(\bar{y})}{\partial \bar{y}^2} = n\tau^{-2} - n^2\tau^{-4}\,\mathrm{var}(\theta \mid \mathbf{y}). \tag{3.4.17}$$

Therefore, general expressions for the posterior mean and variance of θ are

$$E(\theta \mid \mathbf{y}) = \bar{y} + n^{-1}\tau^2 \frac{\partial \log p(\bar{y})}{\partial \bar{y}} \tag{3.4.18}$$

and

$$\mathrm{var}(\theta \mid \mathbf{y}) = n^{-1}\tau^2 + n^{-2}\tau^4 \frac{\partial^2 \log p(\bar{y})}{\partial \bar{y}^2}. \tag{3.4.19}$$

These results relate to the regression of true score on observed score in classical test theory (see Lord and Novick, 1968, p. 50).

Dawid (1973) and Leonard (1974) address the issue that the estimator in (3.4.11), based upon a conjugate prior, is open to discussion on the grounds that as $\bar{y}$ moves a large distance away from μ, it may discredit this prior estimate and therefore should not be shrunk the same proportion ρ of the distance towards μ. They show that prior distributions with thicker tails yield possibly more desirable properties. Dawid recommends a generalized t-prior density, taking the form

$$\pi(\theta) \propto \left[\nu\lambda + (\theta - \mu)^2\right]^{-\frac{1}{2}(\nu+1)} \qquad (-\infty < \theta < \infty), \tag{3.4.20}$$

while Leonard employs an improper prior density, with infinitely thick tails, taking the form

$$\pi(\theta) \propto \exp\left\{-\frac{1}{2\sigma^2}(\theta-\mu)^2\right\} I\big[|\theta-\mu| \le \xi\big] + \exp\left(-\frac{\xi^2}{2\sigma^2}\right) I\big[|\theta-\mu| > \xi\big] \qquad (-\infty < \theta < \infty), \tag{3.4.21}$$

where $I[A]$ denotes the indicator function for the set A.

Under the prior density (3.4.21), the posterior mode of θ is

$$\tilde{\theta} = \begin{cases} \theta^* & \text{for } |\bar{y}-\mu| \le \xi^* \\ \bar{y} & \text{for } |\bar{y}-\mu| > \xi^*, \end{cases} \tag{3.4.22}$$

where θ^* is the posterior mean (3.4.11) under the conjugate prior, and

$$\xi^* = \xi\left(1 + \frac{n\sigma^2}{\tau^2}\right)^{\frac{1}{2}}. \tag{3.4.23}$$

Whenever $\bar{y}$ is within a distance ξ^* of μ, the estimator (3.4.23) reduces just to the usual weighted average θ^*. However, when $\bar{y}$ lies more than a distance ξ^* from μ, the prior estimate μ gets discredited, and $\tilde{\theta}$ refers to the sample estimate $\bar{y}$.

In situations where the parameter space Θ is unbounded, Bayesians are faced with the problem that their estimates are generally quite sensitive to the thickness of the tails of the prior density. However, in practice it is quite difficult to model the thickness of the tails based upon the prior information, for example, how to assess ν in (3.4.20) or ξ in (3.4.22). Therefore, in practice, many Bayesians refer to the entropy criterion (3.4.6) and a conjugate normal prior.

It is worth noting that posterior expectations of bounded functions of an unbounded parameter θ are not as sensitive to the tail behavior of the prior density as the posterior mean of θ. For example, the posterior mean value function of the sampling density (e.g., Exercise 3.4.g) deserves consideration. Sensitivity problems are discussed further by Lavine (1991), and robust estimates of location are considered by Doksum and Lo (1990).

Worked Example 3F: *A Maximum Entropy Prior Distribution*

You specify the prior probabilities

$$p(\theta \le a_j) = \phi_j \qquad (j = 1, 2, \ldots, p)$$

for a parameter θ concentrated on the region $(a_0, a_{p+1}]$, where $a_0 \le a_1 \le a_2 \le a_p \le a_{p+1}$. Find the maximum entropy distribution for θ, subject to these p conditions.

Model Answer 3F:

These conditions are equivalent to

$$E\big[t_j(\theta)\big] = \phi_j \qquad (j = 1, 2, \ldots, p),$$

where

$$t_j(\theta) = I[\theta \le a_j].$$

Therefore, by Theorem 3.4.1, the posterior density of θ is

$$\pi(\theta) \propto \exp\left\{\sum_{j=1}^{p} \lambda_j I[\theta \le a_j]\right\} \qquad (0 \le \theta \le 1),$$

where $\lambda_1, \lambda_2, \lambda_p$ are determined by the p conditions. It is easily checked that $\pi(\theta)$ must equal a histogram, with intervals $(a_0, a_1], (a_1, a_2], \ldots, (a_p, a_{p+1}]$.

Worked Example 3G: *Calculating a Posterior Mean from a Predictive Density*

If $\bar{y} \mid \theta \sim N(\theta, 1)$, suppose that the prior distribution of θ is uniform on $(-1, 1)$. Find the predictive density of $\bar{y}$. Hence find the posterior mean of θ.

Model Answer 3G:

Since

$$p(\bar{y} \mid \theta) = \frac{1}{\sqrt{2\pi}} \exp\left\{-\frac{1}{2}(\bar{y} - \theta)^2\right\},$$

the predictive density of θ is

$$\begin{aligned} p(\bar{y}) &= \int_{-1}^{1} \frac{1}{\sqrt{2\pi}} \exp\left\{-\frac{1}{2}(\theta - \bar{y})^2\right\} d\theta \\ &= \Phi(1 - \bar{y}) - \Phi(-1 - \bar{y}) \qquad (-\infty < \bar{y} < \infty). \end{aligned}$$

Then equation (3.4.18) tells us that the posterior mean of θ is

$$\begin{aligned} E(\theta \mid \bar{y}) &= \bar{y} + \frac{\partial \log p(\bar{y})}{\partial \bar{y}} \\ &= \bar{y} - \frac{\phi(\bar{y} - 1) - \phi(\bar{y} + 1)}{\Phi(1 - \bar{y}) - \Phi(-1 - \bar{y})}, \end{aligned}$$

where $\phi(z)$ denotes a standard normal density.

Worked Example 3H: *The Posterior Expectation of a Poisson Log-Mean*

Let y possess a Poisson distribution with mean θ and let $\pi(\theta)$ denote the posterior density of θ. Show that the posterior mean of $\alpha = \log\theta$ is

$$E(\alpha \mid y) = \Psi(y+1) + \frac{\partial \log p(y)}{\partial y},$$

where $\Psi(z) = \partial\Gamma(z)/\partial z$, and $p(y)$ is the predictive probability mass function of y.

Model Answer 3H:

$$p(y) = \int_0^\infty \frac{e^{-\theta}\theta^y}{\Gamma(y+1)}\pi(\theta)\,d\theta,$$

giving

$$\log p(y) = \log \int_0^\infty e^{-\theta}\theta^y \pi(\theta)\,d\theta - \log p(y+1).$$

Since

$$\frac{\partial(\theta^y)}{\partial y} = (\log\theta)\theta^y,$$

we have

$$\frac{\partial \log p(y)}{\partial y} = \frac{\int_0^\infty (\log\theta) e^{-\theta}\theta^y \pi(\theta)\,d\theta}{\int_0^\infty e^{-\theta}\theta^y\pi(\theta)\,d\theta} - \Psi(y+1).$$

The first term on the left-hand side equals $E[\log\theta \mid y]$, since

$$\pi(\theta \mid y) \propto e^{-\theta}\theta^y\pi(\theta),$$

and the required result follows immediately.

Worked Example 3I: *Posterior Expectations of Poisson Sampling Probabilities*

Suppose that y is Poisson distributed with mean θ and that the prior distribution of θ is Gamma with parameters α and β. Find the posterior means of the sampling probabilities

$$\phi_j = p(y = j) = \frac{e^{-\theta}\theta^j}{j!} \qquad (j = 0, 1, \ldots).$$

Model Answer 3I:

Applying the general results of Section 3.3 (B), the posterior distribution of θ is Gamma with parameters $\alpha + y$ and $\beta + 1$. The prior expectation of ϕ_j is

$$\begin{aligned}E(\phi_j) &= \int_0^\infty \frac{1}{\Gamma(j+1)} e^{-\theta}\theta^j \frac{1}{\Gamma(\alpha)}\beta^\alpha\theta^{\alpha-1}\exp\{-\beta\theta\}\,d\theta \\ &= \frac{\beta^\alpha}{\Gamma(j+1)\Gamma(\alpha)}\int_0^\infty \theta^{\alpha+j-1}\exp\big\{-(\beta+1)\theta\big\}\,d\theta \\ &= \frac{\Gamma(\alpha+j)}{\Gamma(j+1)\Gamma(\alpha)}\frac{\beta^\alpha}{(\beta+1)^{\alpha+j}}.\end{aligned}$$

Therefore, the posterior expectation of ϕ_j is

$$E(\phi_j \mid y) = \frac{\Gamma(\alpha+y+j)}{\Gamma(j+1)\Gamma(\alpha+y)}\frac{(\beta+1)^{\alpha+y}}{(\beta+2)^{\alpha+y+j}}.$$

SELF-STUDY EXERCISES

3.4.a Let $y_1, \ldots, y_n$ denote a random sample from a Poisson distribution with mean θ. Show that the likelihood approximation (3.4.4) holds, for $\lambda = \log\theta$, with $\hat{\lambda} = \log\bar{y}$ and $\vartheta = 1/\bar{y}$.

3.4.b In Exercise 3.4.a, let $\pi(\lambda)$ be a prior density for λ with "positive support," that is, $\pi(\lambda) > 0$ for all $\lambda \in R^1$. Show that as n gets large, with $\hat{\lambda}$ fixed, the corresponding posterior distribution of λ will be approximately normal with mean $\hat{\lambda}$ and variance $n^{-1}\vartheta$. (Formally, $n^{\frac{1}{2}}(\lambda - \hat{\lambda})/\vartheta^{\frac{1}{2}}$ converges in (posterior) distribution to a standard normal random variable, with sampling probability one, that is, whenever $\hat{\lambda}$ remains finite.)

Under consideration of the likelihood approximation (3.4.4), discuss how generally this type of approximation will be valid for posterior distributions. (These aspects are investigated by Johnson, 1967, 1970, and DeGroot, 1970, Ch. 10.)

3.4.c Intuitively compare an exact prior-to-posterior analysis of the sampling model in Exercise 3.4.a when the prior distribution of θ is Gamma, with parameters α and β, with an approximate analysis when the prior distribution of $\lambda = \log\theta$ is instead $N(\mu, \sigma^2)$. Which set of prior assumptions do you find more useful?

3.4.d When $\Theta = \{\theta : 0 < \theta < 1\}$, show that the density $\pi(\theta)$ that maximizes the entropy criterion (3.4.6), subject to the constraints that $E[\log\theta] = a$ and $E[\log(1-\theta)] = b$, is a beta density. Describe two constraints such that the maximum entropy density is instead normal, but for the logit $\lambda = \log\theta - \log(1-\theta)$.

When, instead, $\Theta = \{\theta : \theta > 0\}$, find the maximum entropy distribution subject to $E[\log\theta] = 0$ and $E(\theta) = \mu$.

3.4.e Use (3.4.16) to find the posterior mean of θ when $y_1, \ldots, y_n$ are a random sample from a $N(\theta, \tau^2)$ distribution and when the prior density of θ is the Box–Tiao robustifying density (1.5.26), with specified parameters μ, ϕ, σ_1^2, and σ_2^2. Does this estimate for θ provide sensible properties, as $\bar{y}$ moves away from μ?

3.4.f In Exercise 3.4.e, instead complete the prior-to-posterior analysis, using the prior density (3.4.21). Confirm that (3.4.22) is indeed the posterior mode of θ and show how to calculate the posterior median.

3.4.g Let a normal mean θ possess a $N(\theta^*, v)$ posterior density. Find the posterior mean of the $N(\theta, \tau^2)$ density

$$f(u \mid \theta) = \frac{1}{\sqrt{2\pi\tau^2}} \exp\left\{-\frac{1}{2\tau^2}(u-\theta)^2\right\},$$

for each value of $u \in R$. Show that your posterior mean value function is identical to the predictive density, given $\mathbf{y}$, of a single future observation u. Does this result hold more generally?

3.4.h In the previous exercise, compare your estimate for $f(u \mid \theta)$ with the estimate obtained by replacing θ by its posterior mean. Which estimate would you prefer?

3.4.i In the previous exercise, suppose that you now observe the realization u of your future observation. Find the posterior distribution of θ, given $\mathbf{y}$ and u.

Hint: Treat your previous posterior like a prior.

3.4.j A random vector $\mathbf{Y}$ possesses the exchangeable multivariate normal distribution in Exercise 1.3.d, where μ and σ^2 are specified, but ρ is unknown. Show that the maximum entropy distribution for $\mathbf{Y}$, subject to these restrictions, takes $\rho = 0$. Discuss whether this helps to circumvent the problems relating to lack of identifiability of ρ, discussed in Exercise 1.3.d.

3.4.k Consider a random vector $\mathbf{Y} = (Y_1, \ldots, Y_n)^T$, where the elements Y_i of $\mathbf{Y}$ may be dependent, but where the nature of these dependencies are completely unknown. However, the marginal densities $f_i(y_i)$ of the Y_i are completely specified, for $i = 1, \ldots, n$. Show that the maximum entropy distribution of $\mathbf{Y}$, subject to these constraints, has density

$$f(\mathbf{Y}) = \prod_{i=1}^{n} f_i(y_i).$$

Discuss how this result can be used to justify statistical analyses of data, under the theoretical assumption that random sampling has occurred at the design stage, even though the data might not actually be based upon any randomization. Do you find this justification to be convincing?

Show that these ideas would also justify the assumption that the error terms in a normal model are independent, even when they might be correlated, but where there is no information available to justify a particular correlation structure.

3.4.l A statistic t has sampling density

$$p(t \mid \theta) = \exp\left\{A(\theta) + B(\theta) + C(t)\right\}$$

belonging to the one-parameter exponential family, where θ possesses prior distribution G. Show that the posterior expectation of $B(\theta)$ is

$$E\left[B(\theta) \mid t\right] = \frac{\partial \log p(t)}{\partial t} - \frac{\partial C(t)}{\partial t},$$

where $p(t)$ denotes the prior predictive density of t.

3.4.m An observation y is $N(\theta, 1)$ distributed, but $\theta \geq 0$. Show that the maximum likelihood estimate of θ is $\hat{\theta} = \max(y, 0)$. Consider the prior distribution that is uniformly distributed over all positive values of θ. Show that the posterior mean of θ is $\theta^* = y + M(y)$, where the Mills ratio $M(y) = \phi(y)/\Phi(y)$ is the ratio of a normal density and c.d.f.

3.4.n Let θ possess the normal likelihood (3.4.3), with location $\bar{y}$ and dispersion $n^{-1}\tau^2$. Let θ possess the following two-stage prior distribution:

Stage I: Given ξ, θ is $N(\xi, \tau^2)$.

Stage II: ξ is $N(\mu, \zeta^2)$, but truncated to the region $\xi \geq 0$.

Show that this prior distribution expresses uncertainty in the parameter constraint that $\theta \geq 0$. Find the posterior density of θ and the prior predictive distribution of $\bar{y}$. Use (3.4.18) and (3.4.19) to find the posterior mean and variance of θ. Fully interpret your results (see also O'Hagan and Leonard, 1976; Copas and Li, 1997).

3.4.o Under the same sampling assumptions as described in Exercise 3.4.n, let the prior distribution of θ instead be an equally weighted mixture of two normal distributions with respective means μ_1 and μ_2, and unit variance. Show that the posterior distribution of θ is a mixture of two normal distributions and fully define the parameters of this mixture.

3.5 Practical Examples

(A) *Prior and Posterior Densities for Location Parameters*

Consider the situation discussed in Section 3.3 (B), where θ is the average number of scratches per item. If θ possesses likelihood function (3.3.1), then $\lambda = \log\theta$ possesses likelihood

$$l(\lambda \mid \mathbf{y}) \propto \exp\left\{n\bar{y}\lambda - ne^{\lambda}\right\}$$
$$\propto \exp\left\{317\lambda - 150e^{\lambda}\right\} \qquad (-\infty < \lambda < \infty). \tag{3.5.1}$$

Note that equation (3.4.4) tells us that the exact curve (3.5.1) may be approximated by a $N[\log\bar{y}, n^{-1}\bar{y}]$ curve, that is, a $N(0.748, 0.00315)$ curve. The exact density (3.5.1) is described by curve (a) in Figure 3.5.1, and this is closely approximated by the normal curve (b).

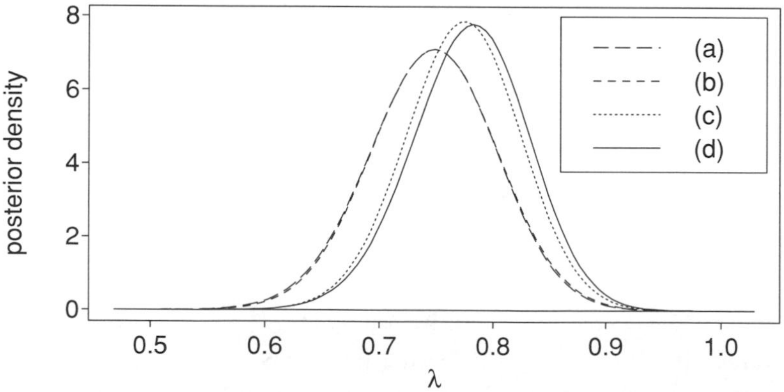

Figure 3.5.1. Posterior inference for a Poisson log-mean: (a) exact likelihood; (b) approximate normal likelihood; (c) exact posterior density (normal prior); (d) exact posterior density (generalized t-prior).

Let the prior mean of θ be 2.5, with standard deviation 0.1. Then, rather than using a Gamma prior for θ, this information could be represented by a normal distribution for λ, with mean μ and variance σ^2 satisfying

$$2.5 = \xi = \exp\left\{\mu + \tfrac{1}{2}\sigma^2\right\} \tag{3.5.2}$$

and

$$0.1 = \xi^2\left(e^{\sigma^2} - 1\right), \tag{3.5.3}$$

that is, $\mu = 0.9084$ and $\sigma^2 = 0.01587$. Note that the generalized t-prior density

$$\pi(\lambda) \propto \left[(\nu - 2)\sigma^2 + (\lambda - \mu)^2\right]^{-\frac{1}{2}(\nu+1)} \tag{3.5.4}$$

will possess the same mean μ and variance σ^2, for any $\nu > 2$.

Curve (c) in Figure 3.5.1 describes the exact posterior density of λ under the proceeding normal prior, and curve (d) describes the exact density under the generalized t-prior (3.5.4) with $\nu = 6$ degrees of freedom. Curves (c) and (d) are quite close, suggesting that (c) is quite robust, in this example, to changes in the tails of the posterior density. We similarly computed the posterior density under different choices of the prior assumptions in Exercise 1.5.j. Again (c) was quite robust. We recommend a similar sensitivity investigation for any observed data set.

(B) *The Regression of True Score on Observed Score*

In the situation described in Section 3.5 (A), we calculated the posterior mean of λ (i.e., the regression of $\lambda = \log\theta$ on $\bar{y}$) under the two sets of prior assumptions: (a) $N(0.9084, 0.01587)$ and (b) generalized t with $\nu = 6$ degrees of freedom, and the same mean and variance as in (a). The two posterior means were calculated for $n = 150$ and for $\bar{y} = 1, 2, \ldots, 10$. The results are summarized in the second and third columns of Table 3.5.1. They suggest that the posterior means can be quite sensitive to the choice between these two prior assumptions when $\bar{y}$ is small (e.g., $\bar{y} = 1$), but not particularly sensitive as $\bar{y}$ gets larger.

(C) *The Posterior Density of a Binomial Logit, Using a Normal Prior*

When y possesses a binomial distribution with probability θ, sample size n, and $\lambda = \text{logit}(\theta)$, the likelihood approximation (3.4.4) may be employed, with $\hat{\lambda} = \log\{p/(1 - p)\}$ and $\varphi = p^{-1} + (1 - p^{-1})$, and $p = y/n$. Let $p = 0.3$. Then curve (a) in Figure 3.5.2 describes the exact likelihood of λ, when $n = 10$, and curve (b) describes approximation (3.4.4). Curve (c) describes a normal prior for λ, which for illustrative purposes is taken to possess unit mean and variance. Curves (d) and (e), respectively, describe the exact posterior density of λ, and the approximate normal posterior density under the likelihood approximation (3.4.4).

Our posterior approximations are open to even more precise validation as n increases. The curves in Figure 3.5.3 possess similar description to those in Figure 3.5.2 except that

Table 3.5.1. *Posterior mean for λ.*

$\bar{y}$	*(a)*	*(b)*
1	0.24794	0.05225
2	0.73143	0.73503
3	1.07608	1.07238
4	1.34075	1.36740
5	1.55438	1.59808
6	1.73291	1.78407
7	1.88598	1.94027
8	2.01977	2.07507
9	2.13850	2.19370
10	2.24516	2.29967

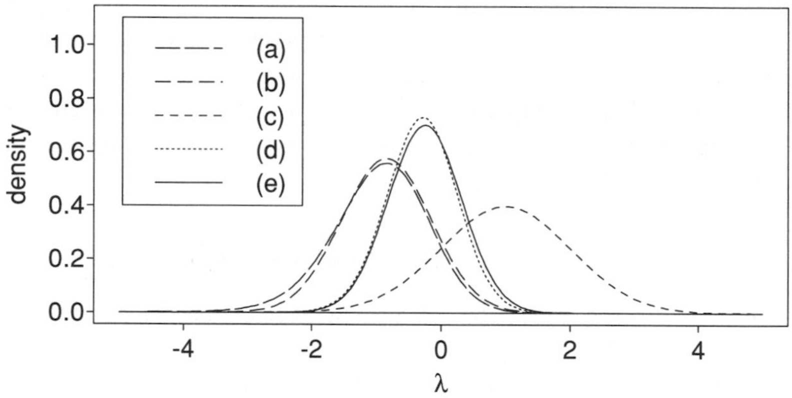

Figure 3.5.2. Posterior inference for a binomial logit (normal prior, $n = 10$): (a) exact likelihood; (b) approximate normal likelihood; (c) normal prior; (d) exact posterior; (e) approximate normal posterior.

$n = 30$, with all other quantities kept fixed. Note that our normal approximations to the posterior densities are even more accurate than the normal likelihood approximations, owing to the effect of the normal prior upon the exact posterior.

(D) *The Posterior Density of a Normal Mean, Using a Prior Mixture of Normals*

A random sample of $n = 20$ observations was generated from a normal distribution with mean $\theta = 3$ and variance $\tau^2 = 36$. The observed sample mean was $\bar{y} = 2.765$. In our prior-to-posterior analysis, the θ was taken to be unknown, and $\tau^2 = 36$ known. The prior distribution of θ was taken to be an equally weighted mixture of two normal distributions, with respective means 0 and 5, and unit variances. The posterior distribution of θ was a mixture of two normal distributions (see Exercise 3.4.c) with respective

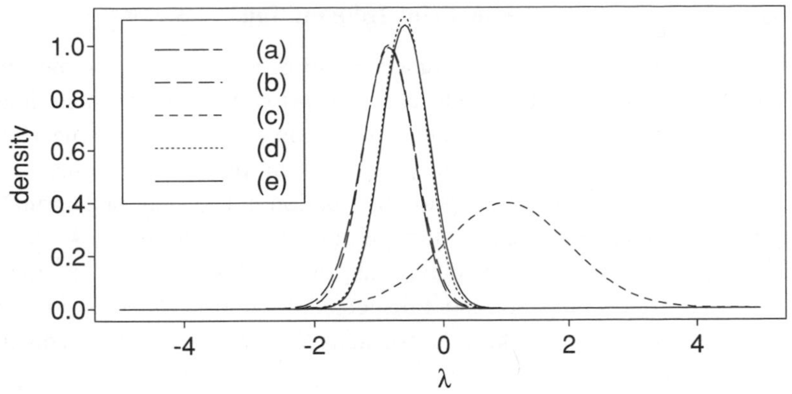

Figure 3.5.3. Posterior inference for a binomial logit (normal prior, $n = 30$): (a) exact likelihood; (b) approximate normal likelihood; (c) normal prior; (d) exact posterior; (e) approximate normal posterior.

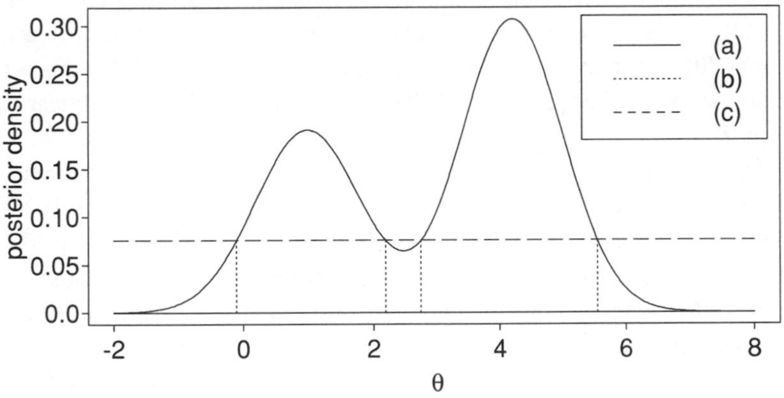

Figure 3.5.4. A posterior mixture of two normal densities: (a) posterior density; (b) boundaries of 90% HPD region; (c) horizontal line for construction of HPD region.

mixing probabilities 0.384 and 0.616, means 0.988 and 4.202, and common variance 0.643.

The posterior density is plotted as curve (a) in Figure 3.5.4. The vertical lines (b) describe the 90% HPD region, which comprises the intervals $(-0.104,\ 2.185)$ and $(2.739,\ 5.544)$. This should be compared with the 90% equal-tailed Bayesian interval $(-0.201,\ 5.624)$. The horizontal line (c) illustrates how to construct the HPD region. Note that the four vertical lines possess equal height and that the intervals contained in the HPD region possess minimized width when compared with other regions with 90% probability coverage. The 95% HPD region is the single interval $(-0.201,\ 5.624)$, while the 95% equal-tailed Bayesian interval is $(-0.226,\ 5.600)$. Furthermore, the 99% HPD region was the interval $(-0.786,\ 6.139)$ compared with the 99% equal-tailed Bayesian interval $(-0.797,\ 6.129)$.

3.6 Vague Prior Information

A scientist is rarely in a state of prior ignorance regarding an unknown parameter θ. The Bayesian paradigm cannot formally handle complete prior ignorance. In such situations, the scientist can use likelihood methods if there is a sampling model available. However, Bayesians can handle situations where the prior information is fairly vague.

First, consider the $N(\mu, \sigma^2)$ normal prior distribution introduced for a normal mean θ, or for appropriate transformations of other parameters, in Section 3.4. A small value of σ^2 indicates your feeling that θ is quite likely to be close to μ. As σ^2 increases, the prior density becomes more and more dispersed, around μ. Then, in the limit, as $\sigma^2 \to \infty$, $\pi(\theta) \to K$, for all θ, where the constant K is arbitrary and does not depend upon θ, that is,

$$\pi(\theta) \propto 1 \qquad (-\infty < \theta < \infty). \tag{3.6.1}$$

The limit (3.6.1) is not a density, since it does not integrate to unity. The prior distribution of θ becomes "improper" as $\sigma^2 \to \infty$. Therefore, (3.6.1) is typically described as an "(improperly) uniform prior density" or a "vague prior density." It cannot be correctly referred to as an "ignorance prior." It indeed represents the specific prior information that θ is equally likely to fall in either of any two intervals $(a, a+w)$, $(b, b+w)$, which are of equal width w, for any possible choices of these intervals, and that this is true for θ rather than for some other parameter ϕ that can be expressed as a nonlinear function $\phi = \phi(\theta)$ of θ.

Under the prior density (3.6.1), the posterior density (3.4.10) for a normal mean θ reduces to a $N(\bar{y}, n^{-1}\tau^2)$ density. We can therefore state that the posterior probability that θ lies in the 95% (frequency-based) confidence interval $(\bar{y}-1.96\tau/n^{\frac{1}{2}}, \bar{y}+1.96\tau/n^{\frac{1}{2}})$ is 0.95. Sir Ronald Fisher tried to justify similar conditional inferences, without assuming a prior, and constructed the "fiducial approach" (e.g., Fisher, 1959, with counterexamples by Lindley, 1958, and Robinson, 1975). It would appear that prior assumptions are needed to justify conditional inferences of this type and that without them the problem does not contain enough intrinsic structure. (See discussion in Section 3.0.)

Under the likelihood approximation (3.4.4) and a uniform prior for λ, we can make the statement that the posterior probability that λ lies in the approximate 95% (frequency-based) confidence interval $(\hat{\lambda} - 1.96\vartheta^{\frac{1}{2}}/n^{\frac{1}{2}}, \hat{\lambda} + 1.96\vartheta^{\frac{1}{2}}/n^{\frac{1}{2}})$ is approximately 0.95. This statement can be made more rigorous, by letting n get large, and for any prior density $\pi(\lambda)$ for λ, with positive support (see Exercise 3.4.b). Under a wide range of regularity conditions, it is true that any $100(1-\epsilon)\%$ Bayesian region will give frequency coverage approaching $100(1-\epsilon)\%$ as n gets large.

Next, consider the choice of vague prior distribution for the probability θ of a binomial distribution. One possibility is to consider the beta family of Section 3.2 and to let the prior sample size γ tend to zero, so that both α and β also approach zero. This gives the vague prior density

$$\pi(\theta) \propto \theta^{-1}(1-\theta)^{-1} \qquad (0 < \theta < 1), \tag{3.6.2}$$

which integrates to infinity and is equivalent to the vague prior density

$$\pi(\lambda) \propto 1 \qquad (-\infty < \lambda < \infty), \tag{3.6.3}$$

for the logit $\lambda = \log\theta - \log(1-\theta)$. Both (3.6.2) and (3.6.3) can also be justified by taking a normal $N(\mu, \sigma^2)$ prior for λ and letting $\sigma^2 \to \infty$.

Under the prior assessment in (3.6.2), the posterior density of θ is beta with parameters y and $n-y$ if neither $y=0$ nor $y=n$, but the posterior distribution is improper if either $y=0$ or $y=n$. If the posterior distribution is proper, then the posterior mean and variance of θ are $p = y/n$ and $p(1-p)/(n+1)$, respectively. It is, however, usually just as reasonable to take the vague prior distribution to be uniform in θ and to possess density

$$\pi(\theta) = 1 \qquad (0 < \theta < 1). \tag{3.6.4}$$

This corresponds to a beta distribution with parameters $\alpha = \beta = 1$. The posterior distribution is then automatically also proper and becomes beta with parameters $y+1$ and $n-y+1$. The posterior mean and mode of θ are $(y+1)/(n+2)$ and y/n, respectively, and you can make posterior probability statements, from (3.2.14), for any choice of y.

Jeffreys' invariance prior (Berger, 1985, p. 390) more generally takes the form

$$\pi(\theta) \propto |F(\theta)|^{\frac{1}{2}}, \tag{3.6.5}$$

where

$$F(\theta) = -\underset{\mathbf{y}|\theta}{E}\left[\frac{\partial^2 \log l(\theta \mid \mathbf{y})}{\partial\theta^2}\right] \tag{3.6.6}$$

denotes Fisher's information for θ. Under a broad range of regularity conditions, it possesses an "invariance" property, that is, if λ is some monotonic nonlinear transformation of θ, the $\pi(\lambda \mid y) \propto |F^*(\lambda)|^{\frac{1}{2}} l(\lambda \mid y)$ automatically incorporates the Jacobian for transforming from $\pi(\theta \mid y) \propto |F(\theta)|^{\frac{1}{2}} l(\theta \mid y)$. Here $F^*(\lambda)$ denotes Fisher's information for λ. In the current situation, the Jeffreys' prior for θ is

$$\pi(\theta) \propto \theta^{-\frac{1}{2}}(1-\theta)^{-\frac{1}{2}} \qquad (0 < \theta < \infty), \tag{3.6.7}$$

which corresponds to a beta distribution, with parameters $\alpha = \beta = \frac{1}{2}$. Furthermore, the Jeffreys' prior for λ is

$$\pi(\lambda) \propto \frac{e^{\frac{1}{2}\lambda}}{1+e^{\lambda}} \qquad (-\infty < \lambda < \infty), \tag{3.6.8}$$

which multiplies (3.6.7) by the Jacobian $\theta(1-\theta)$ of the transformation. Bernardo (1979) provides another justification of (3.6.8) by describing it as a "reference prior" that optimizes a limiting entropy distance criterion, relating to Lindley (1956). Like (3.6.6), his entropy criterion integrates over the sample space and therefore need not satisfy the likelihood principle (see Section 1.5 (C) and Exercises 1.5.e and 1.5.f).

Under Jeffreys' prior, the posterior distribution of θ is beta with parameters $y+\frac{1}{2}$ and $n-y+\frac{1}{2}$, so that the posterior mean $(y+\frac{1}{2})/(n+1)$ is quite appealing when either $y=0$ or $y=n$. However, you are left with the problem of how to choose between different suggestions for the vague prior distribution. One possibility is to demand that the consequent Bayesian intervals should possess frequency properties that are as accurate as possible when the sample size n is finite. In the normal case, you have seen that the uniform prior (3.6.1) yields perfect finite sample frequency properties.

In the binomial case, note from standard properties of the incomplete beta function (e.g., Abramowitz and Stegun, 1968, p. 263) that whenever α is an integer, (3.2.15) satisfies

$$I_{\theta_0}(\alpha, n-\alpha+1) = p(Y \geq \alpha), \tag{3.6.9}$$

where Y denotes a binomial random variable, with probability θ_0 and sample size n. Consequently,

$$p(Y \geq y) = I_{\theta_0}(y, n-y+1) \leq I_{\theta_0}\left(y+\tfrac{1}{2}, n-y+\tfrac{1}{2}\right), \tag{3.6.10}$$

since a beta $(y, n-y+1)$ distribution is "stochastically dominated" by a beta $(y+\frac{1}{2}, n-y+\frac{1}{2})$ distribution. Similarly,

$$p(Y > y) = p(Y \geq y+1) = I_{\theta_0}(y+1, n-y) \geq I_{\theta_0}\left(y+\tfrac{1}{2}, n-y+\tfrac{1}{2}\right). \tag{3.6.11}$$

Therefore, when $\alpha = \beta = \frac{1}{2}$, or indeed α and β are any positive values summing to unity, the Bayesian significance probability in (3.2.17) satisfies

$$p(Y > y \mid \theta = \theta_0) \leq \epsilon_0 \leq p(Y \geq y \mid \theta = \theta_0), \tag{3.6.12}$$

which is a valid definition for a classical significance probability. Therefore, in this particular case, the Jeffreys' prior yields excellent frequency properties. It is similarly possible to demonstrate that the corresponding $100(1-\epsilon)\%$ Bayesian intervals will possess frequency coverage about as close to $100(1-\epsilon)\%$ as can be permitted by the discrete nature of the sampling distribution; see Section 3.7 (B). This is the only formalization available for finite sample confidence intervals for a binomial θ that does not invert a randomized test, by randomizing different intervals, and that is not overconservative.

When using improper distributions in the prior assessment, it is not only necessary to check that the posterior distribution is proper but also that the posterior distribution yields intuitively sensible results. Dawid, Stone, and Zidek (1973) give a number of convincing counterexamples in models involving several parameters.

Worked Example 3J: *Jeffreys' Prior for a One-Parameter Exponential Distribution*

Let $y_1, y_2, \ldots, y_n$ denote a random sample from the exponential distribution with density $\theta e^{-\theta y}$ for $0 < y < \infty$, and with mean $\lambda = 1/\theta$. Find (a) Jeffreys' prior for θ, and (b) Jeffreys' prior for λ. Show that both lead to the same posterior distribution.

Model Answer 3J:

The likelihood of θ is

$$l(\theta \mid \mathbf{y}) = \prod_{i=1}^{n} \theta e^{-\theta y_i} = \theta^n e^{-n\theta \bar{y}} \qquad (0 < \theta < \infty).$$

Consequently,

$$\frac{\partial^2 \log l(\theta \mid y)}{\partial \theta^2} = -\frac{n}{\theta^2},$$

which does not dependent on $\bar{y}$. Therefore $F(\theta) = n/\theta^2$, and Jeffreys' prior for θ is

$$\pi(\theta) \propto |F(\theta)|^{\frac{1}{2}} \propto \theta^{-1}.$$

This gives the posterior density

$$\pi(\theta \mid y) \propto \pi(\theta) l(\theta \mid \mathbf{y}) \propto \theta^{n-1} e^{-\theta \bar{y}} \qquad (0 < \theta < \infty),$$

so that the posterior distribution of θ is Gamma with parameters n and $\bar{y}$. Moreover, the likelihood of λ is

$$l(\lambda \mid \mathbf{y}) = \lambda^{-n} e^{-n\lambda^{-1}\bar{y}} \qquad (0 < \lambda < \infty).$$

Consequently,

$$\frac{\partial^2 \log l(\lambda \mid \mathbf{y})}{\partial \lambda^2} = \frac{n}{\lambda^2} - \frac{2n\bar{y}}{\lambda^3}.$$

Since the expectation of $\bar{y}$ is λ, Fisher's information is $F(\lambda) = n/\lambda^2$, and Jeffreys' prior for λ is

$$\pi(\lambda) \propto |F(\lambda)|^{\frac{1}{2}} \propto \lambda^{-1}.$$

This gives the posterior density

$$\pi(\lambda \mid y) \propto \lambda^{-n-1} e^{-n\lambda^{-1}\bar{y}} \qquad (0 < \lambda < \infty).$$

However, since $\partial\theta/\partial\lambda = \lambda^{-2}$, transforming back gives the same Gamma posterior density for λ, as described above.

Worked Example 3K: *Bayesian Significance Probabilities for the One-Parameter Exponential Distribution*

Under the sampling and prior assumptions of Worked Example 3J, suppose that it is required to investigate the null hypothesis $H_0: \theta = \theta_0$. Express the Bayesian significance probability $p(\theta \leq \theta_0 \mid \bar{y})$ in terms of the normalized incomplete Gamma function $G_\alpha(a)$ in (3.2.26). Show that this is equal to the classical significance probability $p(\bar{Y} \leq \bar{y} \mid \theta = \theta_0)$, where $\bar{Y}$ represents $\bar{y}$ as a random variable.

Model Answer 3K:

Since the posterior density of θ is

$$\pi(\theta \mid \mathbf{y}) = \frac{(n\bar{y})^n}{\Gamma(n)} \theta^{n-1} e^{-n\theta}\, \bar{y} \qquad (0 < \theta < \infty),$$

$$\begin{aligned} p(\theta \le \theta_0 \mid \mathbf{y}) &= \int_0^{\theta_0} \frac{n\theta\bar{y}}{\Gamma(n)} e^{-\theta\bar{y}} d\theta \\ &= \int_0^{n\theta_0\bar{y}} \frac{1}{\Gamma(n)} \xi^{n-1} e^{-\xi} d\xi \\ &= G_{n\theta_0\bar{y}}(n). \end{aligned}$$

Also, given θ, $n\bar{y}$ is Gamma with parameters n and θ, so that

$$\begin{aligned} p\big(\bar{Y} \le \bar{y} \mid \theta = \theta_0\big) &= p\big(n\bar{Y} \le n\bar{y} \mid \theta = \theta_0\big) \\ &= \int_0^{n\bar{y}} \frac{1}{\Gamma(n)} \theta_0^n u^{n-1} \exp\{-\theta_0 u\}\, du \\ &= \int_0^{n\theta_0\bar{y}} \frac{1}{\Gamma(n)} z^{n-1} \exp\{-z\}\, dz \\ &= G_n(n\theta_0\bar{y}) \quad \text{as required.} \end{aligned}$$

Worked Example 3L: *Inference for a Between-Group Variance*

Consider a random effects model with n replications on observations in m groups. Given $z_1, z_2, \ldots, z_m$, the m group means $\bar{y}_1, \bar{y}_2, \ldots, \bar{y}_m$ are independent with respective means $z_1, z_2, \ldots, z_m$ and known variance τ^2/n. However, $z_1, z_2, \ldots, z_m$ are a random sample from the normal distribution with known mean μ and unknown variance σ^2. Noting that unconditionally on the z's, the $\bar{y}$'s are a randomsample from a normal distribution with mean μ and variance $\sigma^2 + n^{-1}\tau^2$,

(a) find the likelihood of the between-group variance σ^2, in terms of

$$U_B = n^{-1} \sum_{i=1}^{n} (\bar{y}_i - \mu)^2,$$

which is assumed to be strictly positive;

(b) show that under an improperly uniform prior for $\log \sigma^2$, the posterior distribution of σ^2 is always improper if $m > 2$;

(c) find and describe the posterior distribution of $\phi = (\sigma^2 + n^{-1}\tau^2)^{-1}$, under an improperly uniform prior for σ^2.

Model Answer 3L:

(a) The likelihood of σ^2, given the $\bar{y}$'s, is equal to the joint density of $\bar{y}_1$, $\bar{y}_2$, $\ldots$, $\bar{y}_m$, as μ and τ^2 are known. This is

$$
\begin{aligned}
l(\sigma^2 \mid \bar{\mathbf{y}}) &= p(\bar{\mathbf{y}} \mid \sigma^2) \\
&\propto (\sigma^2 + n^{-1}\tau^2)^{-\frac{1}{2}m} \exp\left\{-\frac{1}{2}(\sigma^2 + n^{-1}\tau^2)^{-1}\sum_{i=1}^{n}(\bar{y}_i - \mu)^2\right\} \\
&\propto (\sigma^2 + n^{-1}\tau^2)^{-\frac{1}{2}m} \exp\left\{-\frac{1}{2}n^{-1}(\sigma^2 + n^{-1}\tau^2)^{-1}U_B\right\}.
\end{aligned}
$$

Note that $l(\sigma^2 \mid \bar{\mathbf{y}}) > 0$ when $\sigma^2 = 0$.

(b) Since $\pi(\sigma^2) \propto (\sigma^2)^{-1}$,

$$\int_0^{\epsilon} \pi(\sigma^2)\, d\sigma^2 = \infty, \qquad \text{for any } \epsilon > 0,$$

that is, this $\pi(\sigma^2)$ possesses an infinite spike at $\sigma^2 = 0$. Assume, initially, that

$$\pi(\sigma^2 \mid \bar{\mathbf{y}}) = K(\sigma^2)^{-1} l(\sigma^2 \mid \bar{\mathbf{y}})$$

for some constant K. For appropriately small ϵ, there exists a bound M such that

$$0 < M < l(\sigma^2 \mid \bar{\mathbf{y}})$$

for all $\sigma^2 < \epsilon$. For this ϵ, the "posterior probability" that $\sigma^2 \le \epsilon$ is

$$
\begin{aligned}
\int_0^{\epsilon} \pi(\sigma^2 \mid \bar{\mathbf{y}})\, d\sigma^2 &= K\int_0^{\epsilon} (\sigma^2)^{-1} l(\sigma^2 \mid \bar{\mathbf{y}})\, d\sigma^2 \\
&> MK\int_0^{\epsilon} (\sigma^2)^{-1} d\sigma^2 = \infty.
\end{aligned}
$$

As this "posterior probability" exceeds unity, the posterior distribution must be improper.

(c) If $\pi(\sigma^2) \propto 1$, then $\pi(\sigma^2 \mid \bar{\mathbf{y}}) \propto l(\sigma^2 \mid \bar{\mathbf{y}})$, with $\phi = (\sigma^2 + n^{-1}\tau^2)^{-1}$, $\partial\phi/\partial\sigma^2 = -(\sigma^2 + n^{-1}\tau^2)^{-2} = -\phi^2$, so that $\pi(\phi) \propto \phi^{-2}$. Consequently, the posterior density of ϕ is

$$
\begin{aligned}
\pi(\phi \mid \bar{\mathbf{y}}) &\propto \phi^{-2} l(\phi \mid \bar{\mathbf{y}}) \\
&\propto \phi^{\frac{1}{2}(m-2)-1} \exp\left\{-\tfrac{1}{2}n^{-1}\phi U_B\right\} \qquad (0 < \phi < n\tau^{-2}).
\end{aligned}
$$

If $m > 2$, the posterior distribution truncates a Gamma distribution, with parameters $\frac{1}{2}(m-2)$ and $\frac{1}{2}n^{-1}U_B$ to the range $(0, n\tau^{-2})$, and is therefore proper.

SELF-STUDY EXERCISES

3.6.a Suppose that the sampling distribution of nU/ϕ is chi-squared with n degrees of freedom. Find the posterior densities, means, and variances of ϕ, $\xi = \phi^{-1}$, and $\lambda = \log\phi$, given that $U = u$, under the vague prior densities

(a) $\pi(\lambda) \propto 1 \quad (-\infty < \lambda < \infty)$ and
(b) $\pi(\phi) \propto 1 \quad (0 < \phi < \infty)$.

Hint: The posterior density of ξ is a scale multiple of a chi-squared density. The posterior mean and variance of $\lambda = -\log\xi$ can be obtained by noting that this posterior density integrates to unity, and by taking appropriate first and second deviatives under the integral sign. They can be expressed in terms of the digamma function, $\psi(z) = \partial\log\Gamma(z)/\partial z$, and the trigamma function, $\Psi^{(1)}(z) = \partial^2\log\Gamma(z)/\partial z^2$. This device can be used for many other choices of density, with Gamma functions in the denominator.

3.6.b In Exercise 3.6.a, suppose that $\pi(\lambda) \propto 1$. Use the likelihood approximation (3.4.4) to obtain approximations to the posterior mean and variance of λ. Show, from Exercise 3.6.a, how these give approximations to the digamma function and its derivative.

3.6.c In the Poisson random sampling examples (Exercises 3.4.a and 3.4.b), suppose that the prior distribution of θ is Gamma with parameters α and β. Then $2\beta\theta$ possesses a chi-squared distribution with 2α degrees of freedom. The posterior distribution of θ is Gamma with parameters $\alpha^* = \alpha + n\bar{y}$ and $\beta^* + n$. Express the Bayesian significance probability $\epsilon_0 = p(\theta \le \theta_0 \mid \mathbf{y})$ for investigating $H_0\colon \theta = \theta_0$ in terms of

$$\chi_\nu^2(u) = p(U_\nu \le u), \tag{3.6.13}$$

the c.d.f. of a chi-squared variate, with ν degrees of freedom.

3.6.d If α is an integer, then the function in (3.6.13) satisfies

$$\chi_{2\alpha}^2(2n\theta_0) = p(T \ge \alpha), \tag{3.6.14}$$

where T possesses a Poisson distribution, with mean $n\theta_0$.

(See Abramowitz and Stegun, 1968, p. 941.) Show that the Bayesian significance probability ϵ_0 in Exercise 3.6.c, but with $\beta = 0$, satisfies (3.6.14), but with α replaced by $\alpha^* = \alpha + n\bar{y}$. Noting that $T = \sum Y_i$ possesses a Poisson distribution, with mean $n\theta_0$, describe those values of α for which ϵ_0 is also a valid classical significance probability. Describe the corresponding prior densities. Are uniform densities for θ and $\lambda = \log\theta$ included?

3.6.e Let $y_1, \ldots, y_n$ denote a random sample from the exponential distribution, with density

$$f(y \mid \theta) = \lambda\exp\big\{-\lambda(y-\theta)\big\}I[\theta \le y_i]$$
$$(0 < A,\, y_i < \infty,\, 0 < \lambda < \infty), \tag{3.6.15}$$

where $I(A)$ is the indicator function for the set A, θ is unknown, and λ is known. Show that the likelihood of θ, given $\mathbf{y} = (y_1, \ldots, y_n)^T$, is

$$l(\theta \mid \mathbf{y}) \propto \exp\{n\lambda\theta\}I[\theta \le z] \qquad (-\infty < \theta < \infty), \tag{3.6.16}$$

where $z = \min y_i$. Under an improper uniform prior for θ, find the posterior density and mean of θ. Under a normal $N(\mu, \sigma^2)$ prior, show that the posterior distribution is truncated normal and find the posterior mode of θ.

3.6.f Write down a complete expression, in terms of nine Gamma functions, for the predictive probability mass function (3.2.19) of a binomial frequency z, under the Jeffreys' prior $\alpha = \beta = \frac{1}{2}$. An accountant makes 3 errors out of 1,000 entries in an account. Find the predictive mean and standard deviation for the number of errors he will make out of the next 1,000 entries, assuming that his average error rate is constant and that the errors occur independently.

3.6.g Recalculate the predictive mean and standard deviation in Exercise 3.6.f when, instead, $\alpha \to 0$ and $\beta \to 0$. Show that in this case, the predictive probability mass function in (3.2.19) is identical to a modified predictive likelihood, from Exercise 1.5.k, with the exception that some of the Gamma functions are replaced by Stirling's approximation $\Gamma(z) = \sqrt{2\pi} z^{z-\frac{1}{2}} \exp\{-z\}$.

3.6.h Let X posses a binomial $B(\theta, n)$ distribution, where θ possesses a beta prior distribution, with parameters α and β. Consider the Bayes factor

$$B = \frac{p(X = x \mid H_0^*)}{p(X = x \mid H_1^*)}$$

for comparing $H_0^* : \theta \geq \theta_0$ with $H_1^* : \theta < \theta_0$.

(a) Show that B can also be used to compare $H_0 : \theta = \theta_0$ with $H_1 : \theta \neq \theta_0$.

(b) Prove that B is the ratio of posterior odds to prior odds,

$$B = \frac{(1 - \epsilon_0)/\epsilon_0}{(1 - \epsilon)/\epsilon},$$

where ϵ_0 satisfies (3.2.17) and $\epsilon = I_{\theta_0}(\alpha, \beta)$.

(c) Subject to the two propositions

Proposition 1 (Fuzzy ignorance): $\alpha + \beta = 1$,

Proposition 2 (Fairness): $B = 1$ whenever $\epsilon_0 = \frac{1}{2}$ (i.e., whenever θ_0 is posterior median),

show that θ_0 must be the prior median and that

$$B = \frac{1 - \epsilon^*}{\epsilon^*},$$

where $\epsilon^* = I_{\theta_0}(\alpha^* + x, 1 - \alpha^* + n - x)$, and α^* can be determined from

$$I_{\theta_0}(\alpha^*, 1 - \alpha^*) = 0.5.$$

(d) Show that ϵ^* satisfies the inequality for ϵ_0 in (3.6.12) and can therefore be described as a compromise classical significance probability. Noting that the Bayes factor B is just a function of ϵ^*, discuss whether composite null hypotheses can resolve Lindley's paradox (Worked Example 3D).

Note: This methodology was developed, with O. Papasouliotis, to facilitate a fair evaluation of evidence in an investigation for Northumbria Police, Newcastle-upon-Tyne.

3.7 Practical Examples

(A) *Errors in Accounting*

An accountant makes $y = 3$ errors out of $n = 10{,}000$ entries and wishes to draw an inference regarding her underlying error rate θ (when expressed as a proportion). Curve (a) in Figure 3.7.1 describes the beta posterior density under the logistic uniform prior (3.6.2) for θ. Curve (b) describes the posterior density under the Jeffreys' prior (3.6.7), and curve (c) instead refers to the uniform prior (3.6.4). Note that there are substantial differences between the curves under different choices of vague prior.

However, hopefully, the computations of the next section would convince the accountant to refer to curve (b) on the grounds of excellent frequency properties.

(B) *Frequency Coverage of Bayesian Intervals Under Jeffreys' Prior*

Consider the posterior density

$$\pi(\theta \mid y) \propto \theta^{y-\frac{1}{2}}(1-\theta)^{n-y-\frac{1}{2}} \qquad (0 < y < 1)$$

for a binomial probability under a Jeffreys' prior. Then the 95% equal-tailed Bayesian interval for θ is (θ_1, θ_2), where

$$I_{\theta_1}\left(y+\tfrac{1}{2}, n-y+\tfrac{1}{2}\right) = 0.025$$

and

$$I_{\theta_2}\left(y+\tfrac{1}{2}, n-y+\tfrac{1}{2}\right) = 0.975.$$

For any true value θ and fixed n, it is possible to simulate the exact frequency coverage of this interval by generating a large number of independent realizations of y from

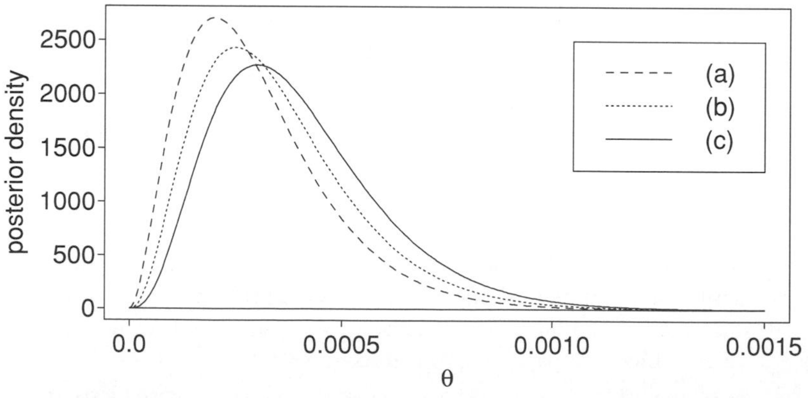

Figure 3.7.1. Posterior densities for accountancy error rate: (a) using logistic uniform prior; (b) using Jeffreys' prior; (c) using uniform prior.

Table 3.7.1. *Frequency coverages for 95% Bayesian intervals.*

θ	Lower tail	Interval	Upper tail
0.10	0.0257	0.9428	0.0315
0.15	0.0330	0.9551	0.0119
0.20	0.0335	0.9507	0.0158
0.25	0.0312	0.9522	0.0166
0.30	0.0268	0.9578	0.0154
0.35	0.0228	0.9468	0.0304
0.40	0.0332	0.9421	0.0247
0.45	0.0242	0.9555	0.0203
0.50	0.0319	0.9358	0.0323
0.55	0.0196	0.9570	0.0234
0.60	0.0255	0.9399	0.0346
0.65	0.0305	0.9462	0.0233
0.70	0.0152	0.9574	0.0274
0.75	0.0164	0.9522	0.0314
0.80	0.0162	0.9510	0.0329
0.85	0.0125	0.9552	0.0323
0.90	0.0324	0.9416	0.0260

a binomial distribution, with probability θ and sample size n, and by computing the proportion of times that

$$0.025 \leq I_\theta\left(y + \tfrac{1}{2}, n - y + \tfrac{1}{2}\right) \leq 0.0975.$$

Results, based upon $M = 100{,}000$ simulations for $n = 50$, are summarized in Table 3.7.1. The figures in the third column denote the frequency coverages for the different values of θ in the first column. Note how remarkably close these are to 0.95. Any differences are caused by the discrete nature of the binomial distribution. The frequency coverages of the tails are described in the second and fourth columns and also seem convincing. Similar results for the 99% equal-tailed Bayesian interval are summarized in Table 3.7.2. Similarly convincing results are available for very low sample sizes, for example, $n = 10$.

3.8 Bayes Estimators and Decision Rules and Their Frequency Properties

Consider an $n \times 1$ vector of observations $\mathbf{Y}$, with specified sampling density $p(\mathbf{y} \mid \theta)$, for $\mathbf{y} \in S$ and $\theta \in \Theta$. Let D denote the decision space. For estimation problems, this comprises the space of all possible estimates for θ, so that typically $D = \Theta$. For other decision-making problems, D denotes the set of all possible decisions. For problems involving the investigation of two hypotheses, set $D = \{H_0, H_1\}$.

Definition 1: A *decision rule* $\delta = \delta(\cdot)$ is a mapping from S to D such that if the observation $\mathbf{y}$ occurs, you will make *decision* $d = \delta(\mathbf{y})$ in D.

Table 3.7.2. *Frequency coverages for 99% Bayesian intervals.*

θ	Lower tail	Interval	Upper tail
0.10	0.0043	0.9914	0.0043
0.15	0.0067	0.9910	0.0023
0.20	0.0082	0.9872	0.0046
0.25	0.0081	0.9867	0.0052
0.30	0.0071	0.9872	0.0057
0.35	0.0057	0.9893	0.0050
0.40	0.0036	0.9921	0.0043
0.45	0.0056	0.9910	0.0034
0.50	0.0028	0.9942	0.0030
0.55	0.0032	0.9914	0.0054
0.60	0.0042	0.9920	0.0039
0.65	0.0055	0.9889	0.0056
0.70	0.0055	0.9878	0.0067
0.75	0.0055	0.9863	0.0082
0.80	0.0045	0.9873	0.0082
0.85	0.0022	0.9911	0.0066
0.90	0.0044	0.9913	0.0044

Definition 2: An *estimator* is a decision rule such that if the observation $\mathbf{y}$ occurs, you will use the decision $d = \delta(\mathbf{y})$ as your *estimate* for θ.

Definition 3: A *loss function* $L(d, \theta)$ is a mapping from $D \times \Theta$ to R' such that if you make decision d, when the true value of the parameter is θ, you will, usually hypothetically, "lose" a real-valued quantity $L(d, \theta)$.

Blattberg and George (1992) consider profit-driven loss functions. However, losses need not be monetary but can involve subjective feelings of sadness or joy at losing or winning. "Lose" is a statistical term, and the "loss" is usually hypothetical. For comparing the two composite hypotheses $H_0 : \theta \le \theta_0$ and $H_1 : \theta > \theta_0$, with $\Theta \subseteq R'$, set $D = \{H_0, H_1\}$. A possible loss function is

$$L(H_i, \theta) = \begin{cases} l_0 I[\theta > \theta_0] & (i = 0) \\ l_1 I[\theta \le \theta_0] & (i = 1), \end{cases} \tag{3.8.1}$$

where $I[A]$ denotes the indicator function for the set A, and the losses l_0 and l_1, under the Type-I and Type-II errors, should be specified subjectively. The traditional Type-I error relates to selecting H_1 when H_0 is true, and the Type-II error relates to selecting H_0 when H_1 is instead true. Note that the decision making is unaffected by instead considering any linear transformation $L^*(d, \theta) = a + bL(d, \theta)$ of the loss function, with $b > 0$. For problems where d is a point estimate of $\theta \in \Theta \subseteq R'$, the class of symmetric loss functions

$$L(d, \theta) = G(|d - \theta|) \qquad (d \in D, \theta \in \Theta) \tag{3.8.2}$$

can be considered, where $G(\cdot)$ is some monotonic increasing function, with $G(0) = 0$. This family includes squared error loss

$$L(d, \theta) = (d - \theta)^2 \tag{3.8.3}$$

and absolute loss

$$L(d, \theta) = |d - \theta|. \tag{3.8.4}$$

The existence of a loss function is a big assumption. Applied statisticians may prefer to use formal inference to summarize the information available, say, by a likelihood or posterior distribution, and then make less formal decisions involving pragmatic judgments, based upon the information thus summarized, together with any further background information. For example, the applied approach to investigating hypotheses outlined in Section 3.2, which combines the concepts of statistical significance and practical significance, may be preferred to a decision-theoretic approach. Also, formal decisions may appeal to a constrained type of rationality (see Chapter 4). However, if a formal decision-theoretic approach is employed, it is important to show that your decisions are reasonable under a broad range of loss functions. This motivates the variety of loss functions considered in this chapter. Roulette provides a motivation for considering frequency properties (see the numerical example of Section 1.2 (B)). If the wheel is fair, then you will, in the long run, lose 1/19 of all money bet. Therefore, the wheel would have to be quite biased to give any Bayesian decision rule a chance of succeeding, and you would need quite a large sample size to detect these biases. More generally, we regard the computation of frequency properties (such as risk functions) of decision rules or estimators as of the most essential importance.

Definition 4: The *risk function* of a decision rule or estimator $\delta = \delta(\cdot)$ is

$$r_\delta(\theta) = \underset{\mathbf{y}|\theta}{E}\, L\{\delta(\mathbf{y}), \theta\}, \tag{3.8.5}$$

where $L(d, \theta)$ is the loss function, and the expectation is with respect to the sampling distribution of $\mathbf{y}$, given θ.

For example, under the quadratic loss function (3.8.3), the risk function is

$$\begin{aligned} r_\delta(\theta) &= \{E(\delta(\mathbf{y}) \mid \theta) - \theta\}^2 + \text{var}(\delta(\mathbf{y}) \mid \theta) \\ &= \textit{bias}^2 + \textit{variance} \\ &= \textit{mean squared error (MSE)} \end{aligned} \tag{3.8.6}$$

whenever the sampling mean and variance exist. This is clearly a frequency criterion.

Now let $\mathcal{D}$ denote the space of all possible decision rules δ.

Definition 5: The decision rule δ is *inadmissible* if there exists $\delta^* \in \mathcal{D}$ such that

$$r^*_\delta(\theta) \leq r_\delta(\theta) \qquad \text{for all } \theta \in \Theta$$

and

$$r^*_\delta(\theta) < r_\delta(\theta) \qquad \text{for some } \theta \in \Theta,$$

in which case δ^* is said to *dominate* δ.

Definition 6: The decision rule $\delta \in \mathcal{D}$ is *admissible* if δ is not inadmissible, that is, if it is not dominated by any other $\delta^* \in \mathcal{D}$.

In most situations where $\mathcal{D}$ is not artificially constrained to be small, there will be a large class of admissible decision rules, including, when $D = \Theta$, the "dogmatic" decision rules

$$\delta(\mathbf{y}) = d_0 \qquad (\mathbf{y} \in D), \tag{3.8.7}$$

which are constant in $\mathbf{y}$, for any specific $d_0 \in D$, and for estimation problems perform remarkably well when indeed $\theta = d_0$. It is therefore important to find frequency-based criteria that help us to choose a particular $\delta \in \mathcal{D}$, which can be regarded as best.

Definition 7: $\delta^* \in \mathcal{D}$ is *minimax* if it satisfies

$$\sup_{\theta \in \Theta} r_{\delta^*}(\theta) = \inf_{\delta \in \mathcal{D}} \sup_{\theta \in \Theta} r_\delta(\theta). \tag{3.8.8}$$

A minimax rule tries to minimize the worst that can happen, and therefore might not fare quite as well when the worst does not happen. Now let π denote some probability distribution on θ. From a frequency viewpoint, π need not denote the prior distribution. It could instead describe a weight function that distinguishes those values of θ for which it is more or less important to make an accurate decision. However, π should denote a proper probability distribution. Definitions 8 and 9 and Lemma 3.8.1 do not apply for improper choices of π.

Definition 8: The *average risk with respect to* π of a decision rule δ is

$$R_\delta(\pi) = \underset{\theta|\pi}{E}\, r_\delta(\theta), \tag{3.8.9}$$

where the expectation is with respect to the probability distribution π for θ.

Definition 9: A decision rule $\delta^* \in \mathcal{D}$ is *Bayes with respect to* π if

$$R_{\delta^*}(\pi) = \inf_{\delta \in \mathcal{D}} R_\delta(\pi), \tag{3.8.10}$$

in which case $R_{\delta^*}(\pi)$ is the *Bayes risk*.

Note that "Bayes with respect to π" is a frequency property and that all procedures resulting from this property will work well in frequency terms when θ lies in a region of Θ to which π attaches high probability. Moreover,

Lemma 3.8.1: *A decision rule $\delta^* \in \mathcal{D}$ that is uniquely Bayes (up to adjustments to δ^* on a space of* $\mathbf{y}$*'s with sampling probability zero) with respect to a particular π and loss function is admissible (for that loss function).*

Proof. Suppose that $\delta^* \in \mathcal{D}$ is uniquely Bayes and not admissible. Then, there exists $\delta' \in \mathcal{D}$, which dominates δ^*. Hence,

$$r_{\delta'}(\theta) \leq r_{\delta^*}(\theta) \quad \text{for all } \theta \in \Theta,$$

so that

$$R_{\delta'}(\pi) \leq R_{\delta^*}(\pi),$$

contradicting the assumption that δ^* is uniquely Bayes. Reducto ad absurdum. ■

In order to find Bayes decision rules, note that whenever $L(d, \theta)$ is bounded below,

$$\begin{aligned} R_\delta(\pi) &= \underset{\theta|\pi}{E}\ \underset{\mathbf{y}|\theta}{E}\ L\big(\delta(\mathbf{y}), \theta\big) \\ &= \underset{\mathbf{y}}{E}\ \underset{\theta|\mathbf{y}}{E}\ L\big\{\delta(\mathbf{y}), \theta\big\} \\ &= \underset{\mathbf{y}}{E}\ q_{\mathbf{y}}\big[\delta(\mathbf{y})\big], \end{aligned} \tag{3.8.11}$$

where

$$q_{\mathbf{y}}(d) = \underset{\theta|\mathbf{y}}{E}\ L(d, \theta) \tag{3.8.12}$$

denotes the posterior expectation of the loss function. For example, under the squared error loss (3.8.3),

$$q_{\mathbf{y}}(d) = \big[d - E(\theta \mid \mathbf{y})\big]^2 + \operatorname{var}(\theta \mid \mathbf{y}), \tag{3.8.13}$$

whenever the posterior mean $E(\theta \mid \mathbf{y})$ and variance $\operatorname{var}(\theta \mid \mathbf{y})$ exist, whereas under absolute loss (3.8.4),

$$q_{\mathbf{y}}(d) = 2d\Lambda_y(d) - 2\int_{-\infty}^{d} \theta\pi(\theta \mid \mathbf{y})\, d\theta + E(\theta \mid \mathbf{y}) - d, \tag{3.8.14}$$

whenever a continuous posterior density $\pi(\theta \mid \mathbf{y})$ for θ exists, where $\Lambda_y(\theta)$ denotes the posterior c.d.f. of θ, and the posterior mean $E(\theta \mid \mathbf{y})$ is assumed finite.

Definition: A *Bayes decision d^* with respect to π* is a member of D satisfying

$$q_{\mathbf{y}}(d^*) = \inf_{d \in D} q_{\mathbf{y}}(d). \tag{3.8.15}$$

Note, from (3.8.13), that the Bayes decision (or Bayes estimate) under squared error loss is the posterior mean $E(\theta \mid \mathbf{y})$ whenever the posterior variance is finite, and from (3.8.14) and a more general proof described by DeGroot (1970, p. 232) that under absolute loss, the Bayes decision is always a posterior median for θ. (Differentiating (3.8.14) with respect to d gives $q_y^{(1)}(d) = 2\Lambda_y(d) - 1$ and $q_y^{(2)}(d) = 2\pi(d \mid y) > 0$.) The proof of the following lemma is straightforward upon considering the representation (3.8.11).

Lemma 3.8.2: *Under the assumption that $L(d, \theta)$ is bounded below, suppose that for each $\mathbf{y} \in D$, $d^* = \delta^*(\mathbf{y})$ is a Bayes decision. Then δ^* is a Bayes decision rule with respect to π.*

These lemmas have important consequences. Suppose, say, that you construct a prior density $\pi(\theta)$ for θ and estimate θ by its posterior mean $\theta^* = \theta^*(\mathbf{y}) = E(\theta \mid \mathbf{y})$. Then $\theta^*(\mathbf{y})$ can possess excellent frequency properties. First, it is admissible by Lemma 3.8.2. Second, the MSE (3.8.13) will be particularly low for θ lying in these regions of Θ to which π assigns high probability. These practical advantages hold even if the choice of prior density cannot be based upon prior information or if the prior density only approximately represents the prior information. It is, of course, important to fully investigate the entire MSE of θ^*, since some admissible estimators can possess poor mean squared error over large regions of the parameter space.

As an example, the estimator (3.4.11), for a normal mean θ, is admissible under quadratic loss for any choice of μ and $\rho \in (0, 1)$, and its MSE will be small when θ lies within some neighborhood of μ (see Exercise 3.8.d). It is therefore useful to consider even if the tails of the prior density are thicker than normal, even though this estimator is not Bayes under a nonnormal prior distribution. The sample mean $\bar{y}$ is also admissible under squared error loss (see Berger, 1985, p. 548). The estimator (3.2.12) for a binomial probability is admissible under squared error loss for any $\alpha, \beta > 0$. The raw proportion y/n is also admissibleadmissible(see Johnson, 1971).

Choosing a Bayesian decision d^* by minimizing the posterior expectation of the loss function, rather than, say, the posterior median of the loss function, is open to justification because of (a) connections with the expected utility hypothesis (see Chapter 4) and (b) the consequent long-run average frequency properties. "Martian Bayes" estimates, which minimize the posterior median of the loss function, are discussed in Exercise 3.10.d.

Under the loss function (3.8.1), the posterior expected loss (3.8.12) satisfies

$$q_{\mathbf{y}}(H_0) = l_0(1 - \epsilon_0) \tag{3.8.16}$$

and

$$q_{\mathbf{y}}(H_1) = l_1\epsilon_0, \tag{3.8.17}$$

where $\epsilon_0 = p(\theta \leq \theta \mid \mathbf{y})$ denotes our Bayesian significance probability. The Bayes decision tells us to prefer H_1 to H_0 whenever $q_{\mathbf{y}}(H_1) < q_{\mathbf{y}}(H_0)$, that is, $\epsilon_0 < l_0/(l_0 + l_1)$. This gives a rationale for judging the smallness of a significance probability in relation to losses subjectively assigned to the Type-I and Type-II errors. Rejection of H_0, when $\epsilon_0 < 0.01$, is only justified when $l_1 = 99l_0$. Finding guilty people innocent appears to be exactly 99 times more serious than finding innocent people guilty.

A "generalized Bayes decision" is a decision minimizing (3.8.12), but under an improper prior distribution for θ. For example, $\bar{y}$ is a "generalized Bayes estimate" of a normal mean θ under quadratic loss, since it is the posterior mean under an improperly uniform prior for θ. The corresponding decision rules and estimators may or may not be admissible. However, Berger (1985, p. 547) tell us that under very broad regularity conditions, any admissible decision rule or estimator must be either Bayes or the limit of a sequence of Bayes decision rules or estimators (under different proper priors that do not converge to a proper limit). This result is known as the complete class theorem. Berger (p. 546) also describes "Stein's necessary and sufficient condition for admissibility," which constrains the way in which the limiting sequence can be chosen. Note that the complete class theorem applies also to decisions involving the choice of sampling model and implies that you should also consider prior distributions over the space of possible sampling models (e.g., Ferguson, 1973; Leonard, 1973a, 1978).

The following three lemmas relate to minimax estimators. The first lemma effectively tells us that a minimax estimator is Bayes or limiting-Bayes against the worst prior possible. The last two help us to use Bayesian procedures to find minimax procedures if we so wish.

Lemma 3.8.3: *Let δ^* be minimax, then δ^* satisfies*

$$\sup_{\pi} R_{\pi}(\delta^*) = \inf_{\delta \in \mathcal{D}} \sup_{\pi} R_{\pi}(\delta^*), \tag{3.8.18}$$

where the supremum should be taken over all possible distributions π on Θ.

Lemma 3.8.4: *Suppose that δ^* is Bayes against some prior π and has risk function $r_{\delta^*}(\theta)$, which is constant in θ. Then δ^* is minimax.*

Lemma 3.8.5: *Suppose that δ^* has risk function $r_{\delta^*}(\theta) = K$, which is constant in θ. Suppose that there exists a sequence of decision rules $\delta_1, \delta_2, \ldots,$ such that $\delta_j(\mathbf{y}) \to \delta^*$ as $j \to \infty$, for each $\mathbf{y} \in S$. Then if the Bayes risks $R_{\delta_1}(\pi_1), R_{\delta_2}(\pi_2), \ldots$ converge to K, as $j \to \infty$, δ^* is minimax.*

Note that minimax procedures may be more convincing when applied to the theory of games (e.g., von Neumann and Morgenstern, 1953; Clayton and Berry, 1985) and to the economic policy of an insurance company.

Worked Example 3M: *Mean Squared Error when Estimating a Range*

Let $y_1, y_2, \ldots, y_n$ denote a random sample from a uniform distribution over the range $(0, \theta]$, noting that the likelihood is $l(\theta \mid \mathbf{y}) = \theta^{-n} I[z \le \theta]$, where z is the maximum of the observations.

(a) Find the c.d.f. of z, and hence find the mean and variance of z.
(b) Find the mean squared error of the estimator $\theta^* = \alpha z$, where α is a constant.
(c) Show that the MSE of θ^* is uniformly minimized, for all $\theta > 0$, when $\alpha = (n+1)/[n - (n+2)^{-1}]$.
(d) Hence show that the maximum likelihood estimator $\hat{\theta} = z$ and the unbiased estimator $\theta^+ = (n+1)\hat{\theta}/n$ are both inadmissible, with respect to squared error loss.

Model Answer 3M:

(a) The c.d.f. of y_i is

$$G_i(y_i \mid \theta) = \begin{cases} 0 & \text{for } y_i \leq 0 \\ \theta^{-1} y_i & \text{for } 0 < y_i \leq \theta \\ 1 & \text{for } y_i > \theta. \end{cases}$$

The c.d.f. of z is therefore

$$\begin{aligned} F(z \mid \theta) &= \prod_{i=1}^{n} G_i(z \mid \theta) \\ &= \begin{cases} 0 & \text{for } z \leq 0 \\ \theta^{-n} z^n & \text{for } 0 < z \leq \theta \\ 1 & \text{for } z > \theta. \end{cases} \end{aligned}$$

Consequently, the density of z is

$$f(z \mid \theta) = n\theta^{-n} z^{n-1} \qquad (0 \leq z \leq \theta),$$

and it is now elementary to demonstrate that the mean and variance of z are $n\theta/(n+1)$ and $c_n\theta^2$, where $c_n = n/(n+1)^2(n+2)$.

(b) As

$$E(\theta^*) = \alpha E(z) = \frac{\alpha n \theta}{n+1},$$

the bias of θ^* is

$$\text{Bias}\,(\theta^*) = \frac{\alpha n \theta}{n+1} - \theta.$$

Consequently,

$$\begin{aligned} \text{MSE}\,(\theta^*) &= \text{Bias}^2(\theta^*) + \text{var}(\theta^*) \\ &= \theta^2 \left[\frac{\{\alpha n - (n+1)\}^2}{(n+1)^2} + \alpha^2 c_n \right]. \end{aligned}$$

(c) The optimal value $\tilde{\alpha}$ of α minimizes

$$\left[\alpha n - (n+1)\right]^2 + \frac{n\alpha^2}{n+2},$$

so that

$$\tilde{\alpha} n - (n+1) + \frac{\tilde{\alpha}}{n+2} = 0,$$

and

$$\tilde{\alpha} = (n+1)\left[n - (n+2)^{-1}\right].$$

This quantity provides a unique minimum.

(d) The estimators $\hat{\theta}$ and θ^+ correspond to the choices $\alpha = 1$ and $\alpha = (n+1)/n$ and therefore must be dominated by

$$\tilde{\theta} = \frac{(n+1)z}{n - (n+2)^{-1}}.$$

They are therefore inadmissible.

Worked Example 3N: *Admissible Estimators of a Range*

In the situation discussed in Worked Example 3M, consider the Pareto prior density

$$\pi(\theta) = \gamma\beta^\gamma\theta^{\gamma-1}I[\beta \le \theta],$$

where $\beta > 0$ and $\gamma > 2$.

(a) Find the prior mean and variance of θ.
(b) Find the posterior mean and variance of θ.
(c) Show that all estimators for θ, taking the form

$$\frac{\gamma+n}{\gamma+n-1}\max(z,\beta) \qquad (\beta > 0, \gamma > 2),$$

are admissible under squared error loss.

Model Answer 3N:

(a) The prior mean is

$$\begin{aligned} E(\theta) &= \int_\beta^\infty \theta\pi(\theta)\,d\theta \\ &= \gamma\beta^\gamma \int_\beta^\infty \theta^{-\gamma}\,d\theta \\ &= \gamma\beta^\gamma \frac{\beta^{-(\gamma-1)}}{(\gamma-1)} \\ &= \frac{\gamma\beta}{\gamma-1}. \end{aligned}$$

Similarly,

$$E(\theta^2) = \gamma\beta^\alpha \frac{\beta^{-(\gamma-2)}}{(\gamma-2)} = \frac{\gamma}{(\gamma-2)\beta^2},$$

so that $\text{var}(\theta) = c_\gamma\beta^2$, with

$$c_\gamma = \frac{\gamma}{(\gamma-2)} - \left(\frac{\gamma}{\gamma-1}\right)^2.$$

(b) The posterior density of θ is

$$\begin{aligned}\pi(\theta \mid \mathbf{y}) &\propto \pi(\theta) l(\theta \mid \mathbf{y}) \\ &\propto \theta^{-\gamma-1} I[z \le \theta] \theta^{-n} I\big[\max(z, \beta) \le \theta\big] \\ &\propto \theta^{-\gamma^*-1} I[\beta^* \le \theta],\end{aligned}$$

where $\gamma^* = \gamma + n$ and $\beta^* = \max(z, \beta)$. This takes the same form as the prior, but with γ and β, respectively, replaced by γ^* and β^*. Consequently,

$$\theta^* = E(\theta \mid y) = \frac{\gamma^* \beta^*}{(\gamma^* - 1)}$$

and

$$\operatorname{var}(\theta \mid y) = c_{\gamma^*} \beta^*.$$

(c) Clearly θ^* in (b) is the unique Bayes estimator under squared error loss, as the posterior variance is also always finite when $\beta > 0$ and $\gamma > 2$. As θ^* is identical to the estimator in part (c), this estimator is therefore admissible.

SELF-STUDY EXERCISES

3.8.a Let y possess a binomial distribution with probability θ and sample size n. Consider the three estimators $\delta_1(y) = y/n$, $\delta_2(y) = (y + \frac{1}{2})/(n + \frac{1}{2})$, and $\delta_3(y) = (y + 1)/(n + 1)$ of θ in relation to the squared error loss function (3.8.3). Prove that δ_2 and δ_3 are uniquely Bayes and admissible estimators of θ. Prove that the MSE of the unbiased estimator δ_1 is equal to its variance. Sketch the MSEs of your three estimators when $n = 10$. (See Section 3.9 (A) for some graphics when $n = 20$.) For which values of θ is Jeffreys' estimator δ_2 the best of the three? Use Lemma 3.8.4 to show that the estimator $\delta_4 = (y + \frac{1}{2}\sqrt{n})/(n + \sqrt{n})$ is minimax and find its constant risk function. For which values of θ would you prefer δ_4 to δ_1? What happens to this comparison as $n \to \infty$?

3.8.b Let $y_1, \ldots, y_n$ constitute a random sample from a normal distribution, with zero mean and unknown variance ϕ, and let $S^2 = \sum y_i^2$. Note that S^2/ϕ possesses a sampling distribution, given ϕ, which is chi-squared with n degrees of freedom, that is, χ_n^2, and possesses mean n and variance $2n$. Show that $\phi_1^* = S^2/n$ is an unbiased estimator of ϕ and possesses minimum variance among all unbiased estimators of ϕ (e.g., upon application of the general result in Section 1.5 (A) for maximum likelihood estimators), and so is, in a traditional sense, "best."

Show that under squared error loss $L(d, \phi) = (d - \phi)^2$, ϕ_1^* is dominated by $\phi_2^* = S^2/(n+2)$, so that the "best" estimator is inadmissible! Compare the biases and variances of ϕ_1^* and ϕ_2^* and try to explain this result.

3.8.c In Exercise 3.8.b, assume that the prior distribution of $\nu\lambda/\phi$ is χ_ν^2. Then λ^{-1} is the prior mean of $\xi = \phi^{-1}$ and ν provides a "prior sample size." Show that the posterior distribution of $(\nu\lambda + S^2)/\phi$ is $\chi_{\nu+n}^2$. Hence, prove that for any $\nu > 0$ and $\lambda > 0$, the estimator

$$\xi^* = \frac{\nu + n}{\nu\lambda + S^2}$$

is admissible under squared error loss for ξ. If $\nu > 2$, then the prior mean of ϕ is $\nu\lambda/(\nu-2)$ and the prior variance is finite if $\nu > 4$. Very carefully describe a class of estimators for ϕ, which are admissible under squared error loss for ϕ.

3.8.d In the normal random sampling situation of Section 3.4, where the mean θ is unknown and the variance τ^2 known, consider the shrinkage estimator θ^* in (3.4.11). Show that the mean squared error of this estimator is

$$r_\delta(\theta) = (1-\rho)^2(\theta-\mu)^2 + n^{-1}\rho^2\tau^2,$$

where $\rho = n/(n+k)$. Show that the shrinkage estimator will possess smaller mean squared error than the sample mean $\bar{y}$ whenever

$$|\theta - \mu| < n^{-\frac{1}{2}}\frac{(1+\rho)\tau}{1-\rho}.$$

Suppose that you require an estimator that will be superior to $\bar{y}$, in terms of mean squared error, whenever θ falls in the interval $(\mu - 3n^{-\frac{1}{2}}\tau, \mu + 3n^{-\frac{1}{2}}\tau)$, but that will be inferior to $\bar{y}$ whenever θ lies strictly outside this interval. Show that $\frac{1}{2}(\bar{y}+\mu)$ is a suitable candidate.

3.8.e In the Poisson sampling situation, described in Exercises 3.4.a–3.4.c, remember that if the prior distribution of θ is Gamma with parameters α and β, the posterior distribution is Gamma with parameters $\alpha + n\bar{y}$ and $\beta + n$. Show that for any α, $\beta > 0$, the estimator $\theta^* = (\alpha + n\bar{y})/(\beta + n)$ is admissible, under squared error loss. Now consider the loss function

$$L(d,\theta) = \frac{(d-\theta)^2}{\theta} \qquad (0 < d < \infty, 0 < \theta < \infty). \tag{3.8.19}$$

Show that the Bayes estimate, under this loss function, is $d^* = 1/E(\theta^{-1} \mid \mathbf{y})$ whenever the posterior mean $E(\theta \mid y)$ is also finite. Find this Bayes estimator for the above Poisson situation and show when it is admissible. Find the risk function and Bayes risk of your Bayes estimator.

Consider instead the "entropy loss function"

$$L(d,\theta) = \frac{d}{\theta} - \log\frac{d}{\theta} - 1 \qquad (0 < d < \infty, 0 < \theta < \infty). \tag{3.8.20}$$

Show that the Bayes estimate, under this loss function, is $d^* = 1/E(\theta^{-1} \mid \mathbf{y})$ whenever the posterior mean $E[\log\theta \mid \mathbf{y}]$ is also finite. Show that any estimate which is Bayes under (3.8.19) is also Bayes under (3.8.20), but not necessarily vice versa.

3.8.f In Exercise 3.8.e, show that the sample mean $\bar{y}$ has constant risk function equal to n^{-1} under loss function (3.8.19). Use Lemma 3.8.5, together with the Bayes estimators of Exercise 3.8.e, to prove that $\bar{y}$ is a minimax estimator of θ under this loss function.

Prove, however, that $\bar{y}$ has infinite risk function, for all $\theta \in (0,\infty)$, under the loss function (3.8.20) and is inadmissible under this loss function. Can $\bar{y}$ be expressed as the limit of a sequence of Bayes estimators?

3.8.g *Linear Prediction* (Ericson, 1970): An observation y has unknown mean θ and variance $v(\theta)$, where the functional form of $v(\cdot)$ is specified. It is required to estimate θ by a "linear predictor" of the form

$$\theta^* = a + by,$$

where a and b do not depend upon y.

(a) Show that the MSE of θ^* is

$$r^*(\theta) = \left[a - (1-b)\theta\right]^2 + b^2 v(\theta).$$

(b) Show that the average risk of θ^* is

$$R^*(\pi) = \left[a - (1-b)\mu\right]^2 + (1-b)^2\sigma^2 + b^2\tilde{v},$$

whenever μ and σ^2, the prior mean and variance of θ, exist, together with $\tilde{v}$, the prior expectation of $v(\theta)$. Show that the average risk is minimized by the estimator

$$\theta^* = \frac{v^{-1}y + \sigma^{-2}\mu}{\tilde{v}^{-1} + \sigma^{-2}}$$

and interpret this weighted average form.

(c) Suppose that the posterior mean is known to take the form $\theta^* = a + by$. Describe a and b.

3.8.h Let Y possess a Gamma distribution whose mean and variance are both equal to a positive integer M. The prior distribution of M is geometric, with probability mass function

$$p(M = m) = (1-\xi)\xi^{m-1} \qquad (m = 1, 2, \ldots),$$

where $0 < \xi < 1$.

(a) Prove that the posterior distribution of $M - 1$, given that $Y = y$, is Poisson, with mean $\lambda = y\xi$.

(b) Under the loss function

$$L(d, m) = \frac{(m-d)^2}{m} \qquad (m = 1, 2, \ldots; 0 < d < \infty),$$

show that the posterior expected loss, for an estimate d of m, is

$$\begin{aligned} q_{\mathbf{y}}(d) &= E(M \mid y) - 2d + E\left(M^{-1} \mid y\right) d^2 \\ &= 1 + \lambda - 2d + \lambda^{-1}\left(1 - e^{-\lambda}\right) d^2. \end{aligned}$$

(c) Find the Bayes estimate d^* of M and show that

$$q_{\mathbf{y}}(d^*) = E(M \mid y) - \left\{E\left(M^{-1} \mid y\right)\right\}^{-1} = 1 - \lambda \sum_{m=1}^{\infty} e^{-m\lambda}.$$

(d) Prove that the marginal distribution of ξY is exponential with mean $\xi/(1-\xi)$. Express the Bayes risk of your Bayes estimator of M as an infinite series.

3.8.i In Exercise 3.8.h, use Lemma 3.8.5 to prove that Y is a minimax estimator of M under the stated loss function.

3.9 Practical Examples

(A) Estimating a Binomial Probability

Consider first the situation described in Exercise 3.8.a. Curves (a), (b), (c), and (d) in Figure 3.9.1 respectively describe the MSEs of $\delta_1(y) = y/n$, $\delta_2(y) = (y+\frac{1}{2})/(n+1)$, $\delta_3(y) = (y+1)/(n+2)$, and $\delta_4(y) = (y+\frac{1}{2}\sqrt{n})/(n+\sqrt{n})$ when $n = 20$. Note that $\delta_1(y)$ is preferred when the true value θ is close to either zero or one. However, all four estimates are admissible. While the minimax criterion would prefer δ_4, a weighting distribution on $\theta \in (0, 1)$, averaging the MSEs, would provide another way of selecting a particular estimator, even if the weighting distribution is not interpreted as comprising part of the prior assessment.

Note that δ_2, δ_3, and δ_4 are all Bayes estimators and all three possess appealing mean squared error properties. A related result is described by Brown, Chow, and Fong (1992), who prove admissibility of the maximum likelihood estimator of $\theta(1-\theta)$.

(B) *Estimating a Normal Variance*

Next, consider the situation described in Exercise 3.8.b. Curves (a) and (b) in Figure 3.9.2 describe the MSEs of $\phi_1^* = S^2/n$ and $\phi_2^* = S^2/(n+2)$ when $n = 20$. Clearly ϕ_2^* dominates ϕ_1^*, so that the UMVU estimator is inadmissible. Curve (c) describes the MSE of

$$\phi_3^* = \frac{150 + S^2}{n+5},$$

which, by Exercise 3.8.c, must be admissible. However, note that ϕ_3^* performs well only for values of ϕ close to 2. Therefore, ϕ_3^* is an example of a proper Bayes and admissible estimator, with poor frequency properties. Curves (a), (b), and (c) in

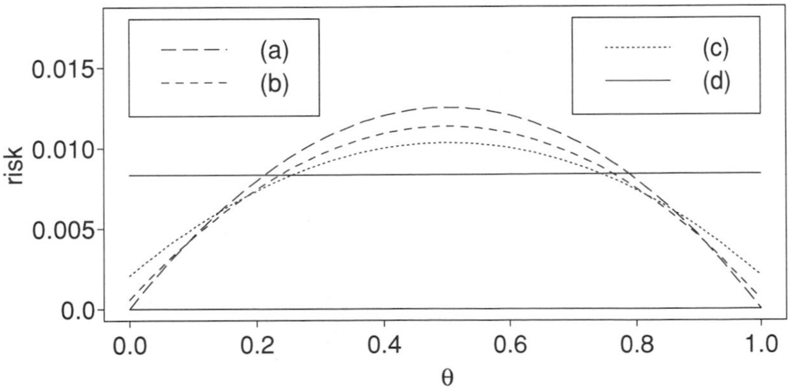

Figure 3.9.1. Mean squared errors for estimators of a binomial probability: (a) δ_1; (b) δ_2; (c) δ_3; (d) δ_4.

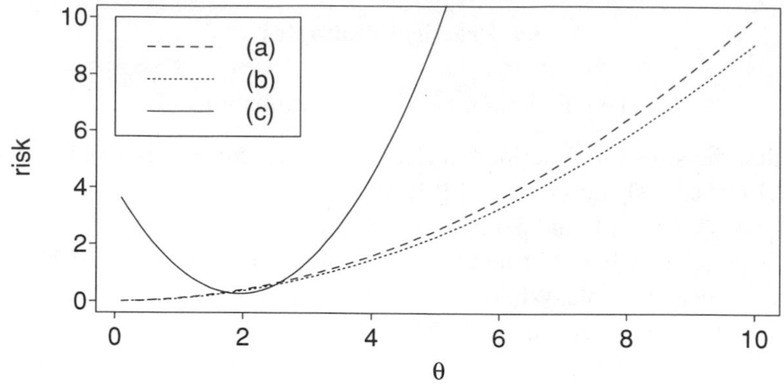

Figure 3.9.2. Mean squared errors for estimators of a normal variance: (a) ϕ_1^*; (b) ϕ_2^*; (c) ϕ_3^*.

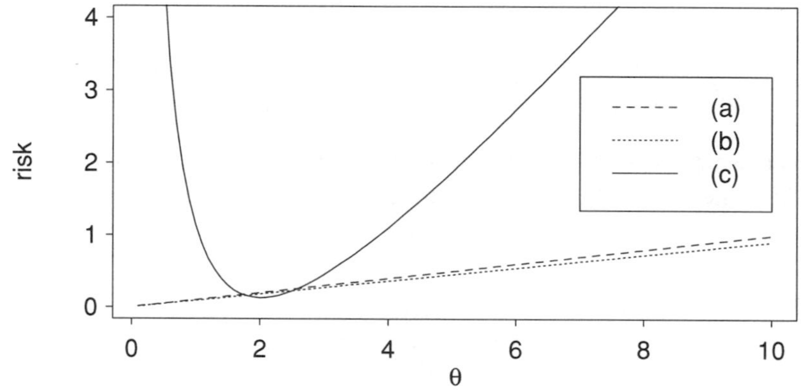

Figure 3.9.3. Risk functions for estimators of a normal variance: (a) ϕ_1^*; (b) ϕ_2^*; (c) ϕ_3^*.

Figure 3.9.3 describe the risk functions of ϕ_1^*, ϕ_2^*, and ϕ_3^* under the loss function

$$L(\phi^*, \phi) = \frac{(\phi^* - \phi)^2}{\phi}.$$

Again ϕ_1^* dominates ϕ_2^*, and ϕ_3^* does not perform well.

(C) *Estimating a Poisson Log-Mean*

Now consider the situation discussed in Section 3.3 (B), where $y_1, \ldots, y_n$ are a random sample from a Poisson distribution with mean θ. Suppose, however, that it is now required to estimate $\lambda = \log\theta$ under squared error loss for λ. Then the maximum likelihood estimate $\hat{\lambda} = \log\bar{y}$ possesses infinite mean squared error, as $\bar{y}$ assumes the value of zero, with positive probability. However, under a Gamma prior distribution for θ, with parameters α and β, the posterior mean of λ is

$$\tilde{\lambda} = \psi(\alpha + n\bar{y}) - \log(\beta + n), \tag{3.9.1}$$

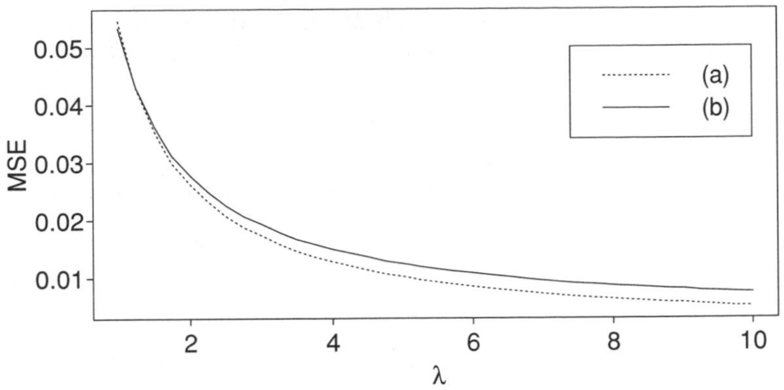

Figure 3.9.4. Mean squared errors for estimators of a Poisson mean: (a) $\alpha = 0.5$, $\beta = 0$; (b) $\alpha = \beta = 1$.

where $\psi(z) = \partial \log \Gamma(z)/\partial z$. A standard series approximation to the digamma function (Abramowitz and Stegun, 1968, p. 259, equation 6.3.18) tells us that $\tilde{\lambda}$ is approximated by

$$\lambda^* = \log(\alpha + n\bar{y}) - \tfrac{1}{2}(\alpha + n\bar{y}) - \log(\beta + n). \tag{3.9.2}$$

The Jeffreys' prior (3.6.5) corresponds to the choices $\alpha = 0.5$ and $\beta = 0$. Curve (a) in Figure 3.9.4 describes the mean squared error of $\tilde{\lambda}$ when $\alpha = 0.5$ and $\beta = 0$, and $n = 20$. This is identical, to three-significant-digit accuracy, to the mean squared error of the corresponding approximation λ^*. Curve (b) in Figure 3.9.4 describes the mean squared error of the Bayes estimator, when instead $\alpha = 1$ and $\beta = 1$. The Jeffreys' prior density works better in this case, unless λ is very small.

3.10 Symmetric Loss Functions

Let a real-valued parameter θ possess continuous posterior density $\pi_y(\theta)$ and posterior c.d.f. $\Lambda_y(\theta)$, for $\theta \in \Theta = (-\infty, \infty)$. Consider, first, symmetric bounded loss functions

$$L(d, \theta) = G(|d - \theta|) \qquad (-\infty < \theta, d < \infty), \tag{3.10.1}$$

for θ. Here $G(\xi)$ possesses the defining properties of a c.d.f., on $(0, \infty)$, that is, it is monotonic increasing in ξ, continuous on the right, and satisfying $G(0) = 0$ and $G(\infty) = 1$. Methods for finding a statistician's choice of G are discussed in Section 4.1 (The Experimental Measurement of Utility) and Exercise 4.2.c.

Note the representation (Smith, 1980)

$$L(d, \theta) = \underset{\xi}{E} L_\xi(d, \theta), \tag{3.10.2}$$

where

$$L_\xi(d, \theta) = \begin{cases} 0 & \text{if } |d - \theta| \le \xi \\ 1 & \text{if } |d - \theta| > \xi \end{cases} \tag{3.10.3}$$

denotes a step loss function, with step width 2ξ, and ξ denotes the random variable with c.d.f. G. Under step loss, the posterior expected loss (3.8.12) becomes

$$q_{\mathbf{y}}(d, \xi) = 1 - \Lambda_y(d + \xi) + \Lambda_y(d - \xi) \qquad (-\infty < d < \infty). \tag{3.10.4}$$

Under more general symmetric bounded loss, the posterior expected loss is the corresponding mixture

$$\begin{aligned} q_{\mathbf{y}}(d) &= \underset{\xi}{E}\, q_{\mathbf{y}}(d, \boldsymbol{\xi}) \\ &= 1 - \underset{\xi}{E}\Lambda_y(d + \xi) + \underset{\xi}{E}\Lambda_y(d - \xi) \qquad (-\infty < d < \infty), \end{aligned} \tag{3.10.5}$$

where the expectations should be taken with respect to a random variable ξ, with c.d.f. G.

The posterior expected loss (3.10.5) can be computed and minimized numerically, for example, using one-dimensional numerical integrations with respect to ξ, to find the Bayes estimate of d under the loss function (3.10.1). However, it is of interest to use the representation (3.10.2) to investigate some properties of the solution. Note, upon differentiating (3.10.4) with respect to d, that any Bayes estimate d_ξ^*, under step loss (3.10.3), must satisfy

$$\pi_y\left(d_\xi^* + \xi\right) = \pi_y\left(d_\xi^* - \xi\right). \tag{3.10.6}$$

The estimated d_ξ^* can be obtained graphically from a sketch of the posterior density $\pi_y(\theta)$. Find a line horizontal to the θ axis that intersects the posterior density at points a distance 2ξ apart, then drop a vertical line to the θ axis from a point halfway between the two points of intersection. The value achieved for θ gives a local optimum of (3.10.4). For unimodal symmetric posterior densities, this point will always be equal to posterior mode $\tilde{\theta}$ and will hence also equal the posterior median and the posterior mean, if the latter exists. For unimodal posterior densities, these will more generally be just one local optimum, and this will provide the Bayes estimate d_ξ^*. For bimodal posterior densities, depending upon the choice of ξ, there will be either one or three local optima. In situations where there is more than one local optimum, the choice should be the one that minimizes (3.10.4).

For the normal mixture example of Section 3.5 (D), consider the loss function (3.10.3), but with $\xi = 0.5$. The three solutions $d_1^* = 0.0992$, $d_2^* = 2.445$, and $d_3^* = 4.200$ of equation (3.10.6) are described by the three vertical lines in Figure 3.10.1. It is necessary to choose the solution, minimizing (3.10.4), that is in the notation of Exercise 3.4.o, minimizing the posterior expected loss

$$\begin{aligned} q_{\mathbf{y}}(d, \xi) = 1 &- \phi\Phi^*\left(d + \xi - \mu_1^*\right) - (1 - \phi)\Phi^*\left(d + \xi - \mu_2^*\right) \\ &+ \phi\Phi^*\left(d - \xi - \mu_1^*\right) + (1 - \phi)\Phi^*\left(d - \xi - \mu_2^*\right), \end{aligned}$$

where $\Phi^*(d) = \Phi(\lambda^{-\frac{1}{2}} d)$. The three solutions d_1^*, d_2^*, d_3^* give respective posterior expected losses 0.821, 0.924, and 0.712, so that d_3^* clearly provides the Bayes estimate.

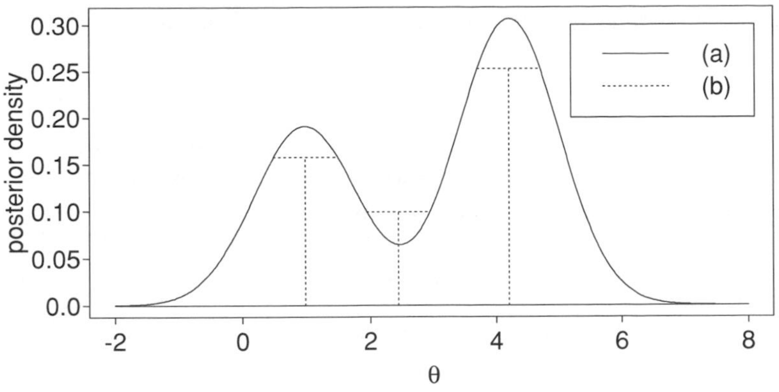

Figure 3.10.1. Bayes estimators under step loss: (a) posterior density; (b) graphical procedure for local optima of loss function.

Note, in general, that as $\xi \to 0$, d_ξ^* will approach the posterior mode $\tilde{\theta}$. However, in the limit, (3.10.3) becomes "zero–one loss"

$$L_0(d, \theta) = \begin{cases} 1 & \text{if } d = \theta \\ 0 & \text{if } d \neq \theta. \end{cases} \tag{3.10.7}$$

Zero–one loss is not generally a sensible loss function, since its posterior expectation, under a continuous posterior density, is always zero. Therefore, unless it happens to coincide with a Bayes estimate under another loss function, the posterior mode $\tilde{\theta}$ is in general neither a Bayes estimate nor admissible under any particular loss function. It can, however, be described as a "limiting Bayes" estimate.

Now let D_B denote the class of estimates that are Bayes estimates, for a particular posterior density $\pi_y(\theta)$, against step loss (3.10.3), for some $\xi \in (0, \infty)$. For symmetric unimodal posterior densities, $D_B = \{\hat{\theta}\}$ contains just the posterior mode or median $\tilde{\theta}$.

Lemma 3.10.1 (Smith's lemma): *If d^* is Bayes against some symmetric bounded loss function of the form* (3.10.1), *then $d^* \in D_B$.*

Proof. A general proof is described by Smith (1980). As an illustration, consider the special case where, for each $\xi \in (0, \infty)$, (3.10.4) has a unique local optimum d_ξ^*. Then $\pi(d_\xi^* + \xi) = \pi_y(d_\xi^* - \xi)$, for $\xi \in (0, \infty)$.

If d^* is a local optimum of (3.10.5), then d^* satisfies

$$\underset{\xi}{E}\, \pi_y(d^* + \xi) = \underset{\xi}{E}\, \pi_y(d^* - \xi).$$

Therefore, by the mean value theorem, there exists $a \in (0, \infty)$ satisfying

$$\pi_y(d^* + a) = \pi_y(d^* - a),$$

so that $d^* = d_a^*$. ■

Next, consider symmetric unbounded loss functions of the form

$$L(d,\theta) = G^*(|d-\theta|) \qquad (-\infty < \theta < \infty, -\infty < d < \infty), \tag{3.10.8}$$

where $G^*(0) = 0$, $G^*(\infty) = \infty$, and $G^*(\cdot)$ is monotonic and increasing on the right. The class defined by (3.10.8) includes absolute loss $L(d,\theta) = |d-\theta|$ and squared error loss $L(d,\theta) = (d-\theta)^2$. Note that the loss function

$$L^A(d,\theta) = 1 - \exp\big\{-AG^*(|d-\theta|)\big\} \tag{3.10.9}$$

is a symmetric bounded loss function for $A > 0$. The class (3.10.9) includes exponential absolute loss $L(d,\theta) = 1 - \exp\{-A|d-\theta|\}$ and exponential squared error loss $L(d,\theta) = 1 - \exp\{-A(d-\theta)^2\}$.

Notice, from (3.10.9), that $G^*(|d-\theta|) = \lim_{A\to 0}[A^{-1}L^A(d,\theta)]$. Smith uses this representation of a symmetric unbounded loss function, as the limit of a sequence of bounded loss functions, to prove the following lemma.

Lemma 3.10.2: *If d^* is Bayes against some symmetric unbounded loss function of the form* (3.10.8), *then $d^* \in D_B^*$, the closure of D_B.*

Corollary 3.10.1: *If $\pi_y(\theta)$ is symmetric, unimodal, and continuous, then the posterior mode or median $\tilde{\theta}$ is uniquely Bayes under any choice of bounded or unbounded symmetric loss function.*

Worked Example 3O: *Asymmetric Step Loss*

Let θ possess a posterior density $\pi_y(\theta)$ and c.d.f. $\Omega_y(\theta)$ for $\theta \in (-\infty,\infty)$ and consider the asymmetric step loss function

$$L(d,\theta) = \begin{cases} 1, & d > \theta + a \quad \text{or} \quad d < \theta - b \\ 0 & \text{otherwise.} \end{cases}$$

(a) Find the posterior expected loss.
(b) Find an equation for the Bayes estimator d^*.

Model Answer 3O:

(a) As the loss function is an indicator function, its posterior expectation equals the corresponding posterior probability, that is,

$$q_{\mathbf{y}}(d) = \Omega_y(d-a) + 1 - \Omega_y(d+b).$$

(b) Upon differentiating, we find that the Bayes estimate d^* satisfies the equation $\pi_y(d^*-a) = \pi_y(d^*+b)$.

SELF-STUDY EXERCISES

3.10.a For the posterior density (3.4.10), for a normal mean θ, show that the posterior mean $\theta^* = (n\tau^{-2}y + \sigma^{-2}\mu)/(n\tau^{-2} + \sigma^{-2})$ is, for any μ, and $\sigma^2 > 0$, uniquely Bayes and hence admissible under exponential squared error loss $L(d,\theta) = 1 - \exp\{-A(d-\theta)^2\}$.

Hint: Use Lemma 3.10.1 and save yourself some algebra!

3.10.b Consider the posterior distribution, in Exercise 3.8.e, for a Poisson mean θ, which is Gamma with parameters $\alpha + n\bar{y}$ and $\beta + n$. Show that the posterior density of $\lambda = \log\theta$ is

$$\pi(\lambda) \propto \exp\left\{\lambda(\alpha + n\bar{y}) - (\beta + n)e^{\lambda}\right\} \qquad (-\infty < \lambda < \infty). \tag{3.10.10}$$

Show that the Bayes estimate λ^* for λ, under an $L_\xi(d,\lambda)$ step loss function for λ, satisfies

$$2\xi(\alpha + n\bar{y}) = (\beta + n)\left(e^{\lambda^*+\xi} - e^{\lambda^*-\xi}\right) - (\beta + n)e^{\lambda^*-\xi}. \tag{3.10.11}$$

Show that the solution to equation (3.10.11) is

$$\lambda^* = \log\theta^* - \log\left(\frac{e^{\xi} - e^{-\xi}}{2\xi}\right), \tag{3.10.12}$$

where $\theta^* = (n\bar{y}+\alpha)/(n+\beta)$. Hence show that the class D_B of all estimates that are Bayes under some symmetric bounded loss function for λ is $D_B = \{\lambda : -\infty < \lambda < \log\theta^*\}$.

Hence prove that $E(\lambda \mid \mathbf{y}) \le \log\theta^*$. Express $E(\lambda \mid \mathbf{y})$ in terms of the digamma function $\Psi(z) = \partial \log\Gamma(z)/\partial z$. (See also Exercise 3.6.a.)

3.10.c Let θ possess a distribution, where $\theta = 1/\xi$ and ξ possesses a $N(0,1)$ distribution. Sketch this density (obtained by differentiating the c.d.f.) and show that two equal modes exist. Find the class D_B.

Hint: The cusp point of the density (at $\theta = 0$) is infinitely differentiable.

3.10.d Consider a loss function of the form $L(d,\theta) = G(|d-\theta|)$, where $G(\xi)$ is continuous and strictly increasing in ξ. You seek a "Martian Bayes estimate" minimizing the posterior median of the loss function rather than the posterior expectation of the loss function. Note that

$$\begin{aligned} p(L(d,\theta) \le a \mid \mathbf{y}) &= p\left\{|d-\theta| \le G^{-1}(a)\right\} \\ &= 1 - \Lambda_y\left\{d + G^{-1}(a)\right\} \\ &\quad + \Lambda_y\left\{d - G^{-1}(a)\right\}. \end{aligned} \tag{3.10.13}$$

Hence the posterior median a^* of the loss function satisfies

$$\Lambda_y\left\{d + G^{-1}(a^*)\right\} - \Lambda_y\left\{d - G^{-1}(a^*)\right\} = \tfrac{1}{2}. \tag{3.10.14}$$

Show that the Martian Bayes estimate d^* does not depend upon G at all and provides the value of d^* satisfying

$$\Lambda_y(d^* + \xi) - \Lambda_y(d^* - \xi) = \tfrac{1}{2}, \tag{3.10.15}$$

such that the distance ξ is the smallest possible. Show when d^* can be interpreted as the center of a 50% highest posterior density interval for θ. Do you think that this is a useful estimate?

3.10.e If the posterior density $\pi_y(\theta)$ is continuous, find an expression, for any fixed d, of the posterior density $\pi_d(L)$ of your loss function. When $L(d, \theta) = (d - \theta)^2$, show how by sketching $\pi_d(L)$ as a function of the loss L and by comparing $\pi_d(L)$, for different choices of d, you can make an inferential decision regarding your choice of d, which takes good account of the uncertainty and shape of your posterior density.

3.10.f Show, in Exercise 3.2.c, that the prior predictive density of $\bar{y}$ is

$$p(\bar{y}) = n^n \frac{\beta^\alpha \Gamma(\alpha + n)}{\Gamma(n)\Gamma(\alpha)} \frac{\bar{y}^{n-1}(\alpha + n)}{(\beta + n\bar{y})} \qquad (0 < \alpha < \infty).$$

(a) Suppose now that you wish to investigate $H_0 : \theta = \theta_0$ versus $H_1 : \theta \neq \theta_0$. You possess prior probabilities ϕ and $1-\phi$ that H_0 and H_1, respectively, are true. Given H_1, let θ possess the Gamma distribution of Exercise 3.2.c, but concentrated on $\theta \neq \theta_0$. Show that your posterior probability of H_0 being true is

$$\phi^* = \frac{\phi R}{\phi R + 1 - \phi},$$

where

$$R = \frac{p_0(\bar{y})}{p(\bar{y})},$$

with

$$p_0(\bar{y}) = \frac{n^n \bar{y}^{n-1}}{\Gamma(n)} \theta_0^n \exp\{-n\bar{y}\theta_0\}.$$

(b) Your decision space $D = \{d_0, d_1\}$ consists of two possible decisions, d_0 : accept H_0 and d_1 : reject H_0. Consider the zero–one loss function defined by

$$L(d_0, \theta) = \begin{cases} 0 & \text{if } \theta = \theta_0 \\ 1 & \text{if } \theta \neq \theta_0 \end{cases}$$

and

$$L(d_1, \theta) = \begin{cases} 1 & \text{if } \theta = \theta_0 \\ 0 & \text{if } \theta \neq \theta_0. \end{cases}$$

Under the prior distribution of part (a), show that d_0 is the Bayes decision whenever $R > K$, where $K = \phi^{-1} - 1$.

(c) Show that increasing α to $\alpha + 1$ multiplies R by $\alpha(n\bar{y} + \beta)/\beta(\alpha + n)$. Discuss whether or not the procedure in (b) is sensitive to the prior assumptions.

3.10.g In the context of Exercise 3.10.f, let $\delta = \delta(\mathbf{y})$ denote some decision rule for mapping the observations to the decision space $D = \{d_0, d_1\}$.

(a) Show that the risk function of δ under our zero–one loss function can be represented as

$$r_\delta(\theta_0) = \textit{size},$$

and

$$r_\delta(\theta) = 1 - \textit{power function} \qquad (\theta \neq \theta_0),$$

where the size is the sampling probability that δ selects H_1 if H_0 is true, and the power function is the sampling probability that δ selects H_1 if H_1 is true.

(b) Hence show that the average risk of δ under the prior distribution of Exercise 3.10.f (part (a)) is

$$R_\delta = \phi\ (size)\ + (1-\phi)(1 - strength),$$

where the strength of δ averages the power function with respect to a Gamma distribution with parameters γ and β. (See Crook and Good, 1982.)

(c) Let δ^* denote the decision rule that accepts H_0 whenever $R > K$, where K is determined by the size of the test. Prove that δ^* maximizes the strength among all decision rules of the same size.

(d) Show how these ideas can be extended to situations where H_0 is also composite. Develop Bayes decision rules that maximize the strength among decision rules with fixed "volume," where "volume" averages a "size function" with respect to the prior distribution under the null hypotheses.

3.10.h Consider the situation described in Exercise 3.4.o, where the posterior distribution is a mixture of two normal distributions. Show that the posterior expected loss (3.10.5) under the exponential squared error loss function

$$L(d,\theta) = 1 - \exp\left\{-A(d-\theta)^2\right\} \tag{3.11.16}$$

reduces to

$$q_{\mathbf{y}}(d) = 1 - (1+2A\lambda)^{-\frac{1}{2}}\Big[\phi \exp\Big\{-\big(A^{-1}+2\lambda\big)^{-1}\big(d-\mu_1^*\big)^2\Big\}$$
$$-(1-\phi)\exp\Big\{-\big(A^{-1}+2\lambda\big)^{-1}\big(d-\mu_2^*\big)^2\Big\}\Big].$$

Investigate the behavior of the posterior expected loss as $A \to 0$.

3.11 Practical Example: Mixtures of Normal Distributions

In the situation investigated in Exercise 3.4.o, Section 3.5 (D), and Exercise 3.10.h, the posterior expected loss is plotted as a function of d, in Figure 3.11.1, for various

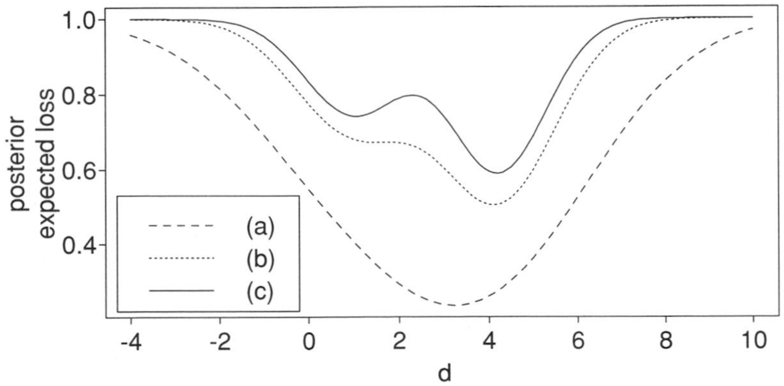

Figure 3.11.1. Posterior expected loss as a function of the decision d: (a) $A = 0.1$; (b) $A = 0.5$; (c) $A = 1.0$.

choices of the quantity A appearing in the loss function (3.10.16). Curves (a), (b), and (c) in Figure 3.11.1 correspond to $A = 0.1, 0.5$, and 1.0, respectively. Note that the posterior expected loss becomes bimodal as A increases.

Bayes decisions can more generally "bifurcate" in this manner either as the loss function changes or as the posterior distribution changes. This provides a possible rational explanation of apparently "flip-flop" decision making, for example, when two local minima of the posterior expected loss suddenly flip in relation to the global minimum property. See Smith (1978).

4

The Expected Utility Hypothesis

4.0 Preliminaries and Overview

In the early 1950s, Maurice Allais and Leonard J. Savage initiated a debate (a) as to how people or corporations should make decisions involving money that are in some sense rational, (b) as to how people in fact make monetary decisions, and whether they can be regarded as rational. Savage believed that subject to a set of axioms, rational monetary decisions must satisfy the expected utility hypotheses. For example, you should possess a utility function on monetary rewards such that you would always seek to maximize the expectation of your utility. An experimental procedure for eliciting your utility function is described in Section 4.1 (B). Allais demonstrated that French workers do not always concur with Savage's criterion, so that this hypothesis does not always provide an appropriate "descriptive theory." However, Savage's normative theory became a cornerstone of microeconomics for the next three decades and influenced an apparent rationality. (See Section 4.2.)

In the late 1970s and early 1980s, McGinness, Machina, Karmarkar, and Chew extended the expected utility hypothesis in the context of financial decision making. In particular, Savage's main axiom, the independence assumption, or sure thing principle, has been relaxed. This has since become a source of considerable further research in the economics area.

In this chapter, we suggest that many people would place premiums on the certain, or virtually certain, components of their possibly random rewards. To compensate for this, we describe "ϵ-adjusted" utility theory (see Section 4.3). An elicitation procedure is described in Section 4.4. In Section 4.5, we demonstrate how a company can make money from individuals who are following our descriptive criterion. This follows suggestions and encouragement by Robert McCullough and Arnold Zellner. In Chapter 4, we consider a total of five paradoxes: the St. Petersburg paradox (Section 4.1 (A)), Allais' paradox and the median utility paradox (Section 4.2), a risk-aversion paradox (Section 4.5), and the Ellsberg paradox (Section 4.6). Our theory can explain all five paradoxes. We parallel the Kahneman–Tversky approach (Kahneman and Tversky, 1979) to prospect theory, which also places greater emphasis on certain rewards, but the results reported here are simpler.

In Section 4.7, a practical case study demonstrates that many individuals would follow our ϵ-adjusted utility criterion, with a positive premium on virtually certain rewards. This is one of several similar studies completed in Madison, Wisconsin, with similar conclusions. Furthermore, apparently risk-averse behavior, under the expected utility hypothesis, can often be largely alternatively explained by a premium on virtually certain rewards.

Throughout this chapter, we carefully separate the construction of subjective probability from the notion of monetary utility. To avoid confusion with utility, subjective probabilities should be judged by reference to an objective auxiliary experiment, for example, the spinning pointer of Exercise 1.1.k, rather than by considering monetary bets (see DeGroot's first five axioms, DeGroot, 1970, Ch. 6). While the fifth axiom and its implications are quite complicated, it does provide us with an excellent prescription for evaluating subjective probabilities, once we are convinced that we are able to represent information in this way. Our separate elicitation scheme for subjective probabilities and utility leaves us free to judge an individual's utility, or ϵ-adjusted utility, by asking the individual to evaluate and compare different lottery tickets. The key theoretical result of the chapter is described in equation (4.4.7), that is, ϵ, the proportionate premium on virtually certain rewards, is equal to $2\phi - 1$, where ϕ denotes the mixing probability for a particular lottery ticket.

While the material in this chapter was independently developed at the University of Wisconsin–Madison during the 1980s, Peter Wakker (personal communication) advises us that we are describing a simple variation of some of the most important practical modifications to expected utility as defined in the economics literature by Bell (1982), extended by Cohen (1992), and axiomatized by Gilboa (1988) and Jaffray (1988). There are also relationships with "rank-dependent utility" (Quiggin, 1982; Tversky and Kahneman, 1992; and Tversky and Wakker, 1995). We are, however, unaware of any previous publication of our very simple experimental elicitation procedure.

4.1 Classical Theory

(A) *The Expected Utility Hypothesis and the St. Petersburg Paradox*

The classical theory of the expected utility hypothesis was developed by Leonard J. Savage (1972) and Friedman and Savage (1948). In the current section, we selectively review existing methodology, including DeGroot (1970, Ch. 7). In Section 4.2, we describe how the classical theory can be updated. A lucid summary of Savage's original axioms is described by Press (1989, pp. 18–19). Note that Press (pp. 10–12) describes as "incoherent" any individual whose subjective conditional probabilities (rather than utilities) do not follow the Rényi axiom system (Rényi, 1970) and argues that "if an incoherent individual is willing to make a sequence of bets using 'incoherent' probabilities, and considers each bet fair or favorable, then the individual will suffer a net loss no matter what happens" (such a bet is called a Dutch book). We regard Press's notion of *incoherence* as a separate issue when compared with Savage's notion of *inconsistency*, a term that can be applied when a person's preferences regarding monetary rewards do not satisfy the Savage axioms.

Let $R = \{r : 0 \le r < \infty\}$ denote the set of nonnegative monetary rewards and let $P = P(\cdot)$ denote a probability distribution defined on subsets of R, such that if you choose P, you will receive a reward x that is the numerical realization of a random variable X with probability distribution P. This reward may, or may not, be payable in return for an amount of money invested for the reward. For example, you may invest \$1,000 for a *portfolio* (i.e., an investment yielding different monetary rewards with different probabilities) that in five years will be worth \$0 with probability 0.01, \$1,500 with probability 0.49, and \$1,600 with probability 0.50.

This situation can be summarized by saying that you are investing \$1,000 for a probability distribution or portfolio P that assigns the probabilities $p_1 = 0.01$, $p_2 = 0.49$, and $p_3 = 0.50$ to the rewards $r_1 = 0$, $r_2 = 1{,}500$, and $r_3 = 1{,}600$, that is, for the random reward $p_1\{r_1\} + p_2\{r_2\} + p_3\{r_3\}$, with expectation \$1,535. Note that P can be a subjective prior distribution, rationalized via the objective auxiliary experiment of Exercise 1.1.k, or it can represent a posterior distribution, based upon some economic data $\mathbf{y}$. In any case, we differ from de Finetti (1937) by keeping our development of utility quite separate from the philosophy of subjective probability distributions. The probabilities should of course follow the Kolmogorov axioms (Press, 1989, p. 8), and any related conditional probabilities should follow the Rényi axioms (Press, 1989, p. 10), but simply because they would not be probabilities or conditional probabilities if they did not satisfy these axioms, and for no deeper reasons.

Assume that you are considering alternative investment strategies relating to a choice among several portfolios. More formally, you wish to make a decision by choosing P from a class $\mathcal{P}$ of those probability distributions on R, or portfolios, that are under consideration. Assume that for any P_1 and P_2 in $\mathcal{P}$, you are able to state one of the preferences $P_1 \preceq P_2$ if P_1 is not preferred to P_2, $P_1 \sim P_2$ if P_1 and P_2 are equally preferred, and $P_2 \preceq P_1$ if P_2 is not preferred to P_1. Use $P_1 \prec P_2$ to denote the complement of $P_2 \preceq P_1$.

Intuitively speaking, a utility function $U(r)$ is a real-valued function measuring the value in real-life terms that you would attach to any definite monetary reward r. For example, if $U(\$3{,}000) = 2U(\$1{,}000)$, then you value a reward of \$3,000 twice as much as a reward of \$1,000. More formally, the existence of a utility function depends upon whether your preferences among probability distributions satisfy the following hypothesis.

The Expected Utility Hypothesis (EUH)

In any specific situation, there exists a utility function $U(r)$, for $r \in R$, such that for any P_1 and P_2 in $\mathcal{P}$, $P_1 \preceq P_2$ if and only if

$$E(U(X) \mid P_1) \leq E(U(X) \mid P_2),$$

whenever the expectations, with respect to the distributions P_1 and P_2 of the random variable X, exist.

Your utility function can vary from situation to situation, for example, it might depend upon your current amount of savings or how much you are prepared to spend on a portfolio. Consider, for example, the dotted curve in Figure 4.1.1, which satisfies

$$U(r) = 1 - e^{-Ar} = 1 - \left(\tfrac{1}{2}\right)^{\tilde{r}},$$

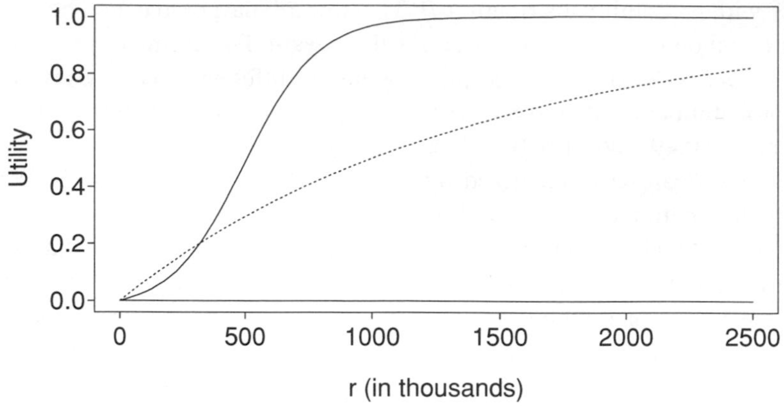

Figure 4.1.1. Two utility functions: dotted curve – a risk-averse utility function; solid curve – a concave/convex utility function.

where $A = (\log 2)/10^6$ and $\tilde{r} = r/10^6$. Then $U(10^6) = \frac{1}{2}$ and $U(\infty) = 1$. Therefore, under this utility function, no amount of money possesses more than twice the subjective value, or utility, than a million dollars. Suppose also that you wish to choose between a certain reward P_1 of \$100,000, with probability one, and the random reward P_2, where you receive either \$80,000 or \$130,000, each with probability $\frac{1}{2}$. Then

$$\begin{aligned} E\big(U(X) \mid P_2\big) &= \tfrac{1}{2}U(80{,}000) + \tfrac{1}{2}U(130{,}000) \\ &= 1 - \tfrac{1}{2}\left(\tfrac{1}{2}\right)^{0.08} - \tfrac{1}{2}\left(\tfrac{1}{2}\right)^{0.13} = 0.070. \end{aligned}$$

Since this is greater than $E[U(X) \mid P_1] = 1 - (\frac{1}{2})^{0.1} = 0.067$, you prefer the random reward. Note that $U(r) = 1 - e^{-Ar}$ is an example of a *convex* utility function. With this utility function, you are *risk averse*, that is, you would always avoid a random reward in favor of receiving the expectation of this random reward with certainty. Consider next a concave/convex utility function of the form

$$U(r) = (1-K)^{-1}\left\{\frac{e^{A(r-r_0)}}{1+e^{A(r-r_0)}} - K\right\}, \tag{4.1.1}$$

with $K = e^{-Ar_0}/(1 + e^{-Ar_0})$. This utility function is *concave* for $r \le r_0$, implying that you are a *risk taker* for small monetary rewards (you will always prefer a random reward in favor of receiving the expectation of this reward with certainty, whenever all components of your random reward are $\le r_0$), whereas you remain risk averse for monetary rewards above r_0. For example, a very wealthy person designates the values $r_0 = 500{,}000$ and $A = (\log 2)/r_0$, in which case,

$$U(r) = \frac{3}{2}\left\{\frac{2^{(r-r_0)/r_0}}{1+2^{(r-r_0)/r_0}}\right\}. \tag{4.1.2}$$

This utility function is sketched as the solid curve in Figure 4.1.1 and can be used to resolve the following famous paradox.

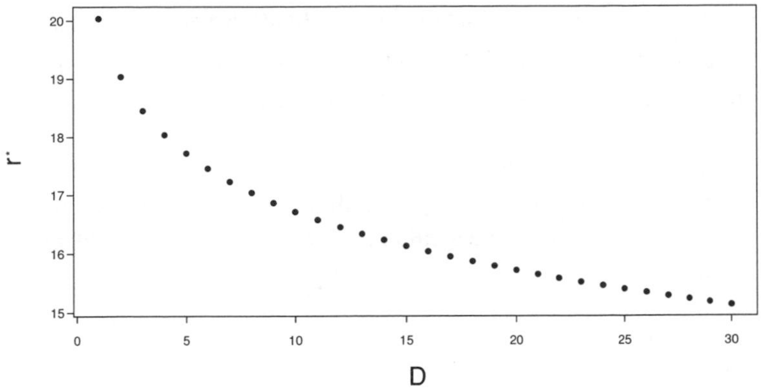

Figure 4.1.2. Equivalent monetary rewards for the St. Petersburg paradox.

The St. Petersburg Paradox

A fair coin is tossed repeatedly until a head is obtained for the first time. If the first head is obtained after n tosses, you receive a reward of 2^n dollars. What certain reward would you accept as equivalent to this random reward?

The "paradox" occurs because your expected winnings are clearly infinite. Therefore, a comparison of expected winnings, or expected linear utility, when $U(r) = a + br$, with $b > 0$, would clearly lead you to refuse any certain finite reward in return for playing this game, even though many people would accept six or eight dollars. However, under EUH, either convex or concave/convex utility would resolve the St. Petersburg paradox (DeGroot 1970, p. 95), since you should be prepared to receive the certain reward r^* satisfying

$$U(r^*) = \sum_{r=1}^{\infty} 2^{-r} U(2^r). \tag{4.1.3}$$

For example, if $U(r) = 1 - e^{-Ar}$, then

$$r^* = -A^{-1} \log\left(\sum_{r=1}^{\infty} 2^{-r} e^{-A2^r} \right). \tag{4.1.4}$$

The dots in Figure (4.1.2) plot r^* as a function of $D = 10^6 A$.

Worked Example 4A: *A Piecewise Linear Utility Function*

Consider the following utility function on monetary rewards, where the reward units are measured in units of a thousand dollars:

$$U(r) = \begin{cases} r/2 & \text{for } 0 < r \leq 1 \\ 0.5 + (r-1)/36 & \text{for } 1 < r \leq 10 \\ 0.75 + (r-10)/3{,}960 & \text{for } 10 < r \leq 1{,}000 \\ 1 & \text{for } r > 1{,}000. \end{cases}$$

Assuming that you follow the expected utility hypothesis,

(a) how much certain reward would you be prepared to receive in place of a lottery ticket that would pay you \$500 with probability 0.99 and \$500,000 with probability 0.01?

(b) You are offered the alternative of \$50,000 cash and an investment, which would pay you \$5,000 with probability $1-\phi$ and a million dollars with probability ϕ. How large a value of ϕ would you need to take the investment?

Model Answer 4A:

(a) Note that $U(0) = 0$, $U(1) = 0.5$, $U(10) = 0.75$, and $U(1{,}000) = 1$. Since the possible rewards are $\frac{1}{2}$ or 500, in terms of thousands of dollars, the utility of the lottery ticket is $\frac{1}{2}/2 = \frac{1}{4}$ with probability 0.99 and $0.75 + (500-10)/3{,}960 = 0.8737$ with probability 0.01. Consequently, the expected utility is

$$0.25 \times 0.99 + 0.8737 \times 0.01 = 0.2562.$$

Since this quantity lies between 0 and 0.5, it corresponds to a certain reward of 2×0.2562 thousand dollars = \$512.5.

(b) The possible rewards are \$5, \$50, or \$1,000. The \$5,000 cash has utility

$$U(5) = 0.5 + (5-1)/36 = 0.6111.$$

The \$50,000 cash has a utility of

$$U(50) = 0.75 + (50-10)/3,960 = 0.7601.$$

Since the million dollars cash has utility $U(1{,}000) = 1$, the investment has utility $0.6111(1-\phi)+\phi$. You would therefore prefer the investment if

$$0.6111(1-\phi)+\phi > 0.7601,$$

that is, if

$$\phi > 0.149/0.3889 = 0.383,$$

so that you have a greater than 38.3% chance of earning a million dollars.

(B) *Measuring an Individual's Utility Function*

Under EUH, your utility function $U(r)$ can be measured experimentally for values of r in any bounded interval, for example, [0, 5,000] by the following elicitation procedure, which lets $\langle r_1, r_2\rangle$ represent the lottery ticket that gives you rewards r_1 and r_2, each with probability $\frac{1}{2}$, and which lets $\{x\}$ denote the certain reward that gives you reward x with certainty.

The Experimental Measurement of Utility

Stage I: Elicit x_1 such that $\{x_1\} \sim \langle 0,\ 5{,}000\rangle$.
Stage II: Elicit x_2 such that $\{x_2\} \sim \langle 0,\ x_1\rangle$.
Stage III: Elicit x_3 such that $\{x_3\} \sim \langle x_1,\ 5{,}000\rangle$.
Stage IV: Elicit x_4 such that $\{x_4\} \sim \langle x_2,\ x_3\rangle$.

For example, Stage I asks you to elicit the certain reward x_1 you would prefer equally to a lottery ticket that pays you only with probability 0.5, in which case your reward is \$5,000. Stage IV asked you to elicit the certain reward x_4 that you would prefer equally to a lottery ticket that pays you x_2 or x_3, each with probability 0.5. Here x_2 and x_3 are your values obtained from Stages II and III.

Since $U^*(r) = a + bU(r)$, with $b > 0$, always yields the same preferences as $U(r)$, you can, without loss of generality, set $U(0) = 0$ and $U(5{,}000) = 1$. It then follows that your utility function satisfies $U(x_1) = \frac{1}{2}$, $U(x_2) = \frac{1}{4}$, and $U(x_3) = \frac{3}{4}$. Your choices should satisfy $x_4 = x_1$, whatever the choice of U, whenever you satisfy EUH. If $x_4 \neq x_1$ and you still believe that you are following EUH, then you can set $U\{(x_1 + x_4/2)\} = \frac{1}{2}$ instead of $U(x_1) = \frac{1}{2}$.

Mr. Jones's Utilities

Further points on your utility curve can be measured by assigning further stages to the elicitation process. You can then, with appropriate rescaling of $U(5{,}000)$, try to fit a concave/convex utility function, such as (4.1.1), to your experimental data. As an artificial example, Mr. Jones gives $x_1 = 1{,}100$, $x_2 = 750$, $x_3 = 1{,}650$, and $x_4 = 1{,}000$. The elicitation based upon three further appraisals of lottery tickets suggests that $U(x_5) = \frac{1}{8}$, $U(x_6) = \frac{7}{8}$, and $U(x_7) = \frac{15}{16}$, where $x_5 = 400$, $x_6 = 2{,}100$, $x_7 = 2{,}500$. Mr. Jones's values give the seven points depicted in Figure 4.1.3 for his utility curve. These points happen to be well fitted by a curve of the form (4.1.1), with $r_0 = 1{,}000$, $A = 1/500$, and $K = 0.1192$, but divided by $U(5{,}000)$, that is, the solid curve in Figure 4.1.3. Under the expected utility hypothesis, Mr. Jones appears to be a risk taker for small amounts of money and quite risk averse for large amounts of money.

Worked Example 4B: *Mrs. Jones's Utilities*

Mrs. Jones gives $x_1 = 1{,}000$, $x_2 = 80$, $x_3 = 1{,}700$, and $x_4 = 900$. Find three points on her utility curve. Are Mrs. Jones's opinions well fitted by a linear utility function?

Model Answer 4B:

With $U(0) = 0$ and $U(5{,}000) = 1$, we have

Stage I: $U(1{,}000) = \frac{1}{2}U(0) + \frac{1}{2}U(5{,}000) = \frac{1}{2}$.

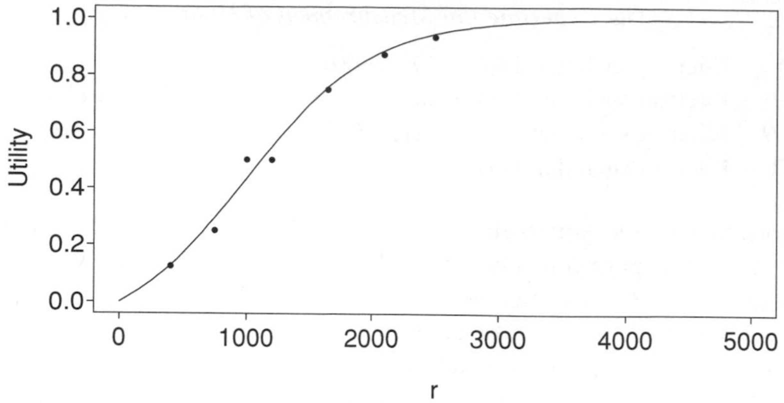

Figure 4.1.3. Mr. Jones's experimental measurement of utility: solid points – experimental observations; solid curve – concave/convex utility function.

Stage II: $U(80) = \frac{1}{2}U(0) + \frac{1}{2}U(1{,}000) = \frac{1}{4}$.

Stage III: $U(1{,}700) = \frac{1}{2}U(1{,}000) + \frac{1}{2}U(5{,}000) = \frac{3}{4}$.

Stage IV: $U(900) = \frac{1}{2}U(80) + \frac{1}{2}U(1{,}700) = \frac{1}{2}$.

As $900 < 1{,}000$, the expected utility hypothesis is not completely satisfied. As $890 = (80 + 1{,}700)/2$, the points $U(80) = \frac{1}{4}$ and $U(1{,}700) = \frac{3}{4}$ lie on the same straight line as $U(890) = \frac{1}{2}$. As the value 890 falls just outside the range (900, 1,000), Mrs. Jones's utility function is slightly different from linear.

4.2 The Savage Axioms

Why should your preferences on the space $\mathcal{P}$ of probability distributions, on a reward space R, satisfy EUH? DeGroot (1970, pp. 101–112) lists a set of six axioms, subject to which your preferences will always satisfy EUH. He refers to *bounded* members of $\mathcal{P}$, which assign all their probabilities to bounded subsets of R, in cases where R is unbounded. For any members P and Q of $\mathcal{P}$ and any probability ϵ, let $\{\epsilon P + (1-\epsilon)Q\}$ denote the unconditional distribution, which refers to P with probability ϵ, and to Q with probability $1 - \epsilon$. Consider, in particular, the following axiom.

Main Axiom of Utility (Independence assumption or sure thing principle): *Let P_1 and P_2 denote any bounded probability distributions in $\mathcal{P}$. Then, for any other bounded member of P of $\mathcal{P}$ and any real number ϵ satisfying $0 < \epsilon < 1$, $P_1 \preceq P_2$ is equivalent to*

$$\{\epsilon P_1 + (1-\epsilon)P\} \preceq \{\epsilon P_2 + (1-\epsilon)P\}.$$

Savage (1972) believes that similar axiom systems define rational or consistent decision making. Moreover, he shows that if your preferences do not satisfy EUH, then it would be possible to place repeated bets with you, which if accepted, in the long run would cause you to be a "sure loser," that is, lose, on average, in the long run. It would be possible to arrange a "Dutch book" against you, assuming that you are prepared to participate. This aspect is discussed further by Lindley (1985).

A Reward Space with Three Elements

Now suppose that $R = \{r_1, r_2, r_3\}$ consists of just three possible monetary rewards and that you have the strict preference ordering $\{r_1\} \preceq \{r_2\} \preceq \{r_3\}$ on these simple rewards. Without loss of generality, set $U(r_1) = 0$ and $U(r_3) = 1$. Let $P = (p_1, p_2, p_3)$, with $p_i \geq 0$ for $i = 1, 2, 3$, and $p_1 + p_2 + p_3 = 1$ denote the probability distribution, in the class $\mathcal{P}$, which assigns probabilities p_1, p_2, and p_3 to the rewards r_1, r_2, and r_3. Then, under EUH, $u = U(r_2)$ is the unique number $u \in [0, 1]$ such that $(u, 0, 1-u) \sim (0, 1, 0)$.

If $P = (p_1, p_2, p_3)$ and $Q = (q_1, q_2, q_3)$ are any members of $\mathcal{P}$, then $E(U \mid P) = p_2 u + p_3$ and $E(U \mid Q) = q_2 u + q_3$. Hence $P \preceq Q$ is equivalent to

$$(p_3 - q_3) + (p_2 - q_2)u \leq 0. \tag{4.2.1}$$

Hence, under EUH, it follows that quite apart from the actual value of your utility u, your preferences will depend only upon the differences $(p_3 - q_3)$ and $(p_2 - q_2)$. Is this real rationality or constrained rationality? DeGroot presents this example without comment. Allais (1953) and Allais and Hagen (1979) show that individuals do not follow EUH. Savage and Lindley think that they should; that is, they propose EUH as a "normative" theory rather than as a "descriptive" theory. As an example of this constrained type of rationality, suppose that your preferences satisfy $\mathbf{Q}_1 \sim \mathbf{Q}_2$, where $\mathbf{Q}_1 = (\frac{1}{3}, \frac{1}{3}, \frac{1}{3})$ and $\mathbf{Q}_2 = (\frac{1}{4}, \frac{7}{12}, \frac{1}{6})$. Then $E(U \mid Q_1) = \frac{1}{3} + \frac{1}{3}u$ and $E(U \mid Q_2) = \frac{1}{6} + \frac{7}{12}u$. Equating these two expectations gives $u = \frac{2}{3}$. Moreover, the essential differences are $\frac{1}{3} - \frac{7}{12} = -\frac{1}{4}$ and $\frac{1}{3} - \frac{1}{6} = \frac{1}{6}$. For example, you are constrained, under EUH, to equally prefer the three random rewards $(\frac{1}{2}, \frac{1}{2}, 0)$, $(\frac{7}{12}, \frac{1}{4}, \frac{1}{6})$, and $(\frac{2}{3}, 0, \frac{1}{3})$. This would appear to stretch credulity. In general, Q_1 and Q_2 imply that $P_1 \sim P_2$, where

$$P_1 = (p_1, p_2, p_3)$$

and

$$P_2 = \left(p_1 - \tfrac{1}{12}, p_2 + \tfrac{1}{4}, p_3 - \tfrac{1}{6}\right),$$

whenever each triple comprises three probabilities, summing to unity.

The triangle in Figure 4.2.1 denotes the space of all possible probability distributions of the form $\mathbf{P} = (p_1, p_2, p_3)$. The vertical distances between any point in the triangle, and the three edges (1), (2), and (3), describe the three probabilities p_1, p_2, and p_3. The four pluses denote all probability distributions that should be equally preferred to $\mathbf{Q}_1$, under EUH. The four dots instead denote all probability distributions that should be equally preferred to the probability distribution $(\frac{1}{4}, \frac{1}{3}, \frac{5}{12})$. The preceding single preference places numerous similar constraints on preferences across this entire triangle of probability distributions.

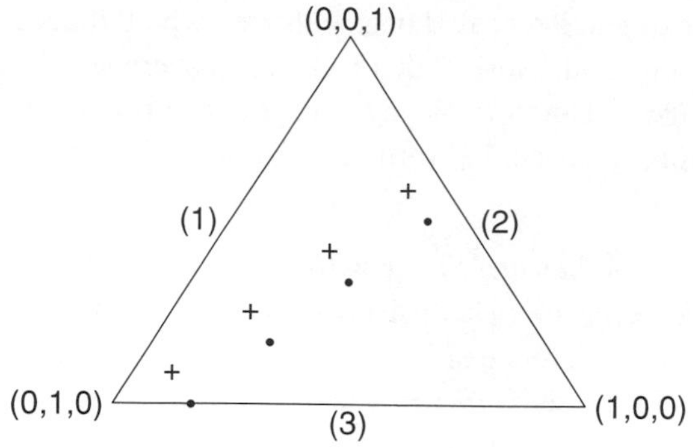

Figure 4.2.1. Equivalent probability distributions (under EUH): pluses – equally preferred to $(\frac{1}{3}, \frac{1}{3}, \frac{1}{3})$; dots – equally preferred to $(\frac{1}{4}, \frac{1}{3}, \frac{5}{12})$.

Allais' Paradox

(a) You have the choice of the following rewards:

$$G_1: \$500{,}000 \text{ with certainty,}$$

$$G_2: \begin{cases} \$2{,}500{,}000 & \text{with probability } 0.10 \\ \$500{,}000 & \text{with probability } 0.89 \\ \$0 & \text{with probability } 0.01. \end{cases}$$

Which gamble out of G_1 and G_2 would you prefer? These give expected rewards \$500,000 and \$695,000.

(b) Now suppose that you instead have the following choices:

$$G_3: \begin{cases} \$500{,}000 & \text{with probability } 0.11 \\ \$0 & \text{with probability } 0.89, \end{cases}$$

$$G_4: \begin{cases} \$2{,}500{,}000 & \text{with probability } 0.10 \\ \$0 & \text{with probability } 0.90. \end{cases}$$

Which gamble out of G_3 and G_4 would you prefer? These give expected rewards \$55,000 and \$250,000.

According to EUH, $G_2 \preceq G_1$ is equivalent to $G_4 \preceq G_3$. There are just three possible certain rewards, \$0, \$500,000, and \$2,500,000, under any choice of utility function. The four gambles G_1, G_2, G_3, and G_4 correspond to the probability distributions (0, 1, 0), (0.01, 0.89, 0.90), (0.89, 0.11, 0), and (0.90, 0, 0.10) on the space of these three rewards. For G_1 and G_2, $p_1 - q_1 = -0.01$ and $p_2 - q_2 = 0.11$, and this is exactly the same for G_3 and G_4. Hence EUH constrains your preferences in the manner stated. However, out of a class of about twenty statistics graduate students, nearly every student possessed the preferences $G_2 \preceq G_1$ and $G_3 \preceq G_4$, primarily because they were keen on the certain reward G_1. When this certainty vanished, they were unconcerned about the difference between 10% and 11%, and keener upon trying for the $2,500,000 under G_4.

Were these students irrational? They certainly didn't think so, even though they were told that they did not satisfy EUH. More generally, why should a person wish to make the optimal average rate of profit in the long run? The main priority for an individual is to take the certain rewards and avoid the catastrophic losses. If a company can model how its investors make monetary decisions, then the company should be able to adjust its policies to increase its own long-term profits. A possible model for individual decision making is described in the next section. It is possible to achieve solutions where companies profit well from individual investments and where the investors still achieve a safe profit, even if this is below average in the long run.

The Median Utility Paradox

For a general $\mathcal{P}$, consider the preference ordering for those $P \in \mathcal{P}$ with unique medians, defined by

$$P_1 \preceq P_2 \text{ whenever Median } (U \mid P_1) \leq \text{ Median } (U \mid P_2), \tag{4.2.2}$$

where Median $(U \mid P)$ is the median of the distribution of $U(X)$ when X has distribution P. Then, for any $U(r)$, which is strictly increasing in r,

$$P_1 \preceq P_2 \text{ whenever Median } (X \mid P_1) \leq \text{ Median } (X \mid P_2), \tag{4.2.3}$$

where Median $(X \mid P)$ is the median of the distribution of the distribution P for X. Note that the criterion (4.2.2) does not depend upon the choice of U at all!

This paradox shows that EUH is not criterion robust. In particular, when $U(r)$ is bounded above, the expected utility will frequently be close to the median utility. However, with this slight change, from mean to median, the preference ordering on P will become the same whether we consider concave or convex or concave/convex utility functions. Moreover, the median utility can give counterintuitive preferences. For example, a random reward of $1,000,001 with probability 0.51 and $0 with probability 0.49 is always preferred to a certain reward of $1,000,000. For the St. Petersburg paradox, certain rewards of either $2 or $4 are acceptable, again an unconvincing solution.

For a variety of applications of the related concept of "subjective expected lexicographic utility," see Lavalle and Fishburn (1991) and Fishburn and Lavalle (1992). The main axiom of utility was considered for multiattribute-expected utility theory by Fishburn (1965, 1974).

Worked Example 4C: *Do You Satisfy EUH?*

You consider two portfolios A and B. Portfolio A gives a reward of $1,000 with probability 0.5 and $2,000 with probability 0.5. Portfolio B gives you a reward of $1,000 with probability 0.3 and a reward of $5,000 with probability 0.7. Show that under EUH, with a strictly increasing utility function, you cannot regard portfolios A and B as equally valuable.

Model Answer 4C:

Without loss of generality, assume that you assign $U(\$1{,}000) = 0$, $U(\$2{,}000) = u$, and $U(\$5{,}000) = 1$, for some $u \in [0, 1]$. Then the expected utility of portfolio A is

$$0.5U(\$1{,}000) + 0.5U(\$2{,}000) = 0.5u,$$

and the expected utility for portfolio B is

$$0.3U(\$2{,}000) + 0.7U(\$5{,}000) = 0.3u + 0.7.$$

These expectations are equal whenever

$$0.5u = 0.3u + 0.7,$$

so that $u = 3.5$. Since this value exceeds utility, EUH with a strictly increasing utility function cannot be satisfied.

SELF-STUDY EXERCISES

4.2.a Use the procedure for the experimental measurement of utility to measure at least twenty points on your utility function for monetary rewards no greater than \$50,000. Do your preferences satisfy EUH? Can you fit a concave/convex curve (e.g., of the type (4.1.1)) to your utility function? Repeat this process for five willing acquaintances.

(In one previous application of this procedure, one subject, a police officer, produced an extremely risk-averse utility function. However, later in the evening, his utility function became equally extremely convex. Therefore, please be careful to complete your survey under reasonable experimental conditions.)

4.2.b If $U(r)$, your utility function on monetary rewards, satisfies $U(0) = 0$ and $U(\infty) = 1$, do you think that $1 - U(-r)$ is a reasonable utility function on monetary losses? Carefully discuss.

4.2.c In Section 3.10, define your "reward" to be $r = -|d - \theta|$, where $|d - \theta|$ is the distance between your point estimate and the true parameter θ. Then, $U(r) = 1 - G(-r)$ can be defined as the statistician's utility function on the reward space for $r \in R = (-\infty, 0)$, and $G(|d - \theta|)$ is a symmetric bounded loss function. Assuming that $G(|d - \theta|) = 1 - \exp\{-A|d - \theta|\}$, show how the value of A can be elicited from the statistician by a single question. (You may assume that the statistician intends to choose d by minimizing the posterior expectation of his loss function.) How would you ask further questions to check that this functional form is appropriate? How would you measure a statistician's entire choice of G when this is not embedded in a particular functional form?

4.3 Modifications to the Expected Utility Hypothesis

Meginniss (1976) proposed comparing different P by comparison of

$$E(U \mid P) + \alpha\omega(P), \tag{4.3.1}$$

where $\omega(P)$ is the entropy associated with the distribution P, and $\alpha > 0$. If P possesses continuous density $p(r)$, then $E(U \mid P) = \int_R U(r)p(r)dr$, and

$$\omega(P) = -\int_R p(r) \log p(r)\, dr. \tag{4.3.2}$$

Meginness also reports an analysis for the situation where R is discrete, and developed an alternative axiom system. Machina (1982, 1983, 1984, 1987), Karmarkar (1978), and Chew (1983) discuss other modifications to EUH. Cyert and DeGroot (1987, Ch. 9) discuss adaptive utility. For example, a company's utility function may evolve during a learning process and hence might give the impression of not satisfying EUH, when in fact the company is referring to the expectations of an evolving utility function.

The following modification to EUH produces a simple attempt to model how individuals actually decide between possibly random monetary rewards, for example, economic portfolios. It is not intended, in the original sense of EUH, to prescribe how individuals should make their decisions.

Associated with any probability distribution $P \in \mathcal{P}$ is a value r_c, the certain component of the reward. For example, if P assigns positive probabilities to k distinct rewards $r_1, r_2, \ldots, r_k$, then $r_c = \min(r_1, r_2, \ldots, r_k)$. An "$\epsilon$-adjusted utility criterion" can be defined by

$$E^*(U \mid P) = E(U \mid P) + \epsilon_{r_c} U(r_c) \qquad (P \in \mathcal{P}), \tag{4.3.3}$$

where, for example, $\epsilon_r = \epsilon$ can be taken to be constant in r. The practical analysis of Section 4.7 suggests that for subjects who are apparently risk averse, under EUH, a typical choice of ϵ_r might be $\epsilon_r = \epsilon = 0.4$ for all monetary rewards in $R =$ (\$500, \$5,000). In other words, the preferences of many individuals can be explained by saying that they put a substantial premium on the certain component of their reward. Also, the behavior of some individuals, who are apparently risk averse under EUH, can alternatively be explained by the "ϵ-adjusted linear utility criterion"

$$E^*(X \mid P) = E(X \mid P) + \epsilon_{r_c} r_c \qquad (P \in \mathcal{P}), \tag{4.3.4}$$

where X denotes the actual reward in R. When $\epsilon_r = \epsilon$ is constant in r, (4.3.3) and (4.3.4) can be respectively replaced by

$$E^*(U \mid P) = E(U \mid P) + \epsilon r_c \tag{4.3.5}$$

and

$$E^*(X \mid P) = E(X \mid P) + \epsilon r_c. \tag{4.3.6}$$

The criteria in (4.3.3) and (4.3.4) satisfy neither the Savage axioms nor the axioms reported by DeGroot (1970, pp. 101–112). They do, however, address the axioms proposed by Gilboa (1988) and Jaffray (1988), together with the broader axiom system described by Kahneman and Tversky (1979). Consider, for example, the linear criterion (4.3.6), with $\epsilon = 0.4$. Let us compare the two rewards

G_1: receive \$1,500 with certainty

and

G_2: receive \$1,200 or \$2,000, each with probability 0.5.

Then

$$E^*(X \mid G_1) = 1{,}500 + 0.4 \times 1{,}500 = 2{,}100$$

and

$$E^*(X \mid G_2) = 0.5 \times (1{,}200 + 2{,}000) + 0.4 \times 1{,}200 = 2{,}080,$$

so that $G_2 \preceq G_1$. However, let

G: receive \$1,000 with certainty,

and consider the random rewards $G_3 = \frac{1}{2}G_1 + \frac{1}{2}G$ and $G_4 = \frac{1}{2}G_2 + \frac{1}{2}G$. Under the criterion in (4.2.6),

$$E^*(X \mid G_3) = 0.5 \times (1{,}500 + 1{,}000) + 0.4 \times 1{,}000 = 1{,}650$$

and

$$E^*(X \mid G_4) = 0.5 \times (1{,}600 + 1{,}000) + 0.4 \times 1{,}000 = 1{,}750,$$

so that $G_3 \preceq G_4$. This demonstrates that the preferences $G_2 \preceq G_1$ and $G_3 \preceq G_4$, which are inconsistent with the main axiom of utility of Section 4.1, can nevertheless be explained by our adjusted criterion (4.3.6).

Just like EUH, the modified criterion (4.3.3) can be used to explain the St. Petersburg paradox of Section 4.1, when instead $R = (0, \infty)$, for example, whenever $U(\infty) = 1$. However, unlike EUH, it also handles Allais' paradox. For example, subject to our adjusted linear criterion in (4.3.6), with $\epsilon = 0.4$, the four gambles in our description in Section 4.2, of Allais' paradox, satisfy

$$\begin{aligned} E^*(X \mid G_1) &= 700{,}000, \\ E^*(X \mid G_2) &= 695{,}000, \\ E^*(X \mid G_3) &= 55{,}000, \\ \text{and} \quad E^*(X \mid G_4) &= 250{,}000. \end{aligned}$$

Therefore $G_2 \preceq G_1$ and $G_3 \preceq G_4$, so that a premium on the uncertain component of the reward can resolve Allais' paradox. It is essential to appreciate that if $\epsilon_r > 0$, a person's utility curve $U(r)$ under our adjusted expected utility criterion (4.3.3) can be very different when compared with the person's utility curve under EUH.

4.4 The Experimental Measurement of ϵ-Adjusted Utility

Consider the criterion in (4.3.5), with general U and constant ϵ. Consider an individual who equally prefers the certain reward r and the equal-chance random reward $\langle r_1, r_2\rangle$, with $r_1 < r_2$. Then (4.3.5) gives the adjusted utilities $(1+\epsilon)U(r)$ for the certain reward r, and $(\frac{1}{2}+\epsilon)U(r_1)+\frac{1}{2}U(r_2)$ for the random reward. Equating these two adjusted utilities gives

$$U(r) = (1-\xi)U(r_1) + \xi U(r_2),$$

with $\xi = 1/2(1+\epsilon)$. Similarly the first four stages of the experimental measurement process of Section 4.1 (B) tell us that

Stage I: $U(x_1) = \xi;$ (4.4.1)

Stage II: $U(x_2) = \xi^2;$ (4.4.2)

Stage III: $U(x_3) = \xi(2-\xi);$ (4.4.3)

Stage IV: $U(x_4) = \xi^2(3-2\xi).$ (4.4.4)

Remember that $U(x_1) = U(x_4)$ whenever $\xi = \frac{1}{2}$ and $\epsilon = 0$, and EUH holds. However, $x_4 < x_1$ is consistent with $\epsilon > 0$. If $\xi \neq \frac{1}{2}$, or $x_4 \neq x_1$, we recommend an addendum to Stage IV, which helps us to measure ϵ.

Stage IVA: This elicitation step can be performed by asking the subject to consider different prespecified choices of the corresponding percentages 100ϕ. Elicit a value of ϕ for which receiving the random reward $(1-\phi)\{x_2\}+\phi\{x_3\}$ is equally preferred to the certain reward $\{x_1\}$.

Calculating an Individual's Premium on Certainty

Under EUH, $\phi = 0.5$, while under our adjusted criterion, with $\epsilon > 0$, we should obtain $\phi > \frac{1}{2}$. For any value of $\phi \in (0, 1)$, (4.3.5) tells us that

$$(1+\epsilon)U(x_1) = (1-\phi)U(x_2) + \phi U(x_3) + \epsilon U(x_2). \tag{4.4.5}$$

Substituting the expressions in (4.4.1), (4.4.2), and (4.4.3) for $U(x_1)$, $U(x_2)$, and $U(x_3)$ in (4.4.5) tells us that ξ satisfies the quadratic equation

$$4\phi\xi^2 - (4\phi+1)\xi + 1 = 0. \tag{4.4.6}$$

This equation has two roots. Note, however, that the root of (4.4.6) satisfying $\xi = \frac{1}{2}$, that is, $\epsilon = 0$, when $\phi = 0.5$, is $\xi = 1/4\phi$. This yields the succinct solution

$$\epsilon = 2\phi - 1. \tag{4.4.7}$$

For example, $\epsilon = 0.4$ whenever $\phi = 0.7$. When $\phi = 1, \epsilon = 1$ and $\xi = \frac{1}{4}$. Therefore, this elicitation procedure tells us that $\epsilon \leq 1$, whenever ϕ can be elicited under our adjusted utility hypothesis, and from (4.4.1), that it is then always true that $U(x_1) \geq \frac{1}{4}$. If, say, $\phi = 0.3$, then $\epsilon = -0.4$, suggesting a strong compulsion for gambling. However, $\epsilon \geq -1$, for any value of ϕ.

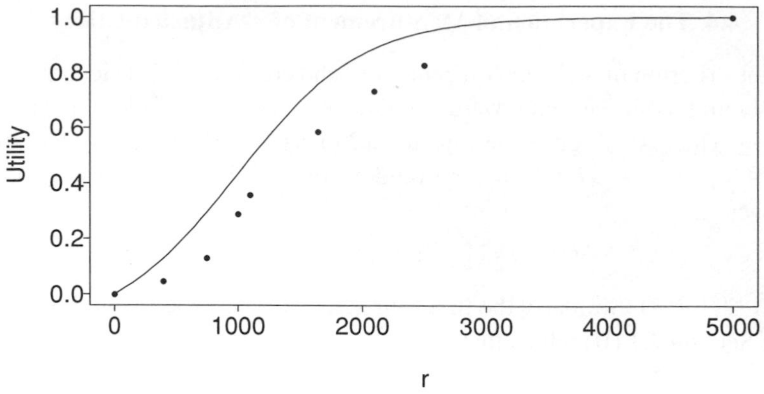

Figure 4.4.1. Mr. Jones's ϵ-adjusted utilities: solid points – modified utilities ($\epsilon = 0.4$); solid curve – utility function under EUH.

Mr. Jones's Preferences

In the Mr. Jones example of Section 4.1 (B), suppose that together with the values $x_1 = 1{,}100$, $x_2 = 750$, $x_3 = 1{,}650$, $x_4 = 1{,}000$, $x_5 = 400$, $x_6 = 2{,}100$, and $x_5 = 2{,}500$, Mr. Jones states, at Stage IVA, that $\phi = 0.7$. Then $\epsilon = 0.4$ and $\xi = 0.3571$. Then equations (4.4.1)–(4.4.5) give $U(1{,}100) = 0.3571$, $U(750) = 0.1276$, $U(1{,}650) = 0.5866$, and $U(1{,}000) = 0.2865$. Furthermore, as $x_5 = 400$ is based upon

$$\{x_5\} = \langle 0, x_2 \rangle,$$

then $(1+\epsilon)U(400) = \frac{1}{2}U(750) = 0.0638$, so that $U(400) = 0.0638/1.4 = 0.0456$. As $x_6 = 2{,}100$ satisfies

$$\{x_6\} \sim \langle x_3,\ 5{,}000 \rangle,$$

then $(1+\epsilon)U(x_6) = (\frac{1}{2}+\epsilon)U(x_3) + \frac{1}{2} = 0.2933(\frac{1}{2}+\epsilon) + \frac{1}{2}$, so that $U(2{,}100) = 0.7321$. Finally, as $x_7 = 2{,}500$ is based upon

$$\{x_7\} \sim \langle x_6,\ 5{,}000 \rangle,$$

then $(1+\epsilon)(U(x_7) = (\frac{1}{2}+\epsilon)U(x_6) + \frac{1}{2} = 0.3252(\frac{1}{2}+\epsilon) + 0.5$, so that $U(2{,}500) = 0.8277$.

The seven points of Mr. Jones's revised utility function under the ϵ-adjusted expected utility hypothesis are described in Figure 4.4.1, together with the two endpoints. This may be compared with the solid curve, which fits his utility function, under the expected utility hypothesis.

Worked Example 4D: *Lucy's ϵ-Adjusted Utility*

Lucy evaluates rewards X (i.e., positive amounts of milk) by the ϵ-adjusted linear utility criterion

$$E^*(X \mid P) = E(X \mid P) + \tfrac{2}{3}r_c,$$

where r_c denotes the certain component of the reward.

(a) Lucy is given the opportunity to receive $\frac{1}{4}$ liter with probability $\frac{1}{3}$ and $\frac{1}{2}$ liter with probability $\frac{2}{3}$. How much certain reward would she accept in return?

(b) Lucy is given the opportunity to receive either 1 liter with certainty, or $\frac{1}{2}$ liter with probability $1 - \phi$ and 2 liters with probability ϕ. How large a value of ϕ would be needed to convince her to take the gamble?

Model Answer 4D:

(a) The expectation of this random reward is

$$E(X \mid P) = \tfrac{1}{4} \times \tfrac{1}{3} + \tfrac{1}{2} \times \tfrac{3}{4} = \tfrac{11}{24}.$$

Therefore the ϵ-adjusted utility is

$$\tfrac{11}{24} + \tfrac{2}{3} \times \tfrac{1}{4} = \tfrac{15}{24} = \tfrac{5}{8}.$$

The ϵ-adjusted utility for a certain reward r is

$$\left(1 + \tfrac{2}{3}\right)r = \tfrac{5}{3}r.$$

Require $\frac{5}{3}r = \frac{5}{8}$, so that $r = \frac{3}{8}$.

(b) The ϵ-adjusted utility for 1 liter with certain reward is

$$1 + \tfrac{2}{3} = \tfrac{5}{3}.$$

The ϵ-adjusted utility for the random reward is

$$\tfrac{1}{2}(1 - \phi) + 2\phi + \tfrac{2}{3} \times \tfrac{1}{2} = \tfrac{5}{6} + \tfrac{3}{2}\phi.$$

Equating this expression with $\frac{5}{3}$ gives $\phi = \frac{5}{9}$.

SELF-STUDY EXERCISES

4.4.a Officer Plod states values $x_1 = 1{,}000$, $x_2 = 200$, $x_3 = 1{,}800$, $x_4 = 520$ and $\phi = 0.6$ at the first four stages of the experimental measurement procedure, for monetary rewards between \$0 and \$5,000. Find choices of $U(200)$, $U(1{,}000)$, and $U(1{,}800)$ that are consistent with these choices of x_1, x_2, x_3, and ϕ, under the criterion (4.3.6). Is Officer Plod's value $x_4 = 520$ consistent with this criterion?

4.4.b If ϵ_r in (4.3.3) is not constant in r, devise an experimental measurement procedure that helps us to model ϵ_r for $r \in (0,\ 5{,}000)$.

4.4.c Like most other Goths, Alaric the Goth, who invaded Rome in 410 A.D., was fond of gambling. For R = the space of monetary rewards between zero and a million gold pieces, he provided, at the first four stages of the experimental measurement procedure, the values $x_1 = 700{,}000$, $x_2 = 490{,}000$, $x_3 = 980{,}000$, $x_4 = 840{,}000$, and $\phi = \frac{5}{14}$. Are these choices consistent with the criterion (4.3.6), with $\epsilon = -\frac{2}{7}$? If so, you may assume that he is following this criterion more generally.

Suppose that for Alaric to leave Rome, the senators offer him the choice between a lottery ticket that would pay him either zero gold pieces or a million gold pieces, each with probability one half, and a bride with a certain dowry of 500,000 gold pieces. How much monetary value would he need to place on the bride in order to prefer the bride and the dowry to the lottery ticket?

4.5 The Risk-Aversion Paradox

Consider an individual whose preferences can be rationalized by noting that for any two fixed rewards r_1 and r_2, with $r_1 \le r_2$, he will always equally prefer the fixed reward

$$G_1 = \{r_1 + \rho(r_2 - r_1)\} \tag{4.5.1}$$

and the equally weighted random reward

$$G_2 = \langle r_1, r_2 \rangle, \tag{4.5.2}$$

for some value of ρ to be determined. We will refer to this individual as possessing "ρ-weighted preferences." This does not, of course, define the individual's preferences for unequally weighted random rewards.

Relationship with EUH

An individual with ρ-weighted preferences will, under the experimental measurement procedure of Section 4.1 (B), for monetary rewards between zero and r^* dollars, always provide values $x_1, x_2, x_3,$ and x_4 at the first four stages of the elicitation procedure, satisfying

$$x_1 = \rho r^*,$$
$$x_2 = \rho^2 r^*,$$
$$x_3 = \rho(2 - p)r^*,$$

and

$$x_4 = \rho^2(3 - 2\rho)r^*. \tag{4.5.3}$$

Now $x_1 \ne x_4$ unless $\rho = \frac{1}{2}$ or 1. For more general values of ρ, this individual is therefore inconsistent with EUH. However, he can nevertheless give the appearance of following EUH. For example, $\rho = 0.45$ and $r^* = \$5{,}000$ give $x_1 = \$2{,}250$, $x_2 = \$1{,}012.5$, $x_3 = \$3{,}487.5$, and $x_4 = \$2{,}126.25$. As $x_1 \le \frac{1}{2}x^*$, $x_2 \le \frac{1}{2}x_1$, and $x_3 \le \frac{1}{2}(x_1 + x^*)$, these values suggest that the subject is risk averse under EUH. Now, however, we demonstrate that this apparent risk aversion, which occurs when $\rho < \frac{1}{2}$, can be entirely untrue, thus providing a risk-aversion paradox. The apparent risk aversion can instead be explained by linear utility, together with a positive premium on the certain component of the reward.

Relationship with ϵ-Adjusted Linear Utility

For general $\rho \in (0, 1)$, the fixed and random rewards in (4.5.1) and (4.5.2) will, under (4.3.6), have respective adjusted expected linear utilities,

$$E^*(X \mid G_1) = (1 + \epsilon)[r_1 + \rho(r_2 - r_1)] \tag{4.5.4}$$

and

$$E^*(X \mid G_2) = \left(\tfrac{1}{2} + \epsilon\right) r_1 + \tfrac{1}{2} r_2. \tag{4.5.5}$$

However, these two expressions are algebraically equivalent whenever $\epsilon = \frac{1}{2}\rho^{-1} - 1$, so that $\rho = \frac{1}{2(1+\epsilon)}$. It follows that the behavior of an individual following ρ-weighted preferences can be rationalized by the adjusted criterion (4.3.6), with $\epsilon = \frac{1}{2}\rho^{-1} - 1$. In our example, $\rho = 0.45$ and $\epsilon = \frac{1}{9}$. The individual will be apparently risk averse whenever $\rho < \frac{1}{2}$, corresponding to a positive premium $\epsilon > 0$ on the certain component of the reward. Our adjusted expected linear utility can also suggest possible preferences for unequally weighted random rewards, which compare sensibly with the individual's equal preferences for (4.5.1) and (4.5.2).

Making a Profit

Now assume that a company is able to demonstrate that 100 of its clients follow the preceding behavior, with, say, $\rho = 0.4$ and $\epsilon = 0.25$, and that they possess portfolios that will sell for \$1,000, \$2,000, and \$3,000, each with probability $\frac{1}{3}$, so that the expected selling price is \$2,000, with standard deviation about \$816. Each client will then, if he understands this rationalization, regard his portfolio as equivalent in value to an amount

$$\frac{2{,}000 + 1{,}000\epsilon}{1+\epsilon} = \$1{,}800,$$

with certainty. If the client doesn't understand the rationalization, then he may regard the offer described below as even more generous.

Given large resources, the company would be wise to offer each client \$1,830 for her or his portfolio. If all 100 deals are contracted, then the clients would make a total of \$3,000 apparent profit as well as avoid the "catastrophe" of receiving only \$1,000 per portfolio. However, the company would earn a profit with expectation \$20,000 and standard deviation \$8,160. If the company's utility function was either linear or slightly concave, in this range of monetary rewards, then this would provide an excellent corporate investment. When repeated for a large number of clients, a company with enough resources to cover unlucky short-term losses will certainly make a large profit in the long run. However, the clients are also well treated, in a "minimax" sense, since they are all protected against the worst possible outcome.

Mrs. Miller's Preferences

As a numerical illustration of the preceding paradox, suppose that Mrs. Miller gives the values $x_1 = \$1{,}666.67$, $x_2 = \$555.56$, $x_3 = \$2{,}777.78$, and $x_4 = \$296.30$ at the first four stages of the experimental procedure of Section 4.1 (B), together with $x_5 = 166.67$, $x_6 = 3{,}519.52$, and $x_7 = 4{,}013.01$, satisfying the preferences $x_5 \sim \langle 0,\ x_2\rangle$, $x_6 \sim \langle x_4,\ 5{,}000\rangle$, and $x_7 \sim \langle x_6,\ 5{,}000\rangle$. This is consistent with apparently risk-averse behavior with $\rho = 0.3$, and with ϵ-adjusted linear utility with $\epsilon = \frac{2}{3}$. However, if it is assumed that Mrs. Miller is under instructions to follow EUH, then

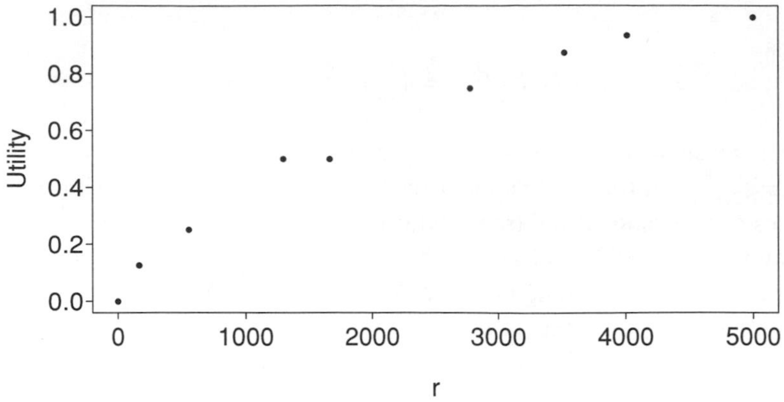

Figure 4.5.1. Mrs. Miller's apparently risk-averse utility (under EUH).

these values suggest points on a utility curve, as indicated in Figure 4.5.1. Hence, Mr. Miller might falsely interpret the convexity by believing that his wife is following conventionally risk-averse behavior.

Worked Example 4E: *Professor MacStat's ϵ-Adjusted Utility*

At the first four stages of the experimental measurement procedure in Section 4.1 (B), with $r^* = \$5{,}000$, Professor MacStat specifies $x_1 = \$500$, $x_2 = \$50$, $x_3 = \$950$, and $x_4 = \$140$.

(a) Show that Professor MacStat does not follow EUH.
(b) Show that he follows ϵ-adjusted linear utility, with $\epsilon = 4$.
(c) How much certain reward would Professor MacStat exchange, under the criterion in (b), for a portfolio that would sell for \$1,000, \$2,000, and \$3,000, each with probability $\frac{1}{3}$?
(d) How much certain reward might a company, following linear utility, offer Professor MacStat for the portfolio, with the objective of keeping everybody satisfied?

Model Answer 4E:

(a) Since $x_4 = \$140 < x_1 = \$500$, the professor does not follow EUH.
(b) x_1, x_2, x_3, and x_4 satisfy (4.5.1) with $\rho = 0.1$ and $r^* = \$5{,}000$. Following the theory of Section 4.5, the professor therefore satisfies ϵ-adjusted linear utility, with $\epsilon = \frac{1}{2}\rho^{-1} - 1 = 4$.
(c) Paralleling Mrs. Miller's example, he would sell for

$$\frac{2{,}000 + 1{,}000\epsilon}{1 + \epsilon} = \$1{,}200.$$

(d) An offer of \$1,600 would give an anticipated advantage of 400 units to both sides.

SELF-STUDY EXERCISES

4.5.a An individual follows the behavior $G_1 \sim G_2$ for all $r_1 \leq r_2$, where

$$G_1: \quad \text{receive } r_1 + \rho_{r_1}(r_1 - r_2) \text{ with certainty,}$$
$$G_2: \quad \text{receive } r_1 \text{ or } r_2, \text{ each with probability } \tfrac{1}{2},$$

where ρ_r depends on r. Show when this behavior is consistent with the criterion in (4.3.4).

4.5.b The experimental measurement procedure, for monetary rewards between \$0 and \$10,000, gives $x_1 =$ \$3,000, $x_2 =$ \$900, $x_3 =$ \$5,100, and $x_4 =$ \$2,160.

Show that under EUH, the subject is apparently risk averse, but that the subject's behavior can be better explained by a linear utility function, together with a premium on the certain component of the reward. Find this premium.

4.6 The Ellsberg Paradox

Ellsberg (1961) considers an urn with 30 red balls and 60 black and yellow balls in unknown proportions. One ball is chosen at random from the urn. Consider choices between

G_1: receive \$100 if ball is red, and \$0 otherwise

and G_2: receive \$100 if ball is black, and \$0 otherwise,

and also between

G_3: receive \$100 if ball is red or yellow, and \$0 otherwise,

and G_4: receive \$100 if ball is black or yellow, and \$0 otherwise.

In Ellsberg's opinion, many people would express the preferences $G_2 \preceq G_1$ and $G_3 \preceq G_4$. Suppose, however, that the unknown proportion of black balls, among the 60 black and yellow balls, is ϕ and that you possess utility function $U(r)$, under EUH, with $U(0) = 0$ and $U(100) = u$. Then the proportions of red, black, and yellow balls are $\frac{1}{3}$, $\frac{2}{3}\phi$, and $\frac{2}{3}(1-\phi)$, respectively, so that

$$E(U \mid G_1) = \tfrac{1}{3}u,$$
$$E(U \mid G_2) = \tfrac{2}{3}\phi u,$$
$$E(U \mid G_3) = \tfrac{1}{3}(3 - 2\phi)u,$$
$$\text{and} \quad E(U \mid G_4) = \tfrac{2}{3}u.$$

Therefore people who judge that $\phi < \frac{1}{2}$ would express the preferences $G_2 \preceq G_1$ and $G_4 \preceq G_3$, and people who judge that $\phi > \frac{1}{2}$ would choose $G_1 \preceq G_2$ and $G_3 \preceq G_4$. However, for many people, $\phi = \frac{1}{2}$ is the only reasonable subjective choice in this ignorance situation. For these people, EUH requires that $G_1 \sim G_2$ and $G_3 \sim G_4$. Ellsberg regards this as a paradox.

The ϵ-adjusted utility criterion (4.3.3) does not immediately resolve this paradox. It can be resolved only if decision makers realize that they should, and probably do, distinguish between objective probabilities (such as those based upon randomization or estimates from substantive data) and subjective probabilities (such as those based upon equality of ignorance) when reaching their decisions. Therefore, whenever $U(0) = 0$, we recommend the "modified ϵ-adjusted utility criterion,"

$$\tilde{E}(U \mid P) = \delta_P E(U \mid P) + \mathcal{E}_{r_c} U_{r_c} \qquad (P \in \mathcal{P}), \tag{4.6.1}$$

where $\delta_P \in (0, 1)$ measures the degree of objectivity of the probability distribution P. Set $\delta_P = 1$ if P is completely objective, and $\delta_P = 0$ if P is a completely random guess.

The modification (4.6.1) resolves the Ellsberg paradox. Suppose, for example, that ϕ is subjectively chosen to be $\phi = \frac{1}{2}$, but that $\delta_P = \frac{1}{2}$ whenever $\mathcal{P}$ is chosen subjectively. We then have

$$\begin{aligned} \tilde{E}(U \mid G_1) &= \tfrac{1}{3}u, \\ \tilde{E}(U \mid G_2) &= \tfrac{1}{6}u, \\ \tilde{E}(U \mid G_3) &= \tfrac{1}{3}u, \\ \text{and} \quad \tilde{E}(U \mid G_4) &= \tfrac{2}{3}u. \end{aligned}$$

Consequently $G_2 \preceq G_1$ and $G_3 \preceq G_4$, thus rationalizing the Ellsberg paradox. An alternative explanation of this paradox is produced by Kadane (1992), who also discusses prospect theory (Kahneman and Tversky, 1979), and prospective reference theory (Viscusi, 1989).

Worked Example 4F: *Portfolio Example*

Consider n portfolios $P_1, P_2, \ldots, P_n$. On portfolio P_i, you receive zero reward with probability $1 - i^{-1}$ and a reward of i dollars with probability i^{-1}. You follow EUH, with linear utility, but with degree of objectivity $\delta_P = \lambda_i$ whenever you assign probability i^{-1} to the reward i.

(a) Show that for $i = 1, 2, \ldots, n$, you would prefer reward P_i to reward P_{i-1} whenever $\lambda_i > \lambda_{i-1}$.

(b) Hence show that P_1 is your optimal portfolio, whenever λ_i is strictly decreasing in i.

Model Answer 4F:

The expected reward for portfolio P_i is

$$\left(1 - i^{-1}\right) \times 0 + i \times 1 = 1.$$

Hence the adjustment in (4.6.1) reduces to λ_i. The results in (a) and (b) follow immediately.

4.7 A Practical Case Study

Eighty-five subjects were interviewed by 17 graduate students, who elicited responses x_1, x_2, x_3, x_4, and ϕ, according to the first four stages of the experimental measurement of utility, described in Section 4.1 (B), and the addendum to the fourth stage, as described in Section 4.4. The reward space R related to monetary rewards between \$0 and \$5,000. Forty-four out of the 85 subjects provided responses to x_1, x_2, and x_3, demonstrating apparently risk-averse preferences, that is, $x_1 < 2{,}500$, $x_2 < \frac{1}{2}x_1$, and $x_3 < x_1 + \frac{1}{2}(5{,}000 - x_1)$. Quite remarkably, 43 out of the 44 subjects reported $x_4 < x_1$. Subject number 18 instead reported that $x_4 = x_1$. Hence our apparently risk-averse subjects are consistently inconsistent with EUH in a very specific way. This study replicated a similar study carried out a couple of years earlier, when very similar conclusions were reached.

The results for the 44 apparently risk-averse subjects in the current study are reported in Table 4.7.1. Out of the 44 apparently risk-averse subjects, 43 reported a value of $\phi > 0.5$ for the betting probability elicited at Stage IVA of Section 4.4. Subject 43 instead reported $\phi = 0.5$. These subjects are therefore largely consistent with the ϵ-adjusted (not necessarily linear) utility criterion in (4.3.3), with a positive value for the premium on certain ϵ. The values of $\epsilon = 2\phi - 1$ are shown in the last column of Table 4.7.1. Note that subject 43 has $\epsilon = 0$ but also $x_4 < x_1$, and so is inconsistent with both criteria.

Further inspection of the data in Table 4.7.1 will show that the 17 starred candidates yield a choice of $U(r)$, under the criterion (4.3.3), via (4.4.1)–(4.4.4), which is reasonably linear in r. The apparently risk-averse behavior of these 17 subjects can be explained by the representations of Section 4.5.

Finally consider the remaining 27 apparently risk-averse subjects, but minus subjects 18 and 43. The choices of U for these 25 people, under the adjusted condition (4.3.3), can be shown to be substantially less concave than the corresponding choices under EUH. It would appear that a large amount of apparently risk-averse behavior can be explained by people placing premiums on the certain, or virtually certain, components of their rewards.

Table 4.7.1. *Results for apparently risk-averse subjects* ($r^* = 5{,}000$).

	Occupation	x_1	x_2	x_3	x_4	ϕ	ϵ
1.	Geneticist	1,000	400	1,400	800	0.66	0.32
2.	Pharmacist	1,000	250	1,500	450	0.85	0.70
3.*	Food scientist	1,700	500	2,700	1,500	0.70	0.40
4.*	Textile worker	1,000	200	2,000	800	0.60	0.20
5.	Philosopher	600	150	1,500	400	0.70	0.40
6.	Educationalist	1,250	500	1,800	1,000	0.65	0.30
7.	Biochemist	1,000	200	2,500	600	0.70	0.40

(continued)

Table 4.7.1 *(continued)*

	Occupation	x_1	x_2	x_3	x_4	ϕ	ϵ
8.	Housewife	500	100	800	200	0.80	0.60
9.	Nurse	2,000	800	2,500	1,500	0.75	0.50
10.*	Food scientist	2,200	1,000	3,200	2,000	0.60	0.20
11.	Food scientist	1,000	300	1,800	700	0.90	0.80
12.*	Food scientist	2,300	1,200	3,400	2,000	0.60	0.20
13.	Occ. therapist	2,000	800	3,200	1,800	0.70	0.40
14.*	Mechanical engineer	1,500	500	2,500	1,200	0.75	0.50
15.*	Forester	1,000	200	1,800	800	0.60	0.20
16.*	Food scientist	800	200	1,500	500	0.90	0.80
17.*	Chemist	1,000	300	2,000	700	0.60	0.20
18.	Electrical engineer	1,500	500	3,000	1,500	0.60	0.20
19.*	Physicist	2,400	1,000	3,600	2,300	0.55	0.10
20.*	Chemist	2,200	1,000	3,300	2,000	0.60	0.20
21.	Musician	1,000	300	2,200	900	0.75	0.50
22.	Agronomist	1,900	900	3,000	1,800	0.60	0.20
23.*	Education psychologist	2,000	700	3,000	1,400	0.80	0.60
24.	Psychologist	2,090	980	2,500	1,480	0.52	0.04
25.	Mechanical engineer	1,200	300	1,300	400	0.58	0.16
26.	Mechanical engineer	1,000	200	1,500	700	0.70	0.40
27.*	Mechanical engineer	1,800	700	2,850	1,500	0.72	0.44
28.	Civil engineer	985	380	2,300	850	0.70	0.40
29.*	Health administrator	2,000	800	3,200	3,760	0.60	0.20
30.	Communication artist	500	50	550	200	0.95	0.90
31.*	Computer engineer	2,450	1,000	3,500	2,000	0.65	0.30
32.	English student	500	100	800	200	0.95	0.90
33.	Electrical engineer	1,000	400	1,500	600	0.80	0.60
34.	Industrial engineer	1,000	300	2,150	700	0.87	0.74
35.	Industrial engineer	2,100	950	3,300	1,800	0.65	0.30
36.	Industrial engineer	1,400	500	2,850	1,350	0.72	0.44
37.*	Industrial engineer	900	200	1,850	550	0.90	0.80
38.	Political scientist	600	75	1,600	500	0.67	0.34
39.*	Graphic designer	1,750	600	3,000	1,400	0.67	0.33
40.	Interior designer	1,500	600	2,000	725	0.80	0.60
41.*	High school kid	1,200	300	2,000	800	0.70	0.40
42.	Chemical engineer	1,800	750	2,500	1,000	0.65	030
43.	Educationalist	900	390	1,800	790	0.50	0.00
44.	Musician	1,250	300	1,650	450	0.68	0.36

* Well-explained by ϵ-adjusted linear utility.

5

Models with Several Unknown Parameters

5.0 Preliminaries and Overview

For models with several parameters, the Bayesian approach can be used to draw inferences about any particular parameter, or function of the parameters, by reference to its marginal posterior density. In many examples, the marginal density can be either (a) calculated analytically, (b) computed using standard numerical integration techniques, for example, Simpson's rule, (c) approximated via a multivariate normal approximation to the joint posterior density, (d) more accurately approximated using conditional maximization with respect to the nuisance parameters, and Laplacian approximations, with saddlepoint accuracy, (e) exactly computed using a variation of the Monte Carlo simulation techniques, known as importance sampling (Section 6.4), or (f) computed by Markov Chain Monte Carlo (MCMC) procedures (Sections 6.6–6.9). The reader can therefore refer to Chapter 6 for fuller technical details of some of the general techniques mentioned in Chapter 5. Chapter 5 concentrates more on the statistical ideas relating to marginalization rather than on simulation procedures that might be feasible when the integrations cannot be performed or approximated analytically.

Once a marginal posterior density has been calculated for a parameter of interest η, then you may apply all the techniques of Chapter 3, relating to point estimation, posterior standard deviation, Bayesian intervals, the investigation of hypotheses, and decision making, to this marginal density. This is a key practical advantage of the Bayesian approach. Note, however, that the modified profile likelihood (1.5.9) does typically closely approximate the marginal posterior density when this exists, under a uniform prior (see Section 5.1 (B)), owing to the relation of profile likelihood to Laplacian methods. Therefore, modified profile likelihood should appeal to many Bayesians.

In Sections 5.1 (F) and 5.1 (G), we describe two key applications. The first concerns marginal inference for the parameters in the Hardy–Weinberg equilibrium model for the OAB blood system. The second concerns a full Bayesian analysis for the two-parameter Weibull model for failure or survival times. In both of these examples, relevant marginal posterior densities can be computed via one-dimensional numerical integrations. In Section 5.1 (I), Laplacian approximations and importance sampling are used to investigate a sampling model generalizing the binomial in the context of multiple-choice testing. For an application of Bayesian marginalization in biology, see Parton et al. (1997).

A number of useful analyses are introduced in Exercises 5.1.a–5.1.g for development by the reader. These include analyses for the normal sampling model, with unknown variance, the two-way random effects ANOVA model (with interactions), and the multinomial distribution. Exercise 5.1.h is concerned with Ferguson's Dirichlet process, which permits Bayesian inference regarding an entire unknown sampling model, without parametric restriction. Exercises 5.1.i and 5.1.j address marginalization problems for nonlinear (parametric) sampling models.

The Bayesian theory of two-way contingency tables is developed in Sections 5.2 (A)–5.2 (C). A number of data sets are analyzed using our proposed methodology. These include the Death Penalty, Engineering Apprentice, Shopping in Oxford, Marine Corps data, and data sets relating to remote sensing, occupational mobility, and transitions between geological layers. In particular, Goodman's full-rank interaction analysis can frequently be used to highlight the key real-life conclusions to be drawn from the data.

Sections 5.2 (D) and 5.2 (E) are concerned with applications in chemistry, and it is shown how complicated nonlinear regression models can be analyzed, for example, using Laplacian approximations. It is again typically possible to draw finitc samplc inferences about any parameter of interest by reference to its marginal posterior density.

In Sections 5.3–5.7, we apply Bayesian marginalization to hierarchical models for time series and forecasting. In Section 5.3 and Exercises 5.3.a (normal case) and 5.3.b (Poisson case), it is shown how the Kalman filter (an updating formula that is appropriate when all variance components in the model are specified) can also be applied when the variance components are unknown and possess marginal posterior densities. Exercise 5.3.b extends the concept of "best linear predictors" introduced in Exercise 3.8.g. In Sections 5.4 and 5.5, these ideas are applied to two real data sets. The first involves an on-line analysis of chemical process readings, and the second involves Poisson observations in a quality control situation. Other forecasting applications are discussed in Sections 5.6 and 5.7.

When constructing a prior distribution for several parameters, it is possible to seek transformations of the parameters such that the transformed parameters possess a multivariate normal prior distribution. This idea helps us to model the prior dependencies between the parameters. Similar parametrizations can be employed whenever applying Laplace's approximation or importance sampling to the marginals, particularly if the likelihood is well approximated by the multivariate normal likelihood (1.3.11). While Section 5.2 (E) includes an example, these ideas also relate to the developments in Chapter 6.

5.1 Bayesian Marginalization

(A) *Marginal Posterior Densities*

Let a $p \times 1$ vector of parameters $\boldsymbol{\theta}$ possess twice-differentiable posterior density

$$\pi_y(\boldsymbol{\theta}) \propto \pi(\boldsymbol{\theta}) l(\boldsymbol{\theta} \mid \mathbf{y}) \qquad \left(\boldsymbol{\theta} \in \Theta = R^P\right), \tag{5.1.1}$$

given an $n \times 1$ vector of observations $\mathbf{y}$, where $\pi(\boldsymbol{\theta})$ denotes the prior density, $l(\boldsymbol{\theta} \mid \mathbf{y}) = p(\mathbf{y} \mid \boldsymbol{\theta})$ denotes the likelihood of $\boldsymbol{\theta}$ given $\mathbf{y}$, and the constant of proportionality should be calculated to ensure that $\int_\Theta \pi_y(\boldsymbol{\theta}) d\boldsymbol{\theta} = 1$. Let $\eta = \theta_1$ denote the first element of $\boldsymbol{\theta} = (\theta_1, \ldots, \theta_p)^T$ and let $\boldsymbol{\xi} = (\theta_2, \ldots, \theta_p)^T$. Then (5.1.1) can be equivalently replaced by

$$\pi_y(\eta, \boldsymbol{\xi}) \propto \pi(\eta, \boldsymbol{\xi}) l(\eta, \boldsymbol{\xi} \mid \mathbf{y}) \qquad \left(\eta \in R, \boldsymbol{\xi} \in R^q\right), \tag{5.1.2}$$

where we will regard η as a parameter of interest and $\boldsymbol{\xi}$ as a $q \times 1$ vector of nuisance parameters, with $q = p - 1$. If you wish to make inferences about η, then just apply straightforward probability theory for continuous random variables. The marginal posterior density of η is

$$\begin{aligned} \pi_y(\eta) &= \int_{R^q} \pi_y(\eta, \boldsymbol{\xi})\, d\boldsymbol{\xi} \\ &\propto \int_{R^q} \pi(\eta, \boldsymbol{\xi}) l(\eta, \boldsymbol{\xi} \mid \mathbf{y})\, d\boldsymbol{\xi} \qquad (\eta \in R), \end{aligned} \tag{5.1.3}$$

where the integrations in (5.1.3) should be performed with respect to $\boldsymbol{\xi} \in R^q$. The marginal posterior density of η summarizes all posterior information about η, assuming the correctness of the prior specification, and should be regarded as a basis for Bayesian inference regarding η. All the techniques of Chapter 1 for a single parameter are now available for η, but are based upon its marginal posterior density. These include posterior probability statements, Bayesian intervals, the posterior median, mean, mode, and Bayes decisions under symmetric loss functions.

Next consider the uniform prior specification $\pi(\eta, \boldsymbol{\xi}) \propto 1$. If $\int_{R^q} l(\eta, \boldsymbol{\xi} \mid \mathbf{y}) d\boldsymbol{\xi} < \infty$, for each $\eta \in R$, then the posterior density, under this uniform prior, reduces to the "integrated likelihood"

$$\pi_y(\eta) \propto \int_{R^q} l(\eta, \boldsymbol{\xi} \mid \mathbf{y})\, d\boldsymbol{\xi} \qquad (\eta \in R), \tag{5.1.4}$$

of η. Dawid, Stone, and Zidek (1973) note that in some examples, the integrated likelihood can ignore valuable information, involving important components of low-dimensional sufficient statistics for $(\eta, \boldsymbol{\xi})$. Therefore, considerable care should be taken when using (5.1.4) rather than the marginal posterior density (5.1.3), based upon a proper prior distribution.

(B) *Laplacian Approximations and Modified Profile Likelihoods*

Where technically feasible, the integrations in (5.1.3) should be performed analytically or by numerical integration. These integrations created considerable difficulties for the Bayesian approach until and during the 1970s. However, (5.1.3) can frequently be represented remarkably accurately, even in the tails, by the Laplacian approximation

$$\pi_y^*(\eta) \propto \pi_y\left(\eta, \boldsymbol{\xi}_\eta\right) \left|\mathbf{R}_\eta^*\right|^{-\frac{1}{2}} \qquad (\eta \in R), \tag{5.1.5}$$

where $\boldsymbol{\xi}_\eta$ conditionally maximizes $\pi_y(\eta, \boldsymbol{\xi})$ with respect to $\boldsymbol{\xi}$, for each fixed η, $\pi_y(\eta, \boldsymbol{\xi}_\eta)$ is the "profile posterior density," and

$$\mathbf{R}_\eta^* = -\left[\frac{\partial^2 \log \pi_y(\eta, \boldsymbol{\xi})}{\partial\left(\boldsymbol{\xi}\boldsymbol{\xi}^T\right)}\right]_{\boldsymbol{\xi}=\boldsymbol{\xi}_\eta}. \tag{5.1.6}$$

See Leonard and Novick (1986), Leonard, Hsu, and Tsui (1989), and others, following Leonard (1982a).

The approximation (5.1.5) will be justified in Section 6.2 and works best for parametrizations for which the conditional posterior distribution of $\pi_y(\boldsymbol{\xi} \mid \eta)$, of $\boldsymbol{\xi}$ given η, is approximately multivariate normal or multivariate t, for each η. (Note that as a function of $\boldsymbol{\xi}$, $\pi_y(\boldsymbol{\xi} \mid \eta) \propto \pi_y(\boldsymbol{\xi}, \eta)$.) Positive definiteness of $\mathbf{R}_\eta^*$ is required. A procedure for handling the nonpositive definite case is described by Hsu (1995). The proportionality constant in (5.1.5) should be calculated via a one-dimensional numerical integration. In this case, Tierney and Kadane (1986) use theory relating to saddle-point approximations (e.g., Barndorff-Nielsen and Cox, 1979; Barndorff-Nielsen, 1983; Reid, 1988) to show that under some conditions, the ratio of the approximation to the true density will behave like $1 + Kn^{-1}$ as n gets large, for some constant K. Note that when $\pi(\eta, \boldsymbol{\xi}) \propto 1$, (5.1.5) reduces to the modified profile likelihood (1.5.10), thus providing a Bayesian justification of this procedure.

An alternative expression is suggested by Tierney, Kass, and Kadane (1989), and their suggestion can be used whenever $\eta = \mathbf{a}^T\boldsymbol{\theta}$ is a linear function of $\boldsymbol{\theta}$. In this case,

$$\pi_y^*(\eta) \propto \pi_y(\boldsymbol{\theta}_\eta)|\mathbf{R}_\eta|^{-\frac{1}{2}}\left(\mathbf{a}^T\mathbf{R}_\eta^{-1}\mathbf{a}\right)^{-\frac{1}{2}}, \tag{5.1.7}$$

where $\boldsymbol{\theta}_\eta$ conditionally maximizes $\pi(\boldsymbol{\theta})$ with respect to $\boldsymbol{\theta}$, and

$$\mathbf{R}_\eta = -\left[\frac{\partial^2 \log \pi_y(\boldsymbol{\theta})}{\partial(\boldsymbol{\theta}\boldsymbol{\theta}^T)}\right]_{\boldsymbol{\theta}=\boldsymbol{\theta}_\eta}. \tag{5.1.8}$$

For an application of (5.1.7) to astrophysics, see Cooke, Espey, and Carswell (1996).

(C) *A General Laplacian Approximation*

Consider now a possibly nonlinear, twice differentiable function

$$\eta = g(\boldsymbol{\theta}) \tag{5.1.9}$$

of $\boldsymbol{\theta}$. The marginal posterior density of η is

$$\pi_y(\eta) = \lim_{\epsilon \to 0} \epsilon^{-1} \int_D \pi_y(\boldsymbol{\theta})\, d\boldsymbol{\theta} \qquad (\eta \in \mathcal{A}), \tag{5.1.10}$$

where

$$D = D(\eta, \epsilon) = \left\{\boldsymbol{\theta} : |g(\boldsymbol{\theta}) - \eta| \leq \epsilon\right\}, \tag{5.1.11}$$

and $\mathcal{A}$ denotes the set of possible values for η.

However, if $\pi_y(\boldsymbol{\theta})$ is approximately multivariate normal for all $\boldsymbol{\theta} \in D(\eta, \epsilon)$, with ϵ arbitrarily small, then a convenient approximation (Hsu, 1995) is available. Note that the conditional maximum $\boldsymbol{\theta}_\eta$ of (5.1.1), with respect to $\boldsymbol{\theta}$, given that $g(\boldsymbol{\theta}) = \eta$, satisfies the equation

$$\frac{\partial \log \pi_y(\boldsymbol{\theta})}{\partial \boldsymbol{\theta}} = \lambda_\eta \frac{\partial g(\boldsymbol{\theta})}{\partial \boldsymbol{\theta}}, \tag{5.1.12}$$

in $\boldsymbol{\theta}$, where the Lagrange multiplier λ_η can be determined, upon noting that $\eta = g(\boldsymbol{\theta}_\eta)$. Then the marginal posterior density in (5.1.3) can be approximated by

$$\pi_y^*(\eta) \propto \pi_y(\boldsymbol{\theta}_\eta)\big|\tilde{\mathbf{R}}_\eta\big|^{-\frac{1}{2}} f\big(\eta \mid \boldsymbol{\theta}_\eta, \tilde{\mathbf{R}}_\eta^{-1}\big) \qquad (\eta \in \mathcal{A}), \tag{5.1.13}$$

where

$$\tilde{\mathbf{R}}_\eta = \mathbf{R}_\eta - \lambda_\eta \left[\frac{\partial^2 g(\boldsymbol{\theta})}{\partial\big(\boldsymbol{\theta}\boldsymbol{\theta}^T\big)}\right]_{\theta=\boldsymbol{\theta}_\eta}, \tag{5.1.14}$$

with $\mathbf{R}_\eta$ satisfying (5.1.8), and where $f(\eta \mid \boldsymbol{\mu}, \mathbf{C})$ denotes the exact density of $\eta = g(\boldsymbol{\theta})$ when $\boldsymbol{\theta}$ possesses a multivariate normal $N(\boldsymbol{\mu}, \mathbf{C})$ distribution, with mean vector $\boldsymbol{\mu}$ and covariance matrix $\mathbf{C}$. In some applications, $f(\eta \mid \boldsymbol{\mu}, \mathbf{C})$ can be obtained analytically. In others, further approximations may be needed. Indeed, in the 1991 correction to Tierney, Kass, and Kadane (1989), the saddle-point approximation

$$\pi_y^*(\eta) \propto \pi_y(\boldsymbol{\theta}_\eta)\big|\tilde{\mathbf{R}}_\eta\big|^{-\frac{1}{2}}\big(\mathbf{b}_\eta^T\tilde{\mathbf{R}}_\eta\mathbf{b}_\eta\big)^{-\frac{1}{2}}$$

is proposed, where

$$\mathbf{b}_\eta = \left[\frac{\partial g(\boldsymbol{\theta})}{\partial\boldsymbol{\theta}}\right]_{\theta=\boldsymbol{\theta}_\eta}.$$

This has excellent finite sample accuracy for many transformations $\eta = g(\boldsymbol{\theta})$ and can be derived from (5.1.13) by taking the f-contribution to denote a normal density. This approximation will be compared with (5.1.13) in Section 6.4 (I), but using a Gamma approximation to the f-contribution. The latter compares extremely closely with an exact result computed by Monte Carlo simulation.

The approximation in (5.1.13) can be justified by expanding the Lagrangian $\log\pi_y(\boldsymbol{\theta}) - \lambda_\eta \log g(\boldsymbol{\theta})$ in a Taylor series expansion, about $\boldsymbol{\theta} = \boldsymbol{\theta}_\eta$, for those $\boldsymbol{\theta}$ lying in the D region (5.1.11), and with ϵ sufficiently small. Neglecting cubic and higher terms in this expansion, together with terms that vanish as $\epsilon \to 0$, and taking exponentials give the second-order Taylor series approximation

$$\pi_y^*(\boldsymbol{\theta}) \propto \pi_y(\boldsymbol{\theta}_\eta)\exp\big\{-\tfrac{1}{2}(\boldsymbol{\theta}-\boldsymbol{\theta}_\eta)^T\tilde{\mathbf{R}}_\eta(\boldsymbol{\theta}-\boldsymbol{\theta}_\eta)\big\} \qquad (\boldsymbol{\theta} \in D). \tag{5.1.15}$$

Marginalizing this local approximation to the posterior density of $\boldsymbol{\theta}$, over the D region, then gives the required result in (5.1.13). The simpler Laplacian approximation in (5.1.5) can be justified by a second-order Taylor series approximation to $\log\pi_y(\eta, \boldsymbol{\xi})$; an alternative derivation is described in Section 6.2. A variety of extensions of the general approximation (5.1.13) are described by Leonard, Hsu, and Ritter (1994), Hsu (1990, 1995), Wong and Li (1992), and Sun et al. (1996). These give increased accuracy in various situations, for example, in the tails of the approximation. An approximation proposed by Leonard, Hsu, and Tsui (1989) can also be applied when $\boldsymbol{\theta}_\eta$ is replaced by another sensible representative $\boldsymbol{\theta}_\eta^*$ of the D region. For example, when $\eta = \mathbf{a}^T\boldsymbol{\theta}$ is a linear function of $\boldsymbol{\theta}$, Leonard et al. modify the approximation in (5.1.7) by

$$\pi_y^*(\eta) \propto \pi_y\big(\boldsymbol{\theta}_\eta^*\big)\exp\big\{\mathbf{l}_\eta^T\mathbf{G}_\eta\mathbf{l}_\eta\big\}|\mathbf{R}_\eta|^{-\frac{1}{2}}\omega_\eta^{-\frac{1}{2}},$$

where $\mathbf{R}_\eta$ is evaluated, from (5.1.8), by setting $\boldsymbol{\theta} = \boldsymbol{\theta}^*_\eta$,

$$\mathbf{l}_\eta = \left[\frac{\partial \log \pi_y(\boldsymbol{\theta})}{\partial \boldsymbol{\theta}}\right]_{\boldsymbol{\theta}=\boldsymbol{\theta}^*_\eta},$$

$$\mathbf{G}_\eta = \mathbf{R}_\eta^{-1} - \omega_\eta^{-1}\mathbf{R}_\eta^{-1}\mathbf{a}\mathbf{a}^T\mathbf{R}_\eta^{-1},$$

and

$$\omega_\eta = \mathbf{a}^T\mathbf{R}_\eta^{-1}\mathbf{a}.$$

Leonard, Hsu, and Ritter (1994), Raftery (1996), and Sun et al. (1996) apply related results when $\boldsymbol{\theta}^*_\eta$ is expressible in either algebraic form or in terms of output from standard computer packages. In such cases, the Laplacian approximation is remarkably simple to calculate. Similar techniques could, for example, be used to replace our methods in Sections 6.2 and 6.3 by an approximate approach that is algebraically explicit in terms of the maximum likelihood vector.

(D) *Predictive Distributions*

Suppose that a vector $\mathbf{z}$ of future observations has sampling density or probability mass function $p(\mathbf{z} \mid \boldsymbol{\theta}, \mathbf{y})$, given $\boldsymbol{\theta} \in \Theta = R^p$, and $\mathbf{y} \in S$, for realization of $\mathbf{z}$ in S^*. Then the "predictive distribution" of $\mathbf{z}$, given $\mathbf{y}$, possesses the density or probability mass function

$$\begin{aligned} p(\mathbf{z} \mid \mathbf{y}) &= \int_\Theta p(\mathbf{z}, \boldsymbol{\theta} \mid \mathbf{y})\, d\boldsymbol{\theta} \\ &= \int_\Theta p(\mathbf{z} \mid \boldsymbol{\theta}, \mathbf{y})\pi_y(\boldsymbol{\theta})\, d\boldsymbol{\theta} \qquad (\mathbf{z} \in S^*), \end{aligned} \tag{5.1.16}$$

where the integrations should be taken over $\boldsymbol{\theta} \in \Theta = R^p$. This distribution possesses mean vector

$$E(\mathbf{z} \mid \mathbf{y}) = \underset{\boldsymbol{\theta}|\mathbf{y}}{E}\left[E(\mathbf{z} \mid \boldsymbol{\theta}, \mathbf{y})\right] \tag{5.1.17}$$

and covariance matrix

$$\mathbf{cov}(\mathbf{z} \mid \mathbf{y}) = \underset{\boldsymbol{\theta}|\mathbf{y}}{E}\left[\mathbf{cov}(\mathbf{z} \mid \boldsymbol{\theta}, \mathbf{y})\right] + \underset{\boldsymbol{\theta}|\mathbf{y}}{\mathbf{cov}}\left[E(\mathbf{z} \mid \boldsymbol{\theta}, \mathbf{y})\right].$$

If the integration in (5.1.16) cannot be completed easily, then you may refer to the Laplacian approximation (Leonard, 1982a)

$$p^*(\mathbf{z} \mid \mathbf{y}) \propto \sup_{\boldsymbol{\theta}|\mathbf{z},\mathbf{y}} p(\mathbf{z}, \boldsymbol{\theta} \mid \mathbf{y})\left|\mathbf{Q}(\mathbf{y}, \mathbf{z})\right|^{-\frac{1}{2}} \qquad (\mathbf{z} \in S^*), \tag{5.1.18}$$

where $\mathbf{Q}(\mathbf{y}, \mathbf{z})$ evaluates $-\partial^2 \log p(\mathbf{z}, \boldsymbol{\theta} \mid \mathbf{y})/\partial(\boldsymbol{\theta}\boldsymbol{\theta}^T)$ at the conditional maximum with respect to $\boldsymbol{\theta}$, given $\mathbf{z}$ and $\mathbf{y}$ of $p(\mathbf{z}, \boldsymbol{\theta} \mid \mathbf{y})$. When $\pi(\boldsymbol{\theta}) \propto 1$, the approximation in (5.1.18) reduces to the modified predictive likelihood (1.5.29). Note, from

Exercises 1.5.k and 3.6.g, that the approximation (5.1.18) can give almost perfect answers. If necessary, the constant of proportionality can be roughly approximated by $(2\pi)^{\frac{1}{2}p}$. Approximations to predictive distributions of nonlinear functions of $\mathbf{z}$ are described by Leonard, Tsui, and Hsu (1990). For more detailed studies of predictive distributions, see Aitchison and Dunsmore (1975), Lindley and Novick (1981), and Geisser (1993).

(E) *Exact Computations*

Rather than using Laplacian approximations, the marginal posterior densities in (5.1.3) and (5.1.10) and the predictive distribution (5.1.16) can frequently be computed exactly by Monte Carlo simulations (see Section 6.4) or importance sampling (e.g., Rubinstein, 1981; Kloek and van Dijk, 1978; Zellner and Rossi, 1984; Geweke 1988, 1989; and Section 6.4). Therefore, from the 1980s onwards, Bayesians have been well able to handle the computations needed for drawing precise marginal inferences. Gelfand and Smith (1990), Geyer (1993), Geyer and Thompson (1992), and Besag and Green (1993) describe more complicated MCMC simulation procedures, for example, involving the Gibbs sampler and the Metropolis algorithm (Metropolis et al., 1953), which can be used to solve quite complex problems. Cowles and Carlin (1996) provide a detailed literature review and also discuss convergence of MCMC procedures.

Note, however, that importance sampling can handle a very broad range of models in quite efficient fashion. The Laplacian approximations discussed in the last paragraph of Section 5.1 (C) can involve negligible computer time. The Gibbs sampler, a version of MCMC described in Sections 6.6 and 6.7, can be reserved for high-level problems that cannot be handled using other techniques. It can, however, converge remarkably quickly in some situations, for example, the linear model with random effects.

(F) *An Example from Genetics*

Following Sen and Singer (1993, p. 267), consider the following table relating to the Hardy–Weinberg equilibrium model for the OAB blood group model:

Blood group	Observed frequency	Probability
O	n_O	$\theta_O = \phi_O^2$
A	n_A	$\theta_1 = \phi_A^2 + 2\phi_O\phi_A$
B	n_B	$\theta_2 = \phi_B^2 + 2\phi_O\phi_B$
AB	n_{AB}	$\theta_3 = 2\phi_A\theta_B$
TOTAL	n	1

If the n observations are based upon a random sample from a much larger population, then a multinomial sampling model is appropriate, in which case the likelihood of the

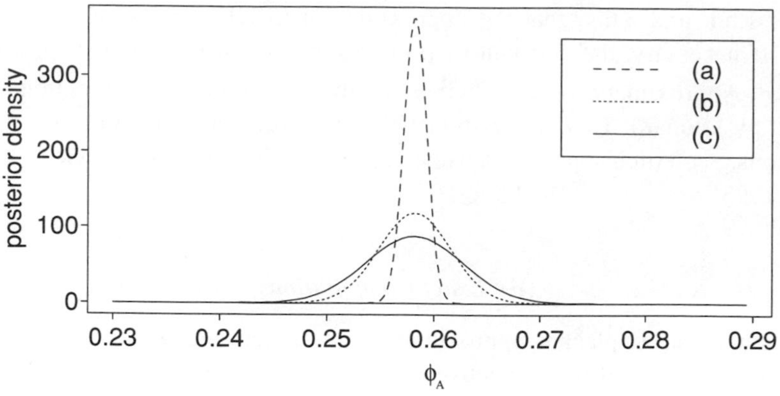

Figure 5.1.1. Posterior densities for ϕ_A: (a) $n = 100{,}000$; (b) $n = 10{,}000$; (c) $n = 5{,}500$.

probabilities ϕ_O, ϕ_A, and ϕ_B is

$$l(\phi_O, \phi_A, \phi_B \mid n_o, n_A, n_B, n_{AB}) \propto \phi_O^{2n_O} \phi_A^{n_{AB}} \phi_B^{n_{AB}} (\phi_A^2 + 2\phi_O\phi_A)^{n_A} (\phi_B^2 + 2\phi_O\phi_B)^{n_B}, \qquad (5.1.19)$$

where $\phi_O + \phi_A + \phi_B = 1$.

Sen and Singer note that iterations are needed to calculate the maximum likelihood estimates. We instead assume the vague Dirichlet prior

$$\pi(\phi_O, \phi_A, \phi_B \mid \mathbf{y}) \propto \phi_O^{-\frac{1}{2}} \phi_A^{-\frac{1}{2}} \phi_B^{-\frac{1}{2}}.$$

Then the entire posterior density of, say, ϕ_A can be calculated via the one-dimensional numerical integration

$$\pi(\phi_A \mid n_o, n_A, n_B, n_{AB}) = \int_0^{1-\phi_n} \pi(\phi_A, \phi_B \mid n_o, n_A, n_B, n_{AB})\, d\phi_B$$

$$\propto \int_0^{1-\phi_A} \phi_0^{2n_0-\frac{1}{2}} \phi_A^{n_{AB}-\frac{1}{2}} \phi_B^{n_{AB}-\frac{1}{2}} (\phi_A^2 + 2\phi_O\phi_A)^{n_A} (\phi_B^2 + 2\phi_O\phi_B)^{n_B} d\phi_B,$$

where $\phi_O = 1 - \phi_A - \phi_B$. This permits precise finite sample inferences regarding ϕ_A. In Figure 5.1.1, we report the marginal posterior density of ϕ_A when the sample proportions are 0.45, 0.41, 0.10, and 0.04, for the O, A, B, and AB blood groups. The three curves (a), (b), and (c) correspond to the sample sizes $n = 100{,}000$, $n = 10{,}000$, and $n = 5{,}500$. The posterior density of ϕ_B is similarly reported in Figure 5.1.2.

(G) *Bayesian Analysis for Failure or Survival Times*

You observe n independent failure times $y_1, \ldots, y_n$ of a machine, or survival times of patients, and take $y_1, \ldots, y_n$ to be a random sample from the Weibull distribution with

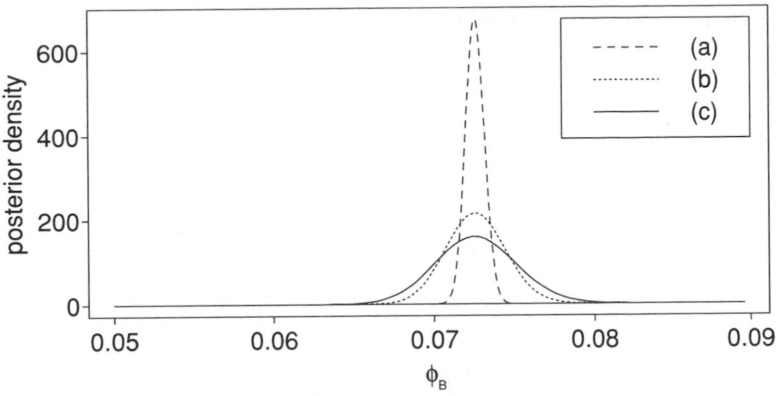

Figure 5.1.2. Posterior densities for ϕ_B: (a) n = 100,000; (b) n = 10,000; (c) n = 5,500.

parameters γ and λ. This distribution possesses c.d.f.

$$F(y) = 1 - \exp\{-\lambda y^{\gamma}\} \qquad (0 < y < \infty, 0 < \lambda, \gamma < \infty) \tag{5.1.20}$$

and can be obtained via a power transformation from the exponential distribution. You can judge how well it fits the data by comparing your fitted c.d.f. with the empirical c.d.f. $F_n(y)$. It is likely to represent the data well whenever $\log(1 - F(y))$ is approximately linear in $\log y$ for all values of y somewhat greater than zero. Otherwise a mixture of Weibull distributions (e.g., Leonard et al., 1994b) might give a better fit. Since the sampling density, under our Weibull assumption, is

$$f(y \mid \lambda, \gamma) = \lambda\gamma y^{\gamma-1}\exp\{-\lambda y^{\gamma}\} \qquad (0 < y < \infty), \tag{5.1.21}$$

the likelihood of λ and γ, given $y_1, \ldots, y_n$, is

$$l(\lambda, \alpha \mid \mathbf{y}) \propto \lambda^n\gamma^n \exp\left\{\gamma t - \lambda\sum_{i=1}^{n} y_i^{\gamma}\right\} I[0 < \lambda, \gamma < \infty] \qquad (-\infty < \lambda, \gamma < \infty), \tag{5.1.22}$$

where $t = \sum_{i=1}^{n} \log y_i$. The preceding general theory can be applied even though the likelihood function is not differentiable, when either $\lambda = 0$ or $\gamma = 0$.

When choosing a prior distribution, algebraically equivalent results may be obtained by assuming a Gamma distribution for either λ or for $\lambda\gamma$. In the current section, we investigate the consequences of taking λ and γ to be independent in the prior assessment. Take γ to possess an arbitrary density $\pi(\gamma)$ (e.g., a Gamma density) that is twice differentiable for $\gamma \in (0, \infty)$. Take λ to possess a Gamma distribution with parameters α and β, mean $\lambda_0 = \alpha/\beta$, and variance $\alpha/\beta^2 = \lambda_0/\beta$. Then the posterior density of λ and γ is

$$\pi(\lambda, \gamma \mid \mathbf{y}) \propto \pi(\gamma)\lambda^{n+\gamma-1}\gamma^n \exp\left\{-\gamma t - \lambda\beta - \lambda\sum_{i=1}^{n} y_i^{\gamma}\right\} \qquad (0 < \lambda, \gamma < \infty). \tag{5.1.23}$$

The marginal posterior density of γ is

$$\begin{aligned}\pi(\gamma \mid \mathbf{y}) &= \int_0^\infty \pi(\lambda, \gamma \mid \mathbf{y})\, d\lambda \\ &\propto \pi(\gamma)\gamma^n \exp\{-\gamma t\} \int_0^\infty \lambda^{n+\alpha-1} \exp\left\{-\lambda\left(\beta + \sum_{i=1}^n y_i^\gamma\right)\right\} d\lambda,\end{aligned} \tag{5.1.24}$$

where $0 < \gamma < \infty$, which by the integral result (3.2.24) reduces to

$$\pi(\gamma \mid \mathbf{y}) \propto \pi(\gamma)\gamma^n \exp\{-\gamma t\} \left(\beta + \sum_{i=1}^n y_i^\gamma\right)^{-(n+\alpha)} \qquad (0 < \gamma < \infty). \tag{5.1.25}$$

The marginal posterior density of λ cannot be obtained analytically, but may be arranged in the form

$$\pi(\lambda \mid \mathbf{y}) = \int_0^\infty \pi(\lambda \mid \gamma, \mathbf{y})\, \pi(\gamma \mid \mathbf{y})\, d\gamma \qquad (0 < \lambda < \infty), \tag{5.1.26}$$

where the first contribution to the integrand is the conditional density of λ, given γ, that is, a Gamma density with parameters $\alpha + n$ and $\beta + \sum y_i^\gamma$. The proportionality constant in (5.1.25) can be computed via a one-dimensional numerical integration; then the density in (5.1.26) can be evaluated by further one-dimensional numerical integrations. Note further that the posterior mean of λ can be computed directly from

$$\lambda_0^* = E(\lambda \mid \mathbf{y}) = \int_0^\infty E(\lambda \mid \gamma, \mathbf{y})\, \pi(\gamma \mid \mathbf{y})\, d\gamma, \tag{5.1.27}$$

and the posterior variance of λ from

$$\begin{aligned}\mathrm{var}(\lambda \mid \mathbf{y}) &= \int_0^\infty \left[E(\lambda \mid \gamma, \mathbf{y})\right]^2 \pi(\gamma \mid \mathbf{y})\, d\gamma - \left(\lambda_0^*\right)^2 \\ &\quad + \int_0^\infty \mathrm{var}(\lambda \mid \gamma, \mathbf{y})\, \pi(\alpha \mid \mathbf{y})\, d\gamma,\end{aligned} \tag{5.1.28}$$

where

$$E(\lambda \mid \gamma, \mathbf{y}) = (\alpha + n) \Big/ \left(\beta + \sum_{i=1}^n y_i^\gamma\right)$$

and

$$\mathrm{var}(\lambda \mid \gamma, \mathbf{y}) = E(\lambda \mid \gamma, \mathbf{y}) \Big/ \left(\beta + \sum_{i=1}^n y_i^\gamma\right).$$

However, a key parameter of interest is the survival or failure probability

$$\eta = 1 - F(u) = \exp\left\{ -\lambda u^\gamma \right\}, \tag{5.1.29}$$

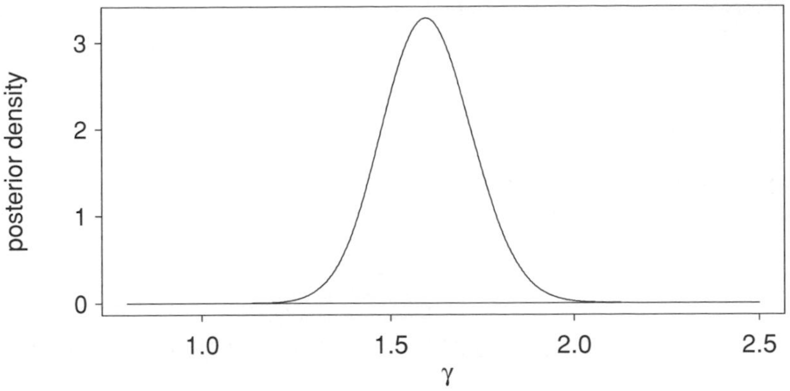

Figure 5.1.3. Posterior density of power parameter γ.

for any fixed $\gamma \in (0, \infty)$. The posterior expectation of η, given γ, is

$$E(\eta \mid \gamma, \mathbf{y}) = \left(\frac{\beta + \sum_{i=1}^{n} y_i^{\gamma}}{\beta + \sum_{i=1}^{n} y_i^{\gamma} + u^{\gamma}} \right)^{\alpha+n}. \tag{5.1.30}$$

Therefore, the posterior mean of η can be obtained by a one-dimensional numerical integration of (5.1.30) with respect to the distribution of γ, given $\mathbf{y}$, with density in (5.1.25). The resultant curve can be plotted as a function of u. This curve gives both the posterior mean value function of the survivor function $1 - F(u)$ and the predictive c.d.f. of the survival or failure time for a further randomly chosen patient or machine. Furthermore, given γ, the survival or failure probability (5.1.29) itself possesses c.d.f.

$$p(\eta \le \eta^* \mid \gamma, \mathbf{y}) = p(\lambda \ge u^{-\gamma}\kappa^* \mid \gamma, \mathbf{y}) \qquad (0 < \eta^* < 1), \tag{5.1.31}$$

where $\kappa^* = -\log \eta^*$.

By an application of (3.2.25), the conditional probability (5.1.31) becomes

$$p(\eta \le \eta^* \mid \gamma, \mathbf{y}) = 1 - G_{\alpha+n}\left[\left(\beta + \sum_{i=1}^{n} y_i^{\gamma} \right) u^{-\gamma}\kappa^* \right]. \tag{5.1.32}$$

The incomplete Gamma function $G_\alpha(a)$ in (3.2.26) can be calculated as a standard function on most computer systems. The unconditional posterior c.d.f. of η should be calculated, for each u, by averaging (5.1.32) numerically with respect to the posterior density (5.1.25) of γ. This permits the exact computation of 95% Bayesian intervals for the failure probability η in (5.1.29), for each value of u. The prior density of γ can be taken to be Gamma with parameters κ_0 and κ_1. The choices $\gamma = \frac{1}{2}$, $\beta = 0$, $\kappa_0 = \frac{1}{2}$, and $\kappa_1 = 0$ are then appealing in vague prior situations.

The preceding procedures were straightforward to apply to the yarn data previously analyzed in Section 2.4 (A). Under the prior assumptions of the previous paragraph for our Weibull parameters γ and λ, the posterior densities of γ and λ are described in Figures 5.1.3 and 5.1.4. Figure 5.1.5 describes the posterior mean value function of

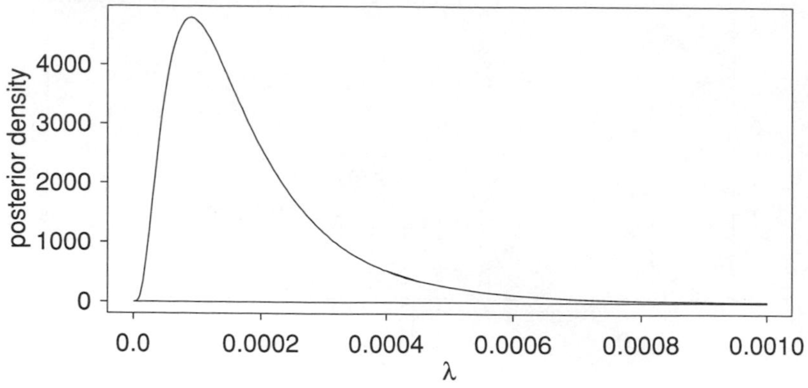

Figure 5.1.4. Posterior density of scale parameter λ.

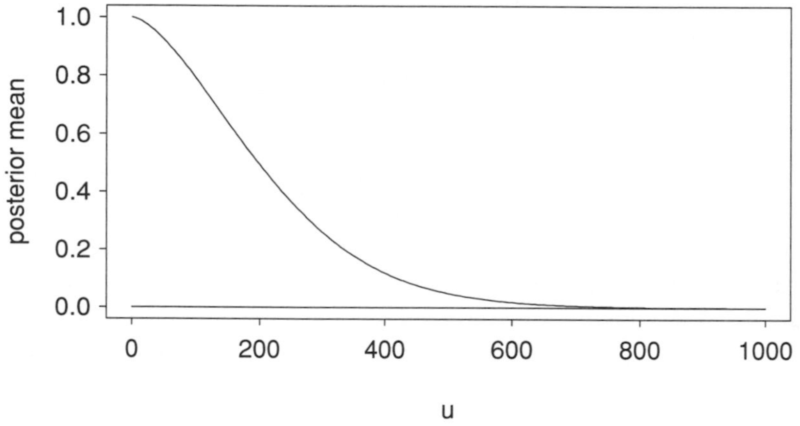

Figure 5.1.5. Posterior mean function of survivor function.

the survivor function. This figure can be used to predict the probabilities of failure for further samples.

(H) *The Linear Logistic Model*

Consider the ingot data of Section 1.4 (B). The dotted curves in Figures 5.1.6 and 5.1.7 describe the Laplacian approximation (5.1.5) when applied to the posterior densities of the parameters β_1 and β_2 in the linear logistic model (1.4.1). These are virtually identical to the exact posterior densities as simulated via importance sampling and using the techniques of Section 6.4. The Laplacian approximations are also identical, to three-decimal-place accuracy, to the exact posterior densities obtained by numerical integration. Note that Figure 5.1.7 demonstrates that β_2 is almost certainly greater than zero. Zellner and Rossi (1984) pioneered similar techniques for more general forms of the logistic model, with several economic applications, including pricing where the explanatory variables denoted prices and the dependent variables denoted numbers of

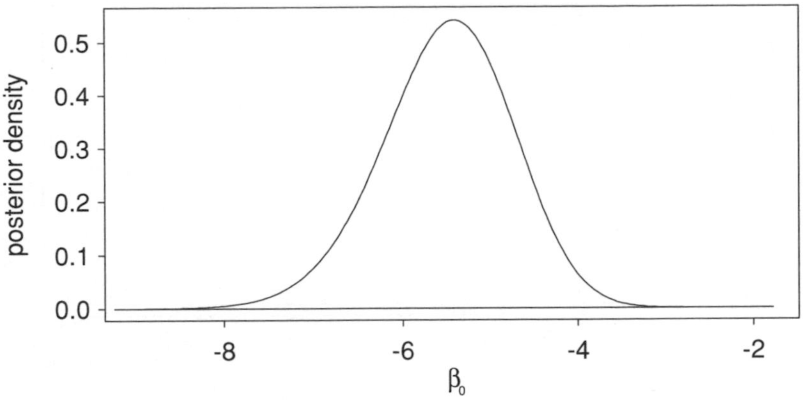

Figure 5.1.6. Posterior density of β_1.

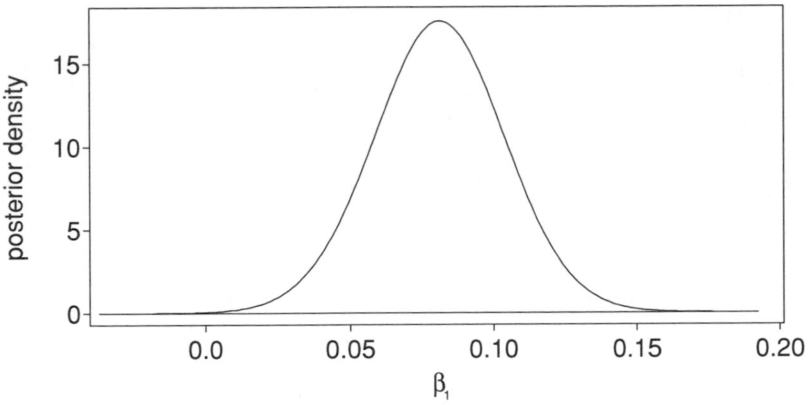

Figure 5.1.7. Posterior density of β_2.

sales. However, Kloek and van Dijk (1978) introduced importance sampling into the Bayesian literature.

It is also useful to apply the approximation (5.1.18) to the predictive probabilities of a further binomial frequency y with sample size n, the same unknown β_1 and β_2, and value x for the explanatory variable. The approximations in Figures 5.1.8–5.1.11 are again virtually identical to the exact results. They describe the predictive distribution for y, under the choices (7, 55), (14, 157), (27, 159), and (51, 16) for (x, n). These predictive distributions can be used to perform an analysis of residuals, which is rather more precise than standard asymptotic analyses. For the ingot data (Table 1.4.3), the observed y's $(y = 0, 2, 7, 3)$ should be contrasted with their corresponding predictive distributions in Figures 5.1.8–5.1.11. For example, the high predictive probability that $y = 0$, when $x = 7$ and $n = 55$, is not refuted by the observed value. Similarly, $y = 2$ lies close to the median of the distribution in Figure 5.1.9, and $y = 7$ and $y = 3$ do not lie in the tails of the respective predictive distributions in

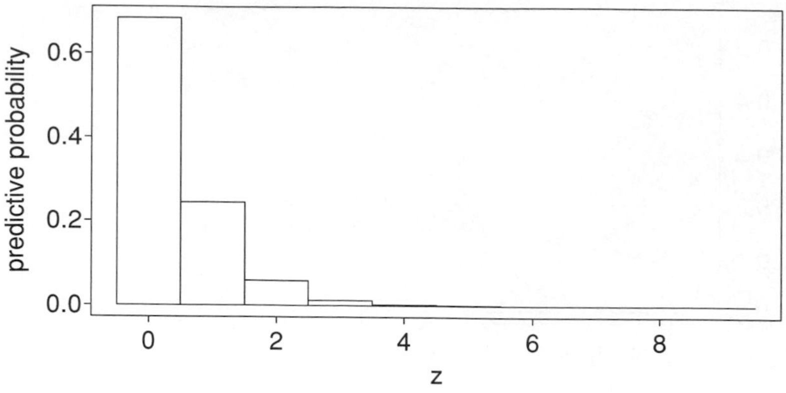

Figure 5.1.8. Predictive distribution of y ($x = 7, n = 55$).

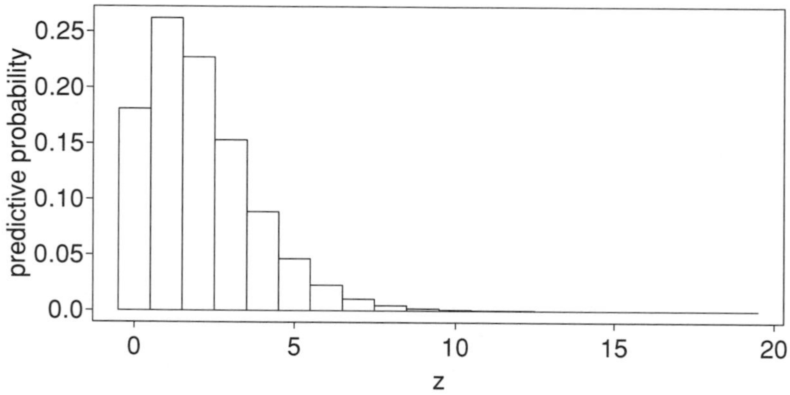

Figure 5.1.9. Predictive distribution of y ($x = 14, n = 157$).

Figures 5.1.10 and 5.1.11. Therefore, the ingot data do not refute the linear logistic assumption (1.4.1).

Leonard (1977b) discusses an informative Bayesian analysis for the linear logistic model, in the context of the Bradley–Terry model for paired comparisons. A biological application, including an analysis of the squirrel monkey data, is described.

(I) *A Discrete Exponential Family Model With Applications to Multiple-Choice Testing*

Consider a discrete random variable y assigning respective probabilities $\phi_0, \phi_1, \ldots, \phi_m$ to the integers $0, 1, \ldots, m$. Consider logits $\gamma_0, \gamma_1, \ldots, \gamma_m$ satisfying

$$\phi_j = \frac{e^{\gamma_j}}{\sum_{g=0}^{m} e^{\gamma_g}} \qquad (j = 0, 1, \ldots, m), \tag{5.1.33}$$

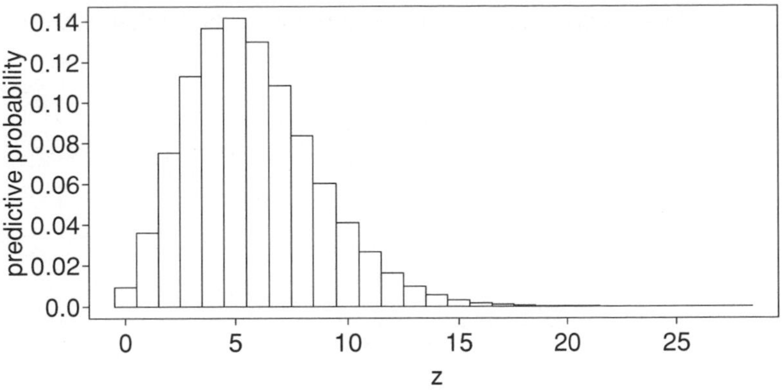

Figure 5.1.10. Predictive distribution of y ($x = 27, n = 159$).

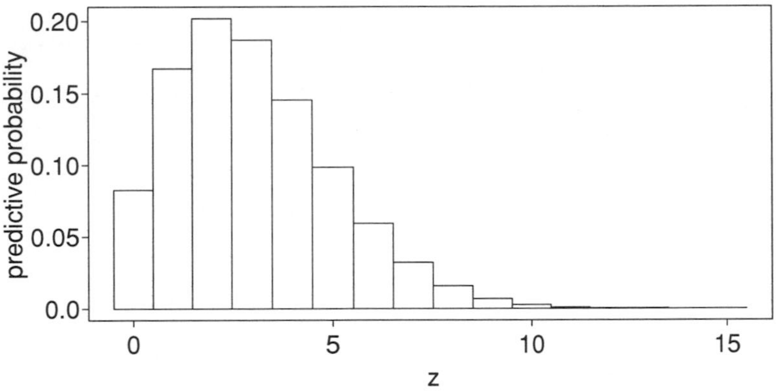

Figure 5.1.11. Predictive distribution of y ($x = 51, n = 16$).

together with the p-parameter exponential family model where

$$\gamma_j = \log {}^m C_j + \theta_1 b_1(j) + \theta_2 b_2(j) + \cdots + \theta_p b_p(j), \tag{5.1.34}$$

with specified basis functions $b_1, b_2, \ldots, b_p$, $p < m$, and ${}^m C_j = m!/j!(m-j)!$. When $p = 1$ and $b_1(j) = j$, this provides a binomial distribution with sample size m and probability $e^{\theta_1}/(1 + e^{\theta_1})$. However, the choices

$$b_k(j) = \left(\frac{j}{m}\right)^k \qquad (k = 1, \ldots, p) \tag{5.1.35}$$

lead to convenient generalizations of the binomial distribution. Such distributions may be appropriate for the number of items y out of m correctly answered by a candidate responding nonindependently to m dissimilar items on a test. See Hsu, Leonard, and Tsui (1991).

Given independent results $y_1, \ldots, y_n$ for n examinees, the likelihood of $\theta_1, \ldots, \theta_p$ is

$$l(\theta_1, \ldots, \theta_p \mid \mathbf{y}) \propto \exp\left\{\sum_{j=1}^{m} \theta_j t_j - n\Lambda(\boldsymbol{\theta})\right\}, \qquad (5.1.36)$$

where

$$t_k = \sum_{j=1}^{m} n_j \left(\frac{j}{m}\right)^k \qquad (k = 1, \ldots, p) \qquad (5.1.37)$$

and

$$\Lambda(\boldsymbol{\theta}) = \log \sum_{j=1}^{m} {}^m C_j \exp\left\{m^{-1}\theta_1 j + m^{-2}\theta_2 j^2 + \cdots + m^{-p}\theta_p j^p\right\}. \qquad (5.1.38)$$

Parameters of interest include

(a) the average true score

$$\begin{aligned} \eta_1 &= \sum_{j=1}^{m} j\phi_j \\ &= \sum_{j=1}^{m} j\, {}^m C_j \exp\left\{m^{-1}\theta_1 j + \cdots + m^{-p}\theta_p j^p - \Lambda(\boldsymbol{\theta})\right\} \end{aligned} \qquad (5.1.39)$$

and

(b) the probability of passing

$$\begin{aligned} \eta_2 &= p(y \geq a) \\ &= \sum_{j=a}^{m} \phi_j \\ &= \sum_{j=a}^{m} {}^m C_j \exp\left\{m^{-1}\theta_1 j + \cdots + m^{-p}\theta_p j^p - \Lambda(\boldsymbol{\theta})\right\}, \end{aligned} \qquad (5.1.40)$$

which comprise quite complicated functions of the parameters $\theta_1, \ldots, \theta_p$.

Consider the Duncan data (Duncan, 1974, p. 55; Morrison and Brockway, 1979, p. 439) and confine attention to their second sample ($n = 150, m = 20$). Under our preceding sampling assumptions, both AIC and BIC, of Section 1.1, were minimized when $p = 2$. The maximum likelihood estimates of θ_1 and θ_2 were $\hat{\theta}_1 = -18.90$ and $\hat{\theta}_2 = 23.77$, with respective estimated standard errors 3.22 and 2.64.

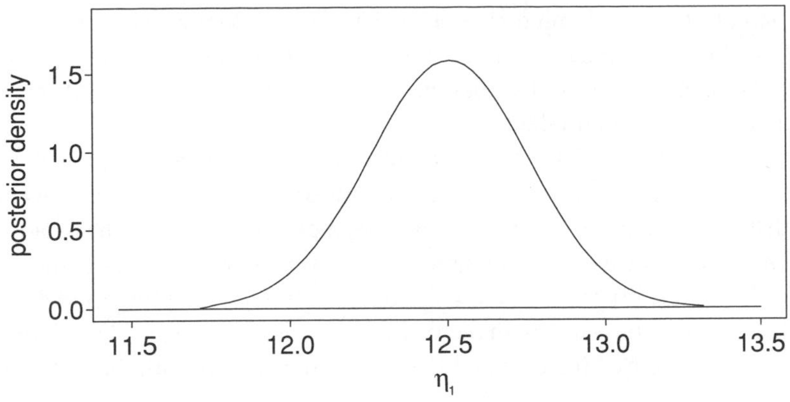

Figure 5.1.12. Posterior density of average true score: Laplacian approximation and exact result.

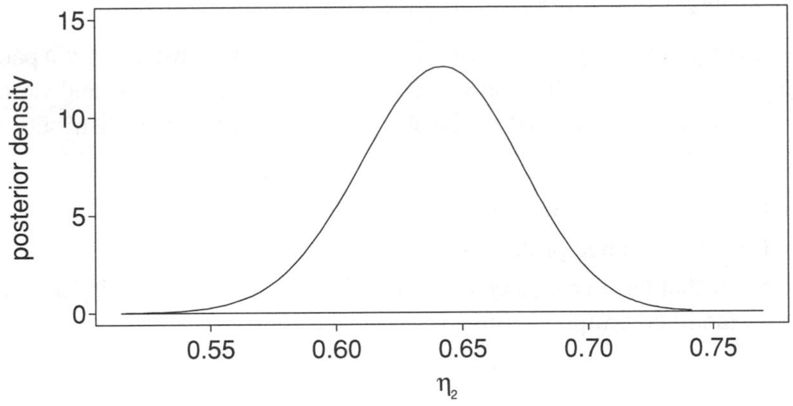

Figure 5.1.13. Posterior density of the probability of passing: Laplacian approximation and exact result.

The Laplacian approximation (5.1.13) was used, for the posterior distributions of η_1 and η_2, under improperly uniform priors for θ_1 and θ_2. In both cases, the "f-contribution" was further approximated by a log-normal distribution with correct first two moments. The curve in Figure 5.1.12 denotes our Laplacian approximation to the posterior density of the average true score η_1. This is virtually identical to exact density obtained by importance sampling. In Figure 5.1.13, we describe the posterior density for the probability η_2 of passing. Hsu, Leonard, and Tsui also report the posterior distribution of a "regression of true score on observed score."

The preceding sampling assumptions can be compared with other generalizations of the binomial distribution via computer simulations, as reported by Hsu et al. (1991, p. 332). For example, $M = 500$ simulations were performed, each generating $n = 200$ observations from a true distribution, which comprised a mixture of a binomial distribution with probability 0.3 and sample size $m = 10$ and a binomial distribution with probability 0.6 and sample size $m = 10$, with respective mixing probabilities 0.4 and 0.6. On each simulation, a beta-binomial distribution was compared with the preceding exponential family distribution, with $p = 4$ and $m = 10$. All parameters were estimated by maximum likelihood.

Our current sampling model yielded a much lower average simulated value for the normalized chi-squared statistic $\overline{\chi}_6^2 = \chi_6^2/6$, with six degrees of freedom, when compared with the statistic $\overline{\chi}_8^2 = \chi_8^2/8$, with eight degrees of freedom, for the beta-binomial distribution, on each of the 500 simulations. The average $\overline{\chi}_6^2 = 8.604$ compared with an average $\overline{\chi}_8^2 = 41.501$ for the beta-binomial model. Our sampling model does also perform better when compared with the various modifications to the beta-binomial distribution summarized by Morrison and Brockway. It contrasts with the overdispersion models introduced by Williams (1975, 1982).

Worked Example 5A: *Marginalization for Linear Regression*

Observations y_i are independent, for $i = 1, 2, \ldots, n$, given two unknown parameters β and σ^2, and normally distributed with respective means βx_i and variance σ^2, where the x_i are specified explanatory variables. The prior distribution of $(\beta, \log \sigma^2)$ is uniform over R^2.

(a) Find the marginal posterior density of σ^2.
(b) Find the modified profile likelihood of σ^2.
(c) Show that the marginal posterior density of $\gamma = \log \sigma^2$ is proportional to its modified profile likelihood.

Model Answer 5A:

(a) The likelihood of β and σ^2 is

$$\begin{aligned} l(\beta, \sigma^2 \mid \mathbf{y}) &\propto (\sigma^2)^{-\frac{n}{2}} \exp\left\{-\frac{1}{2\sigma^2}\sum_{i=1}^{n}(y_i - \beta x_i)^2\right\} \\ &\propto (\sigma^2)^{-\frac{n}{2}} \exp\left\{-\frac{1}{2\sigma^2}S_R^2 - \frac{B}{2\sigma^2}(\beta - \hat{\beta})^2\right\}, \end{aligned}$$

where

$$\hat{\beta} = \sum_{i=1}^{n} x_i y_i \Big/ \sum_{i=1}^{n} x_i^2, \qquad B = \sum_{i=1}^{n} x_i^2,$$

and

$$S_R^2 = \sum_{i=1}^{n}(y_i - \hat{\beta}x_i)^2 = \sum_{i=1}^{n} y_i^2 - \left(\sum_{i=1}^{n} x_i y_i\right)^2 \Big/ \sum_{i=1}^{n} x_i^2.$$

Since $\pi(\beta, \sigma^2) \propto (\sigma^2)^{-1}$, the joint posterior density of β and σ^2 is

$$\pi\left(\beta, \sigma^2 \mid \mathbf{y}\right) \propto \left(\sigma^2\right)^{-\frac{1}{2}(n+2)} \exp\left\{-\frac{1}{2\sigma^2} S_R^2 - \frac{B}{2\sigma^2}\left(\beta - \hat{\beta}\right)^2\right\}.$$

Noting that

$$\int_{-\infty}^{\infty} \exp\left\{-\frac{B}{2\sigma^2}\left(\beta - \hat{\beta}\right)^2\right\} d\beta = \left(\frac{2\pi\sigma^2}{B}\right)^{\frac{1}{2}} \propto \sigma,$$

we find that

$$\pi\left(\sigma^2 \mid \mathbf{y}\right) = \int_{-\infty}^{\infty} \pi\left(\beta, \sigma^2 \mid \mathbf{y}\right) d\beta \propto \left(\sigma^2\right)^{-\frac{1}{2}(n+1)} \exp\left\{-\frac{1}{2\sigma^2} S_R^2\right\}$$
$$(0 < \sigma^2 < \infty),$$

so that S_R^2/σ^2 has a chi-squared distribution with $n-1$ degrees of freedom.

(b) The likelihood of β and σ^2 is maximized, for each fixed σ^2, when $\beta = \hat{\beta}$. Also,

$$-\frac{\partial^2 \log l\left(\beta, \sigma^2 \mid \mathbf{y}\right)}{\partial \beta^2} = \frac{B}{\sigma^2}.$$

Following the definition in (1.5.9), the modified profile likelihood of σ^2 is

$$l_p\left(\sigma^2 \mid \mathbf{y}\right) = \left[\frac{l\left(\beta, \sigma^2 \mid \mathbf{y}\right)}{\left(B/\sigma^2\right)^{-\frac{1}{2}}}\right]_{\beta=\hat{\beta}} \propto \left(\sigma^2\right)^{-\frac{1}{2}(n-1)} \exp\left\{-\frac{1}{2\sigma^2} S_R^2\right\}.$$

(c) If $\gamma = \log \sigma^2$, $\sigma^2 = e^\gamma$, so that

$$l_p(\gamma \mid \mathbf{y}) \propto \left[l_p\left(\sigma^2 \mid \mathbf{y}\right)\right]_{\sigma^2 = e^\gamma} \propto \exp\left\{-\tfrac{1}{2}(n-1)e^\gamma - \tfrac{1}{2}e^{-\gamma} S_R^2\right\}.$$

This is identical to the posterior density of γ obtained by the Jacobian transformation of the chi-squared density in part (a).

Worked Example 5B: *Prediction for Linear Regression*

Suppose, in Worked Example 5A, that σ^2 is instead known, and β has a uniform prior on R. We are about to observe a further independent observation y_{n+1}, which, given β, is normally distributed with mean βx_{n+1} and specified variance σ^2. Find the predictive mean and variance of y_{n+1}.

Model Answer 5B:

In this case, the posterior density of β is proportional to the likelihood

$$\pi(\beta \mid \mathbf{y}) \propto \left(\sigma^2\right)^{-\frac{1}{2}n} \exp\left\{-\frac{1}{2\sigma^2} S_R^2 - \frac{B}{2\sigma^2}\left(\beta - \hat{\beta}\right)^2\right\},$$

where $B = \sum_{i=1}^{n} x_i^2$. As a function of β, we have considerable cancellation, giving

$$\pi(\beta \mid \mathbf{y}) \propto \exp\left\{-\frac{B}{2\sigma^2}\left(\beta - \hat{\beta}\right)^2\right\}.$$

Consequently, the posterior distribution of β is $N(\hat{\beta}, B^{-1}\sigma^2)$. Since

$$E(y_{n+1} \mid \beta) = \beta x_{n+1},$$

the predictive mean of y_{n+1} is

$$E(y_{n+1} \mid \mathbf{y}_n) = \underset{\beta \mid \mathbf{y}_n}{E}\left[E(y_{n+1} \mid \beta)\right] = x_{n+1} E(\beta \mid \mathbf{y}_n) = \hat{\beta} x_{n+1},$$

where $\mathbf{y}_n = (y_1, y_2, \ldots, y_n)^T$. The predictive variance of y_{n+1} is

$$\begin{aligned}
\operatorname{var}(y_{n+1} \mid \mathbf{y}_n) &= \underset{\beta \mid \mathbf{y}_n}{\operatorname{var}}\left[E(y_{n+1} \mid \beta)\right] + \underset{\beta \mid \mathbf{y}_n}{E}\left[\operatorname{var}(y_{n+1} \mid \beta)\right] \\
&= x_{n+1}^2 \operatorname{var}(\beta \mid \mathbf{y}_n) + \sigma^2 \\
&= \left(B^{-1} x_{n+1}^2 + 1\right)\sigma^2.
\end{aligned}$$

This follows the procedure described in Section 5.1 (D).

Worked Example 5C: *Counterexample to Maximum Likelihood*

Observations y_{ij} are independent ($i = 1, 2, \ldots n$; $j = 1, 2$), given unknown parameters $\theta_1, \theta_2, \ldots, \theta_n$ and σ^2, and each y_{ij} is normally distributed with mean θ_i and variance σ^2. The θ_i and $\log \sigma^2$ are a priori independent, and each is uniformly distributed over R.

(a) Find the posterior density of σ^2 and the posterior mode of σ^2.

(b) Find the maximum likelihood estimate of σ^2.

(c) Investigate which of your two very different estimates you would prefer, by considering their limiting behavior, as $n \to \infty$.

Hint: If S^2 denotes the within-group sum of squares, then the sampling distribution of S^2/σ^2 is chi-squared with n degrees of freedom.

Model Answer 5C:

The likelihood of $\theta_1, \theta_2, \ldots, \theta_n$ and σ^2 is

$$l(\theta_1, \theta_2, \ldots, \theta_n, \sigma^2 \mid \mathbf{y}) \propto \left(\sigma^2\right)^{-n} \exp\left\{-\frac{1}{2\sigma^2}\sum_{i=1}^{n}\sum_{j=1}^{2}(y_{ij}-\theta_i)^2\right\}$$

$$\propto \left(\sigma^2\right)^{-n} \exp\left\{-\frac{1}{2\sigma^2}S^2 - \frac{1}{\sigma^2}\sum_{i=1}^{n}(\theta_i - \bar{y}_i)^2\right\},$$

where $\bar{y}_i = (y_{i1} + y_{i2})/2$ and S^2 is the residual sum of squares

$$S^2 = \sum_{i=1}^{n}\sum_{j=1}^{2}(y_{ij} - \bar{y}_i)^2 = \frac{1}{2}\sum_{i=1}^{n}(y_{i1} - y_{i2})^2.$$

From the standard theory for linear models, note that the sampling distribution of S^2/σ^2 is chi-squared with n degrees of freedom, so that

$$E\left(S^2 \mid \sigma^2\right) = n\sigma^2 \quad \text{and} \quad \mathrm{var}\left(S^2 \mid \sigma^2\right) = 2n\sigma^4.$$

(a) The joint posterior density of σ^2 and the θ_i is, since $\pi(\sigma^2) \propto (\sigma^2)^{-1}$,

$$\pi\left(\sigma^2, \boldsymbol{\theta} \mid \mathbf{y}\right) \propto \left(\sigma^2\right)^{-(n+1)} \exp\left\{-\frac{1}{2\sigma^2}S^2 - \frac{1}{\sigma^2}\sum_{i=1}^{n}(\theta_i - \bar{y}_i)^2\right\}.$$

Integrating out each θ_i has the effect of replacing θ_i by $\bar{y}_i$ and multiplying the whole expression by σ. Performing n such integrations, we find that

$$\pi\left(\sigma^2 \mid \mathbf{y}\right) \propto \left(\sigma^2\right)^{-\frac{1}{2}(n+2)} \exp\left\{-\frac{1}{2\sigma^2}S^2\right\} \qquad (0 < \sigma^2 < \infty).$$

This expression is maximized when $\sigma^2 = \tilde{\sigma}^2$, where

$$\tilde{\sigma}^2 = \frac{S^2}{n+2}.$$

(b) The preceding likelihood is maximized when $\theta_i = \bar{y}_i$ and $\sigma^2 = \hat{\sigma}^2$, where

$$\hat{\sigma}^2 = \frac{S^2}{2n}.$$

(c) Since $E(S^2 \mid \sigma^2) = n\sigma^2$ and $\mathrm{var}(S^2 \mid \sigma^2) = 2n\sigma^4$, our two estimators possess sampling mean and variance

$$E\left(\tilde{\sigma}^2\right) = \frac{n}{n+2}\sigma^2 \longrightarrow \sigma^2 \quad \text{as } n \to \infty,$$

$$\mathrm{var}\left(\tilde{\sigma}^2\right) = \frac{2n}{(n+2)^2}\sigma^4 \longrightarrow 0 \quad \text{as } n \to \infty,$$

$$E\left(\hat{\sigma}^2\right) = \frac{1}{2}\sigma^2,$$

$$\mathrm{var}\left(\hat{\sigma}^2\right) = \frac{1}{2n}\sigma^4 \longrightarrow 0 \quad \text{as } n \to \infty.$$

Consequently, $\tilde{\sigma}^2$ is said to be "consistent" for σ^2, while $\hat{\sigma}^2$ is consistent for $\frac{1}{2}\sigma^2$ and inconsistent for σ^2. Clearly, the posterior mode $\tilde{\sigma}^2$ is preferable to the maximum likelihood estimator $\hat{\sigma}^2$ of σ^2. Similar phenomena occur in many such situations where the number of unknown parameters in the model increases with the number of observations.

Worked Example 5D: *Marginal Inferences for the Negative Binomial Distribution*

Observations $m_1, m_2, \ldots, m_n$ are a random sample from the negative binomial distribution with p.m.f.

$$p(m \mid \theta, r) = {}^{m-1}C_{r-1}\theta^r(1-\theta)^{m-r} \qquad (m = r, r+1, \ldots).$$

In the prior assessment, the unknown parameters θ and r are independent, θ possesses a beta distribution with parameters α and β, and r possesses p.m.f. $q(r)$ for $r = 1, 2, \ldots$. Find the marginal posterior p.m.f. of r.

Model Answer 5D:

The likelihood of θ and r is

$$l(\theta, r \mid \mathbf{m}) = \left\{\prod_{i=1}^{n} {}^{m_i-1}C_{r-1}\right\} \theta^{nr}(1-\theta)^{n\bar{m}-nr}$$
$$(0 < \theta < 1, r = 1, 2, \ldots, m^*),$$

where $\mathbf{m} = (m_1, m_2, \ldots, m_n)^T$, $m^* = \min(m_1, m_2, \ldots, m_n)$, and $\bar{m} = (m_1 + m_2 + \cdots + m_n)/n$. Consequently, the posterior density/probability mass function of θ and r is

$$\pi(\theta, r \mid \mathbf{m}) \propto \pi(\theta)\, d(r)\theta^{nr}(1-\theta)^{n\bar{m}-nr} \qquad (0 < \theta < 1, r = 1, 2, \ldots, m^*),$$

where

$$d(r) = \frac{q(r)}{\{\Gamma(r)\}^n \prod_{i=1}^n \Gamma(m_i - r + 1)}$$

and

$$\pi(\theta) \propto \theta^{\alpha-1}(1-\theta)^{\beta-1}.$$

Therefore,

$$\pi(\theta, r \mid m) \propto d(r)\theta^{\alpha+nr-1}(1-\theta)^{\beta+n\bar{m}-nr-1}$$
$$(0 < \theta < 1, r = 1, 2, \ldots, m^*).$$

Since

$$\int_0^1 \theta^{\alpha-1}(1-\theta)^{\beta-1} d\theta = \frac{\Gamma(\alpha)\Gamma(\beta)}{\Gamma(\alpha+\beta)},$$

for any $\alpha, \beta > 0$, the marginal p.m.f. of m is

$$\begin{aligned} p(r \mid \mathbf{m}) &\propto d(r)\Gamma(\alpha + nr)\Gamma(\beta + n\bar{m} - nr) \\ &\propto q(r)\frac{\Gamma(\alpha + nr)\Gamma(\beta + n\bar{m} - nr)}{\{\Gamma(r)\}^n \prod_{i=1}^n \Gamma(m_i - r + 1)} \qquad (r = 1, 2, \ldots, m^*). \end{aligned}$$

Worked Example 5E: *Laplacian Approximations for the Negative Binomial Distribution*

In Worked Example 5D,

(a) find a Laplacian approximation to the marginal posterior p.m.f. of r, under the stated parametrization for θ.
(b) Also find this Laplacian approximation, but under the reparametrization to $\boldsymbol{\gamma}$ satisfying $\theta = e^\gamma/(1 + e^\gamma)$, so that $\gamma = \log\theta - \log(1 - \theta)$.
(c) Which of these approximations compares better with the exact result?

Model Answer 5E:

Note that

$$\pi(\theta, r \mid \mathbf{m}) \propto d(r)\theta^{\alpha^*-1}(1 - \theta)^{\beta^*-1} \qquad (0 < \theta < 1, r = 1, 2, \ldots, m^*)$$

and

$$p(r \mid \mathbf{m}) \propto d(r)\Gamma(\alpha^*)\Gamma(\beta^*) \qquad (r = 1, 2, \ldots, m^*),$$

where $\alpha^* = \alpha + nr$ and $\beta^* = \beta + n\bar{m} - nr$.

(a) $\pi(\theta, r \mid \mathbf{m})$ is maximized as a function of θ for fixed r, when $\theta = \theta^*$, where $\theta^* = (\alpha^* - 1)/(\alpha^* + \beta^* - 2)$. Also,

$$-\frac{\partial^2 \log \pi(\theta, r \mid \mathbf{m})}{\partial\theta^2} = (\alpha^* - 1)\frac{\partial^2 \log(1 - \theta)}{\partial\theta^2} + (\beta^* - 1)\frac{\partial^2 \log(1 - \theta)}{\partial\theta^2},$$

so that

$$-\frac{\partial^2 \log \pi(\theta, r \mid \mathbf{m})}{\partial\theta^2} = \frac{\alpha^* - 1}{\theta^2} + \frac{\beta^* - 1}{(1 - \theta)^2}.$$

Consequently,

$$-\frac{\partial^2 \log \pi(\theta, r \mid \mathbf{m})}{\partial\theta^2} = (\alpha^* + \beta^* - 2)^2\left(\frac{1}{\alpha^* - 1} + \frac{1}{\beta^* - 1}\right).$$

Consequently, and noting that α^* and β^* do not depend upon r, Laplace's approximation to $\pi(r \mid \mathbf{m})$ is

$$\begin{aligned} \pi_L(r \mid \mathbf{m}) &\propto \left(\frac{1}{\alpha^* - 1} + \frac{1}{\beta^* - 1}\right)^{-\frac{1}{2}} [\pi(\theta, r \mid \mathbf{m})]_{\theta=\theta^*} \\ &\propto d(r)\left(\frac{1}{\alpha^* - 1} + \frac{1}{\beta^* - 1}\right)^{-\frac{1}{2}} (\alpha^* - 1)^{\alpha^*-1}(\beta^* - 1)^{\beta^*-1}. \end{aligned}$$

(b) Let $\theta = e^{\gamma}/(1+e^{\gamma})$. Then

$$\frac{d\theta}{d\gamma} = \frac{e^{\gamma}}{(1+e^{\gamma})^2} = \theta(1-\theta).$$

Consequently,

$$\begin{aligned}\pi(\gamma, r) &\propto d(r)\left[\theta^{\alpha^*-1}(1-\theta)^{\beta^*-1}\frac{d\theta}{d\gamma}\right]_{\theta=\frac{e^{\gamma}}{1+e^{\gamma}}}\\ &\propto d(r)\left[\theta^{\alpha^*}(1-\theta)^{\beta^*}\right]_{\theta=\frac{e^{\gamma}}{1+e^{\gamma}}}\\ &\propto d(r)\frac{e^{\alpha^*\gamma}}{(1+e^{\gamma})^{\alpha^*+\beta^*}} \qquad (-\infty < \gamma < \infty, r = 1, 2, \ldots, m^*).\end{aligned}$$

This is maximized, for fixed r, when $\gamma = \gamma^*$, where $\gamma^* = \log(\alpha^*/\beta^*)$. Since

$$\begin{aligned}-\frac{\partial^2 \log \pi(\gamma, r \mid \mathbf{m})}{\partial \gamma^2} &= (\alpha^* + \beta^*)\frac{d^2}{d\gamma^2}\log(1+e^{\gamma})\\ &= \frac{(\alpha^*+\beta^*)e^{\gamma}}{(1+e^{\gamma})^2},\end{aligned}$$

we have

$$\left[-\frac{\partial^2 \log \pi(\gamma, r \mid \mathbf{m})}{\partial \gamma^2}\right]_{\gamma=\gamma^*} = \frac{\alpha^*\beta^*}{\alpha^*+\beta^*}.$$

Again, noting that $\alpha^* + \beta^*$ does not depend on r, our new Laplacian approximation is

$$\begin{aligned}\pi_L^*(r \mid \mathbf{m}) &\propto \frac{\left[\pi(\gamma, r \mid \mathbf{m})\right]_{\gamma=\gamma^*}}{(\alpha^*\beta^*)^{\frac{1}{2}}}\\ &\propto d(r)(\alpha^*\beta^*)^{-\frac{1}{2}}\left[\frac{\exp\{\alpha^*\gamma\}}{(1+e^{\gamma})^{\alpha^*+\beta^*}}\right]_{\gamma=\gamma^*}\\ &\propto d(r)(\alpha^*)^{\alpha^*-\frac{1}{2}}(\beta^*)^{\beta^*-\frac{1}{2}} \qquad (r = 1, 2, \ldots, m^*).\end{aligned}$$

(c) Compared with the exact result

$$\pi(r \mid \mathbf{m}) \propto d(r)\Gamma(\alpha^*)\Gamma(\beta^*),$$

the approximation in (b) will be very accurate, since Stirling's approximation

$$\Gamma(\alpha^*) \approx \frac{1}{\sqrt{2\pi}}(\alpha^*)^{\alpha^*-\frac{1}{2}}e^{-\alpha^*}$$

is known to be very accurate, even for small α^*, and gives approximately

$$\Gamma(\alpha^*)\Gamma(\beta^*) = \frac{1}{2\pi}(\alpha^*)^{\alpha^*-\frac{1}{2}}(\beta^*)^{\beta^*-\frac{1}{2}}e^{-(\alpha^*+\beta^*)}$$
$$\propto (\alpha^*)^{\alpha^*-\frac{1}{2}}(\beta^*)^{\beta^*-\frac{1}{2}},$$

as required. The approximation in (a) is less convincing, since the denominator

$$\left[\frac{1}{\alpha^*-1}+\frac{1}{\beta^*-1}\right]^{\frac{1}{2}}$$

does not factorize into components depending upon α^* and β^* alone.

SELF-STUDY EXERCISES

5.1.a *A Posterior Analysis for a Normal Mean and Variance*:

(a) Observations $y_1, \dots, y_n$ constitute a random sample from a normal distribution with unknown mean θ and unknown variance ϕ. By reference to equation (3.4.2), show that the likelihood of θ and ϕ is

$$l(\theta, \phi \mid \mathbf{y}) \propto \phi^{-\frac{1}{2}n} \exp\left\{-\frac{1}{2\phi}S^2 - \frac{n}{2\phi}(\theta-\bar{y})^2\right\},$$

where $\bar{y}$ is the sample mean and $S^2 = \sum(y_i - \bar{y})^2$. Find the profile likelihoods of θ and ϕ. Find the modified profile likelihoods of θ and $\alpha = \log\phi$, both under this parametrization.

(b) Under the prior $\pi(\theta, \phi) \propto \phi^{-1}$, that is $\pi(\theta, \alpha) \propto 1$, show that

$$\pi(\theta \mid \mathbf{y}) \propto \left[S^2 + n(\theta-\bar{y})^2\right]^{-\frac{1}{2}n} \qquad (-\infty < \theta < \infty),$$

that the posterior density of $n^{\frac{1}{2}}(\theta - \bar{y})/V^{\frac{1}{2}}$ is t with $n-1$ degrees of freedom, where $V = S^2/(n-1)$, that

$$\pi(\phi \mid \mathbf{y}) \propto \phi^{-\frac{1}{2}(n+1)} \exp\left\{-\frac{1}{2\phi}S^2\right\} \qquad (0 < \phi < \infty),$$

and that the posterior density of S^2/ϕ is chi-squared with $n-1$ degrees of freedom.

(c) Show that the posterior density of $\gamma = \log\phi$ is

$$\pi(\gamma \mid \mathbf{y}) \propto \exp\left\{-\tfrac{1}{2}(n-1)\gamma - \tfrac{1}{2}e^{-\gamma}S^2\right\} \qquad (-\infty < \gamma < \infty)$$

and that the posterior density of $\xi = \phi^{-1}$ is

$$\pi(\xi \mid \mathbf{y}) \propto \xi^{\frac{1}{2}(n-3)} \exp\left\{-\frac{1}{2\xi}S^2\right\} \qquad (0 < \xi < \infty).$$

Report the posterior means, modes, and medians of ϕ, γ, and ξ.

(d) Compare your marginal posterior densities with the profile likelihoods of Section 1.5 (D), and the Laplacian approximation (5.1.5), under the (θ, α) parametrization. How well does the Laplacian approximationindexSULaplacian approximation work in this case? Do you think that "modified profile likelihood" is reasonable under the Bayesian philosophy?

(e) Show that any $100(1-\epsilon)\%$ Bayesian interval for θ or ϕ possesses exactly the correct frequency coverage.

(f) Find the predictive density of a further independent $N(\theta, \phi)$ observation y_{m+1}. Use your result to describe the posterior mean value function of the common $N(\theta, \phi)$ sampling density.

5.1.b In Exercise 5.1.a, use (5.1.13) to develop a Laplacian approximation to the posterior density of $\eta = \theta - \gamma$.

5.1.c *A Conjugate Analysis for a Normal Mean and Variance:*

(a) In Exercise 5.1.a, assume instead that in the prior assessment, the distribution of θ, given ϕ, is $N(\mu, \xi^{-1}\phi)$, where μ is the prior mean and ξ is the prior sample size for θ. Assume, furthermore, that $\nu\lambda/\phi$ possesses a chi-squared distribution, with ν degrees of freedom, where λ^{-1} is the prior mean of ϕ^{-1} and ν is the prior sample size for ϕ. Show that the prior densities satisfy

$$\pi(\phi) \propto \phi^{-\frac{1}{2}(\nu+2)} \exp\left\{-\frac{\nu\lambda}{2\phi}\right\} \qquad (0 < \phi < \infty),$$

$$\pi(\theta, \phi) \propto \phi^{-\frac{1}{2}(\nu+3)} \exp\left\{-\left[\nu\lambda + \xi(\theta-\mu)^2\right]\right\}$$

$$(-\infty < \theta < \infty, 0 < \phi < \infty),$$

and

$$\pi(\theta) \propto \left[\nu\lambda + \xi(\theta-\mu)^2\right]^{-\frac{1}{2}(\nu+1)} \qquad (-\infty < \theta < \infty),$$

and show that $(\xi/\lambda)^{\frac{1}{2}}(\theta-\mu)$ possesses a t-distribution with ν degrees of freedom and that μ is the unconditional prior mean of θ.

Hint: Note the integral result, $\int_0^\infty \phi^{-\frac{1}{2}(\nu+2)} \exp\{-\beta/2\phi\}\, d\phi = \Gamma(\frac{1}{2}\nu)/\beta^{\frac{1}{2}\nu}$ for any positive ν and β.

Note: The prior assumptions of Exercise 1.1.a correspond to the unusual choices $\xi = 0$, $\nu = -1$, and $\lambda = 0$. These choices parallel the standard frequency approach.

(b) Show that the posterior density of θ and ϕ satisfies

$$\pi(\theta, \phi \mid \mathbf{y})$$
$$\propto \phi^{-\frac{1}{2}(\nu+n+3)} \exp\left\{-\frac{1}{2\phi}\left[\nu\lambda + S^2 + \xi(\theta-\mu)^2 + n(\theta-\bar{y})^2\right]\right\}$$
$$(-\infty < \theta < \infty, 0 < \phi < \infty).$$

By an application of Lemma 3.4.1, show that this assumes exactly the same form as the prior density $\pi(\theta, \phi)$, but with ν, ξ, μ, and λ respectively replaced by ν^*, ξ^*, μ^*, and λ^*, where $\nu^* = \nu + n$, $\xi^* = \xi + n$,

$$\mu^* = \frac{n\bar{y} + \xi\mu}{n + \xi},$$

and

$$\lambda^* = (\nu + n)^{-1}\left[\nu\lambda + S^2 + \left(\xi^{-1} + n^{-1}\right)^{-1}(\mu - \bar{y})^2\right].$$

Without further integration, show that μ^* is the unconditional posterior mean of θ and that $(\lambda^*)^{-1}$ is the unconditional posterior mean of ϕ^{-1}.

(c) Show that

$$\mu^* = \rho\bar{y} + (1 - \rho)\mu$$

and

$$\lambda^* = \zeta\left[\rho S^2 + (1 - \rho)S_\mu^2\right] + (1 - \zeta)\lambda,$$

where $\rho = n\mu/(n + \xi), \zeta = n/(n + \nu)$, and $S_\mu^2 = \sum(y_i - \mu)^2 = S^2 + n(\mu - \bar{y})^2$. Carefully interpret this result. Following Leonard and Hsu (1992), the "phenomenon of the expansion of the Bayes estimate of the variance" can occur, that is, since $S_\mu^2 \geq S^2$, it follows that $\rho S^2 + (1 - \rho)S_\mu^2$ can be much greater than S^2. Does this concern you?

(d) Stein (1964) shows that any estimate for ϕ that is a scale multiple of S^2 is inadmissible under squared error loss for ϕ, and that admissible estimators for ϕ should depend upon both S^2 and $\bar{y}$. Show that $S^2/(n + 1)$ dominates $V = S^2/(n - 1)$ under squared error loss for ϕ. Investigate conditions on ζ, ρ, μ, and λ, such that λ^* is both proper Bayes and admissible under

(A) squared error loss for $\xi = \phi^{-1}$,

(B) squared error loss for ϕ,

(C) the loss function (3.8.19) for ϕ,

(D) the loss function (3.8.20) for ϕ.

(e) Without further integration, show that the posterior distribution of $(\xi^*/\lambda^*)^{\frac{1}{2}}(\theta - \mu^*)$ is t with ν^* degrees of freedom. Hence describe a Bayesian significance probability for investigating $H_0 : \theta = \theta_0$.

(f) Without further integration, show that the posterior distribution of $\nu^*\lambda^*/\phi$ is chi-squared with ν^* degrees of freedom. Hence describe a Bayesian significance probability for investigating $H_0 : \phi = \phi_0$.

5.1.d Suppose that in the posterior assessment, $\theta_1, \ldots, \theta_p$ happen to possess independent standard normal distributions. Show that approximation (5.1.13) to the posterior distribution of $\eta = \sum_{i=1}^p \theta_i^2$ gives the correct answer, that is, the chi-squared distribution with p degrees of freedom.

Note: This example can be used for comparison purposes with other approximations. See Leonard, Hsu, and Tsui (1989). The approximation suggested by Tierney, Kass, and Kadane (1989, corrected 1991) also works well.

5.1.e *Two-Way Layout:* Consider observations $y_{ij} (i = 1, \ldots, r;\ j = 1, \ldots, s)$, which are $N(\theta_{ij}, \tau^2)$ distributed, independently and given the θ_{ij}, where τ^2 is specified. In the prior assessment, let the θ_{ij} be conditionally independent and $N(\mu_{ij}, \sigma^2)$ distributed, given the μ_{ij}, where σ^2 is specified. Find the posterior distribution of the θ_{ij}, conditional on the μ_{ij}. Note that, given only the μ_{ij}, the y_{ij} are independently $N(\mu_{ij}, \tau^2 + \sigma^2)$ distributed.

Now let

$$\mu_{ij} = \mu + \alpha_i + \beta_j \qquad (i = 1, \ldots, r;\ j = 1, \ldots s),$$

where, in the prior assessment, μ is uniformly distributed, the α_i are $N(0, \sigma_A^2)$ and the β_j are $N(0, \sigma_B^2)$, with μ, the α_i, and the β_j all independent.

(a) By noting that for a multivariate normal distribution, the means are identical to the modes, find the posterior means of μ, the α_i, the β_j, the μ_{ij}, and the θ_{ij}, expressing your results in convenient weighted average forms.

(b) *Challenge research level question:* Find the posterior distribution of the interaction effect $\eta = \theta_{ij} - \theta_{i.} - \theta_{.j} + \theta_{..}$ by in particular finding its posterior variance. Here the dot notation indicates averaging with respect to the corresponding subscript.

(c) Investigate the limiting behavior of your results, as σ_A^2 and σ_B^2 both tend to infinity, and σ^2 remains fixed.

5.1.f *Conjugate Analysis for Multinomial Cell Probabilities:* Let $y_1, \ldots, y_p$ possess a multinomial distribution with cell probabilities $\theta_1, \ldots, \theta_p$ falling in the unit simplex S_U, and sample size $n = \sum y_i$. Suppose that the prior distribution of $\boldsymbol{\theta} = (\theta_1, \ldots, \theta_p)^T$ is Dirichlet with parameters $\alpha\xi_1, \ldots, \alpha\xi_p$, as described in Exercise 3.1.e. Show that the posterior distribution of $\boldsymbol{\theta}$ is Dirichlet with parameters $\alpha\xi_1 + y_1, \ldots, \alpha\xi_p + y_p$. Without further integrations, show that the posterior mean and variance of θ_j are, respectively,

$$\xi_j^* = E(\theta_j \mid \mathbf{y}) = \frac{np_j + \alpha\xi_j}{n + \alpha}$$

and

$$\mathrm{var}(\theta_j \mid \mathbf{y}) = \frac{\xi_j^*(1 - \xi_j^*)}{\alpha + n + 1},$$

with $p_j = y_j/n$.

5.1.g Suppose that

$$\theta_j = \frac{U_j}{\sum_{k=1}^p U_k} \qquad (j = 1, \ldots, p),$$

where $U_1, \ldots, U_p$ are independent chi-squared variates, with respective degrees of freedom $2\alpha_1, \ldots, 2\alpha_p$. Then it is well known that $\boldsymbol{\theta} = (\theta_1, \ldots, \theta_p)^T$ possesses a Dirichlet distribution with parameters $\alpha_1, \alpha_2, \ldots, \alpha_p$. Use this result to show that in Exercise 5.1.f, the marginal posterior distribution of θ_j is beta with parameters $\alpha\xi_j + y_j$ and $\alpha(1 - \xi_j) + n - y_j$.

5.1.h *The Dirichlet Process* (Ferguson, 1973): Suppose that $p \times 1$ observation vectors $\mathbf{Y}_1, \ldots, \mathbf{Y}_n$ are a random sample from a distribution with unknown c.d.f. $F(\mathbf{y})$ for $\mathbf{y} \in \Re^p$, which induces a probability $\lambda = p(A)$ for any event in R^p.

Definition: The prior distribution of F follows a $\mathcal{D}(\alpha, F_0)$ *Dirichlet process* with sample size α and mean value function F_0 if for any choice of events $A_1, \ldots, A_q$, which partition $\Re^p$, the probabilities $p_1 = p(A_1), \ldots, p_q = p(A_q)$ possess a Dirichlet distribution with parameters $\alpha\xi_1, \ldots, \alpha\xi_q$, where $\xi_1 = p_0(A_1) \ldots, \xi_q = p_0(A_q)$ are the probabilities on $A_1, \ldots, A_q$, induced by F_0.

If the notation in this definition describes the prior distribution for F, show that the posterior distribution for F follows a Dirichlet process with (posterior) sample size $\alpha + n$ and (posterior) mean value function $(\alpha F_0 + nF_n)/(\alpha + n)$, where F_n denotes the empirical c.d.f. Hence show that for any $\mathbf{t} \in R^p$, the posterior distribution of $F(\mathbf{t})$ is beta with parameters $\alpha F_0(t) + nF_n(t)$ and $\alpha\{1 - F_0(t)\} + n\{1 - F_n(t)\}$.

(A procedure for empirically estimating α from the current data is described by Antoniak, 1974. Rubin, 1981, proposes a "Bayesian bootstrap" simulation procedure relating to the Dirichlet process. Some other properties of the Dirichlet process are summarized by Leonard, 1996. Note the unusual property that α cannot be identified empirically unless there are ties in the data.)

5.1.i *Random Sampling from an Exponential Distribution with Unknown Scale and Location:* Observations $y_1, \ldots, y_n$ are a random sample from the two-parameter exponential distribution, with density

$$p(y \mid \theta, \lambda) = \lambda \exp\{-\lambda(y-\theta)\} I[\theta \le y]$$
$$(-\infty < y < \infty, -\infty < \theta < \infty, 0 < \lambda < \infty),$$

where $I(A)$ is the indicator function, for the set A.

(a) Show that the likelihood of θ and λ, given $y = (y_1, \ldots, y_n)^T$, is

$$l(\theta, \lambda \mid \mathbf{y}) = \lambda^n \exp\{-n\lambda(\bar{y}-\theta)\} I[\theta \le z]$$
$$(-\infty < \theta < \infty, 0 < \lambda < \infty),$$

where $z = \min y_i$. In the prior assessment, θ has an $N(\mu, \xi^{-1}\lambda^{-1})$ distribution, conditional on λ, and the unconditional distribution of λ is Gamma, with parameters α and β.

(b) Show that the posterior density of λ is

$$\pi(\lambda \mid \mathbf{y}) \propto \lambda^{\alpha+n-1} \exp\{-\kappa\lambda\} \Phi \left\{ \frac{(z-\mu)^*}{\xi^{\frac{1}{2}} \lambda^{\frac{1}{2}}} \right\},$$

where $\kappa = \beta + n(\bar{y} - \mu - \frac{1}{2}\xi^{-1})$ and $\mu^* = \mu + \xi^{-1} n$, and Φ denotes the cumulative normal distribution function. Show that the Laplacian approximation (5.1.5) to $\pi(\lambda \mid \mathbf{y})$ is equivalent to replacing $\Phi(x)$ by K if $x > 0$, and by $K \exp\{-\frac{1}{2}x^2\}$ if $x < 0$, for some constant K. Carefully investigate the properties of the whole posterior density for different choices of the prior parameters. Is the Laplacian approximation likely to work well in this case?

(c) Show that the Laplacian approximation (5.1.5), under the transformation $\alpha = \log \lambda$, to the marginal posterior density of θ gives the exact result. Suggest a family of prior distributions that yields algebraically explicit expressions for the posterior means of λ and θ.

5.1.j In the situation discussed in Section 1.9 (least squares regression with serially correlated errors), assume a uniform prior $\pi(\boldsymbol{\beta}, \gamma_1, \gamma_2) \propto 1$. Find the marginal posterior density of γ_2.

5.2 Further Methods and Practical Examples

(A) *The Death Penalty Data and Measures of Association for 2 × 2 Contingency Tables*

Let $\{y_{ij};\ i = 1, 2 \text{ and } j = 1, 2\}$ denote the cell frequencies in a 2×2 contingency table. Under appropriate randomization at the design stage, it is reasonable to take the y_{ij} to possess a four-cell multinomial distribution with respective cell probabilities

$\{\phi_{ij}; i = 1, 2 \text{ and } j = 1, 2\}$, summing to unity, and sample size $n = \sum y_{ij}$. Following Altham (1969), take the θ_{ij} to possess a Dirichlet prior distribution with parameters $\{\alpha_{ij}; 1 = 1, 2 \text{ and } j = 1, 2\}$ (see Exercise 5.1.f). In vague prior situations, the choices $\alpha_{ij} \equiv \frac{1}{2}$ are frequently employed.

The posterior distribution of the θ_{ij} is Dirichlet with parameters $\{\alpha_{ij} + y_{ij}; i = 1, 2$ and $j = 1, 2\}$. Consider the log measure of association

$$\lambda = \log \theta_{22} + \log \theta_{11} - \log \theta_{12} - \log \theta_{21}. \tag{5.2.1}$$

Then the representation in Exercise 5.1.g tells us that λ is distributed in the posterior assessment as

$$\log \chi^2_{\nu_{22}} + \log \chi^2_{\nu_{11}} - \log \chi^2_{\nu_{12}} - \log \chi^2_{\nu_{21}}, \tag{5.2.2}$$

where χ^2_ν denotes a chi-squared random variable with ν degrees of freedom, and the four chi-squared variates in (5.2.2) are independent, with $\nu_{ij} = 2(\alpha_{ij} + y_{ij})$, $(i = 1, 2;\ j = 1, 2)$. Consequently λ is distributed as

$$\log \frac{(\alpha_{22} + y_{22})(\alpha_{11} + y_{11})}{(\alpha_{12} + y_{12})(\alpha_{21} + y_{21})} + \log F^{\nu_{22}}_{\nu_{12}} + \log F^{\nu_{21}}_{\nu_{22}}, \tag{5.2.3}$$

where $F^{\nu_1}_{\nu_2}$ denotes an F-variate, with ν_1 and ν_2 degrees of freedom, and the two F variates in (5.2.3) are independent. Note that an F-variate with ν_1 and ν_2 degrees of freedom possesses density

$$f(y \mid \nu_1, \nu_2) = C_{\nu_1,\nu_2} y^{\frac{\nu_1}{2}-1} (\nu_2 + \nu_1 y)^{-\frac{\nu_1+\nu_2}{2}} \qquad (0 < y < \infty),$$

where

$$C_{\nu_1,\nu_2} = \frac{\Gamma\left(\frac{\nu_1+\nu_2}{2}\right)}{\Gamma\left(\frac{\nu_1}{2}\right)\Gamma\left(\frac{\nu_2}{2}\right)} \nu_1^{\frac{\nu_1}{2}} \nu_2^{\frac{\nu_2}{2}}.$$

We can alternatively refer to the representation

$$F^{\nu_2}_{\nu_1} = \frac{\chi^2_{\nu_1}/\nu_1}{\chi^2_{\nu_2}/\nu_2},$$

where $\chi^2_{\nu_1}$ and $\chi^2_{\nu_2}$ are independent chi-squared variates with respective degrees of freedom ν_1 and ν_2. The expression (5.2.3) for λ tells us that the posterior density of λ may be computed by a one-dimensional numerical integration for a convolution of two independent F-variates. If the α_{ij} are integers, Altham represents the posterior probability that $\lambda \le 1$ in terms of the hypergeometric series. If $\alpha_{11} = \alpha_{22} = 0$ and $\alpha_{12} = \alpha_{21} = 1$, this reduces to a significance probability for Fisher's exact test, thus providing an important relation between Bayesian and frequency procedures.

Consider the death penalty data of Section 1.2 (H). Curve (a) in Figure 5.2.1 describes the posterior density of the log measure of association between color of defendant and the death penalty variable when attention is constrained to cases with white victims.

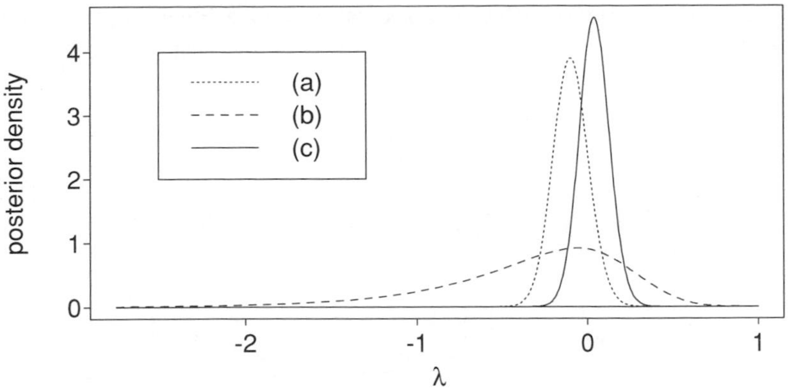

Figure 5.2.1. Posterior densities of measure of association: (a) white victim; (b) black victim; (c) overall.

There is some slight evidence of a negative association. Curve (b) again suggests some slight evidence of a negative association when attention is constrained to cases with black victims. However, curve (c) apparently suggests evidence of a positive association when all cases are pooled. Note, however, that it is inappropriate to report curve (c), since the underlying multinomial assumptions for the overall table are invalidated by the clear inequality of the split tables for cases with white victims and for cases with black victims.

(B) *Goodman's Full-Rank Interaction Analysis and the Engineering Apprentice, Shopping in Oxford, and Marine Corps Data*

Now let $\{y_{ij}; i = 1, \ldots, r \text{ and } j = 1, \ldots, s\}$ denote the cell frequencies in an $r \times s$ contingency table. We proceed under the assumption that the y_{ij} are independent Poisson variates with respective means ξ_{ij}. It then follows that conditional on $\sum y_{ij} = n$, the y_{ij} possess a multinomial distribution with respective cell probabilities $\theta_{ij} = \xi_{ij}/\sum \xi_{kg}$ and sample size n. In the Poisson situation, consider the $\gamma_{ij} = \log \xi_{ij}$ together with the *full-rank interaction model*

$$\gamma_{ij} = \mu + \lambda_i^A + \lambda_j^B + \lambda_{ij}^{AB} \qquad (i = 1, \ldots, r; \quad j = 1, \ldots, s), \tag{5.2.4}$$

introduced by Goodman (1964), and also considered by Leonard and Novick (1986) and Leonard, Hsu, and Tsui (1989). Here μ is the overall effect, λ_i^A the ith row effect, λ_j^B the jth column effect, and λ_{ij}^{AB} denote the interaction effect for the (i, j)th cell. Assume the constraints $\lambda_{\cdot}^A = \lambda_{\cdot}^B = \lambda_{i\cdot}^{AB} \equiv \lambda_{\cdot j}^{AB} \equiv 0$, where the dot notation indicates averaging with respect to the corresponding subscript. Then algebraic manipulations can be used to show that

$$\lambda_{ij}^{AB} = \gamma_{ij} - \gamma_{i\cdot} - \gamma_{\cdot j} + \gamma_{\cdot\cdot}, \tag{5.2.5}$$

so that the (i, j)th interaction effect is a special case of a linear contrast, taking the form

$$\eta = \sum_{k=1}^{r} \sum_{j=1}^{s} a_{kg} \gamma_{kg}, \tag{5.2.6}$$

where $\sum a_{kg} = 0$. Under the prior $\pi(\boldsymbol{\xi}) \propto \prod \xi_{ij}^{-\frac{1}{2}}$, the posterior density of the γ_{ij} is product log-Gamma, that is,

$$\pi(\boldsymbol{\gamma} \mid \mathbf{y}) \propto \exp\left\{ \sum_{i=1}^{r} \sum_{j=1}^{s} \gamma_{ij} y_{ij}^* - \sum_{i=1}^{r} \sum_{j=1}^{s} e^{\gamma_{ij}} \right\}, \tag{5.2.7}$$

where $y_{ij}^* = y_{ij} + \frac{1}{2}$. Lindley (1964) shows that the posterior distribution of η is approximately normal, with mean $\sum_{k=1}^{r} a_{kg} \log y_{kg}^*$ and variance $\sum_{g=1}^{s} a_{kg}^2 / y_{kg}^*$, and some adjustments are investigated by Bloch and Watson (1967). To develop a Laplacian approximation to the posterior density of η, first maximize (5.2.7) but subject to the constraint that $\sum a_{kg} \gamma_{kg} = \eta$. The maximizing values are

$$\tilde{\gamma}_{ij}^{\eta} = \log\left(y_{ij}^* + \lambda a_{ij}\right), \tag{5.2.8}$$

where λ satisfies

$$\eta = \sum_{i=1}^{r} \sum_{j=1}^{s} \log\left(y_{ij}^* + \lambda a_{ij}\right). \tag{5.2.9}$$

Note that the Lagrange multiplier λ can be computed for each fixed η by numerically inverting the function (5.2.9). Then the Laplacian approximation (5.1.7) conveniently reduces to

$$\pi(\eta \mid \mathbf{y}) \propto \frac{\prod_{i=1}^{r} \prod_{j=1}^{s} \left(y_{ij}^* + \lambda a_{ij}\right)^{y_{ij} + \lambda a_{ij} - \frac{1}{2}}}{\left[\sum_{i=1}^{r} \sum_{j=1}^{s} a_{ij}^2 \left(y_{ij}^* + \lambda a_{ij}\right)^{-1}\right]^{\frac{1}{2}}} \qquad (-\infty < \eta < \infty). \tag{5.2.10}$$

Table 5.2.1 summarizes the engineering apprentice data, which cross-classify $n = 492$ apprentices according to the section head's assessment and a written test result. Curve (a) in Figure 5.2.2 describes Lindley's normal approximation (mean 0.279 and standard deviation 0.224) to the posterior density of the $(4, 4)$th interaction effect, under our Jeffreys' prior. This is remarkably close to the exact result in curve (b) in Figure 5.2.2. Indeed this curve represents both our Laplacian approximation and the exact result, based upon Monte Carlo simulation from our sixteen independent log-Gamma distributions for the γ_{ij}. Similarly appealing results were obtained for the posterior densities of the further fifteen interaction effects. Lindley's approximation tells us that the posterior distribution of λ_{ij}^{AB} is approximately normal, with mean $\mu_{ij}^{AB} = \gamma_{ij}^* - \gamma_{i.}^* - \gamma_{.j}^* + \gamma_{..}^*$ and variance

$$\begin{aligned} \epsilon_{ij}^{AB} = {} & \left(1 - 2r^{-1}\right)\left(1 - 2s^{-1}\right) u_{ij} + s^{-1}\left(1 - r^{-1}\right) u_{i.} \\ & + r^{-1}\left(1 - s^{-1}\right) u_{.j} + s^{-1} r^{-1} u_{..}, \end{aligned} \tag{5.2.11}$$

Table 5.2.1. *Interaction analysis for Engineering Apprentice data.*

Section head's assessment		Written test result A	B	C	D
Excellent	y_{ij}	26	29	21	11
	μ_{ij}	0.288	0.073	−0.096	−2.640
	$\sqrt{\epsilon_{ij}}$	0.161	0.151	0.163	0.199
	b_{ij}	1.785	0.482	−0.589	−1.328
Very good	y_{ij}	33	43	35	20
	μ_{ij}	0.096	0.036	−0.020	−1.112
	$\sqrt{\epsilon_{ij}}$	0.147	0.133	0.141	0.166
	b_{ij}	0.656	0.267	−0.145	−0.671
Average	y_{ij}	47	71	72	45
	μ_{ij}	−0.144	−0.057	0.104	0.096
	$\sqrt{\epsilon_{ij}}$	0.133	0.117	0.120	0.138
	b_{ij}	−1.084	−0.488	0.872	0.697
Needs to improve	y_{ij}	7	12	11	9
	μ_{ij}	−0.240	−0.051	0.012	0.279
	$\sqrt{\epsilon_{ij}}$	0.237	0.201	0.208	0.224
	b_{ij}	−1.014	−0.256	0.060	1.244

Note: LRS = 23.871, d.f. = 9.

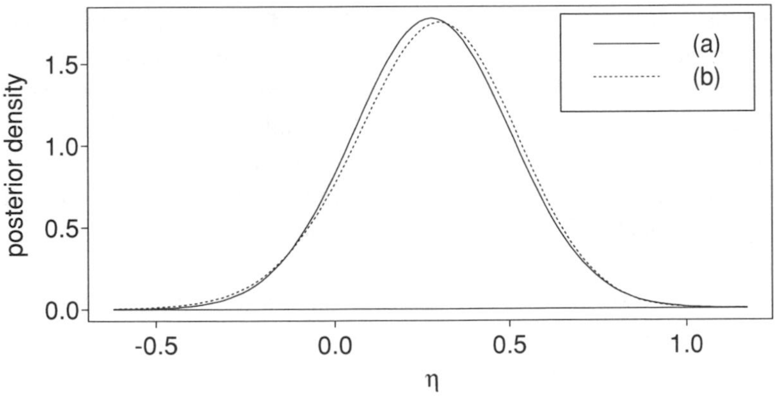

Figure 5.2.2. Posterior densities of (4,4)th interaction effect $\eta = \lambda_{44}^{AB}$: (a) Lindley's normal approximation; (b) Laplacian approximation and exact result.

where $\gamma_{ij}^* = \log(y_{ij} + \frac{1}{2})$ and $u_{ij} = (y_{ij} + \frac{1}{2})^{-1}$. The approximation is likely to be quite accurate whenever each of the y_{ij} is at least 5. These results will also be developed in Exercise 6.2.f, under multinomial sampling assumptions.

Some very simple calculations therefore facilitate the extraction of real-life conclusions from many $r \times s$ contingency tables. Calculate the μ_{ij}^{AB} and ϵ_{ij}^{AB} for each cell, together with the quantities

$$b_{ij} = \mu_{ij}^{AB} / \sqrt{\epsilon_{ij}^{AB}}, \tag{5.2.12}$$

which approximate the distances of the posterior means of the interaction effects from zero when measured in terms of numbers of standard deviations. Therefore the b_{ij} may be used to investigate whether any particular λ_{ij}^{AB} is substantially different from zero, for example, compare $|b_{ij}|$ with the critical normal values 1.960 and 2.576. Note that setting all interaction effects equal to zero in (5.2.4) corresponds to a hypothesis H_0 of independence of rows and columns of the table. This overall hypothesis H_0 may be investigated by reference to the *likelihood ratio statistic*

$$LRS = -2n \sum_{i=1}^{r} \sum_{j=1}^{s} p_{ij} \log \left(\frac{p_{i*} p_{*j}}{p_{ij}} \right), \tag{5.2.13}$$

where the star notation denotes summation with respect to that subscript. For example, refuting H_0, whenever LRS $> 2(r-1)(s-1)$, is equivalent to preferring the full-rank model (5.2.4) to the independence hypothesis H_0, whenever AIC (see equation (1.1.5)) for the full-rank model is greater than AIC under the independence model.

The results of the full-rank interaction analysis for the engineering apprentice data ($n = 492$) are summarized in Table 5.2.1. The overall hypothesis H_0 of independence is not refuted as LRS $= 7.407$ on 9 d.f. Below each cell frequency, we report the corresponding estimate μ_{ij}^{AB} of the interaction effect with, in brackets, the corresponding approximate posterior standard deviation $\sqrt{\epsilon_{ij}^{AB}}$. Below these two numbers appears the value of the ratio b_{ij}, defined in (5.2.12). None of the b_{ij} are greater than 1.960, suggesting that there are no substantive individual interactions in the table.

Note, however, that those b_{ij} largest in magnitude, namely (1.785, −1.328, −1.014, and 1.244) are in the four corners of the table, confirming that in the sample, the section head's assessment is somewhat positively associated with the written test result. Therefore our interaction analysis highlights a key conclusion to be drawn from the data. This type of conclusion often occurs when both margins of the table are ordinal; it will be referred to as a "four corner effect."

The Shopping in Oxford data ($n = 680$ residents of Oxford, England) reported by Fingleton (1984, p. 10), and compiled from data originally analyzed by Bowlby and Silk (1982), are summarized in Table 5.2.2, together with the results of our interaction analysis. The row variable is car availability (none, some, full). The column variable measures the response to the question "Do you find it tiring getting to grocery shops?" on an ordinal scale (the first column corresponds to "strongly disagree" and the last column to "strongly agree"). The number of the b_{ij} for the (1,1), (1,5), (5,1), and (5,5) cells were −2.238, 1.958, −3.020, and 0.138. This gives some indication of a negative association between rows and columns, that is, that as car availability increases, people may find grocery shopping less tiring. This is confirmed by the

Table 5.2.2. *Shopping in Oxford data.*

Car availability		Response to grocery shop question 1	2	3	4	5
None	y_{ij}	55	11	16	17	100
	μ_{ij}	−0.247	−0.070	−0.063	−0.042	0.197
	$\sqrt{\epsilon_{ij}}$	0.110	0.203	0.168	0.168	0.101
	b_{ij}	−2.238	0.346	−0.372	0.250	1.958
Some	y_{ij}	101	7	18	23	103
	μ_{ij}	0.233	−0.480	−0.071	0.214	0.103
	$\sqrt{\epsilon_{ij}}$	0.104	0.225	0.165	0.159	0.102
	b_{ij}	2.247	−2.129	−0.431	1.341	1.022
Full	y_{ij}	91	20	25	16	77
	μ_{ij}	0.014	0.410	0.134	−0.256	−0.301
	$\sqrt{\epsilon_{ij}}$	0.100	0.181	0.153	0.167	0.100
	b_{ij}	0.138	0.258	0.876	−1.528	−3.020

Note: LRS = 23.871, d.f. = 8.

value LRS = 23.871 for the statistic in (5.2.13) on 8 degrees of freedom. Note that LRS is greater than twice the degrees of freedom, thus refuting a hypothesis of zero association. However, perhaps the main feature in the data is that $b_{31} = 0.138$ is small in magnitude. Therefore, a full four corner effect does not occur, as might have been anticipated. It is indeed well known that many residents cycle to shops in Oxford, whereas residents with cars generally drive to a shopping center outside Oxford. It would appear from the data that some residents with full car availability may still find grocery shopping tiring for this reason. This example demonstrates how Goodman's full-rank interaction analysis can more generally be combined with background scientific or social information to mirror and extract the key real-life conclusion from the data.

The Marine Corps data in Table 5.2.3 were reported by Leonard and Novick (1986) and cross-classify $n = 5{,}648$ examinees by school and grade on a military aptitude test. Grade 1 is highest and grade 8 is lowest, and the first three grades correspond to an above-average score. As there are zero frequencies in the data, either an exact Bayesian analysis or an analysis using Laplacian approximations would appear useful. Leonard and Novick perform the latter, but using a complicated hierarchical prior distribution. Based upon the pictures of the posterior densities for the 96 interaction effects in the table, they drew the conclusions summarized in Table 5.2.4.

Again, patterns in the interactions (residuals from the independence model) can be used to detect real-life conclusions. In this case, the interaction analysis suggests that (a) the columns might be usefully collapsed by combining the first three columns and also

Table 5.2.3. *Cross-classification of marines by school and aptitude grade on a military aptitude test.*

	Grade level (high to low)							
School	1	2	3	4	5	6	7	8
A	20	179	276	316	123	27	10	4
B	10	80	112	112	6	5	0	1
C	25	293	390	337	126	66	32	18
D	3	32	55	51	17	7	3	2
E	2	41	46	43	12	4	0	0
F	10	81	138	242	145	32	6	2
G	9	131	270	263	100	39	19	12
H	3	35	57	64	21	7	3	2
I	2	38	69	45	15	8	1	2
J	1	28	49	81	32	28	10	4
K	1	29	51	56	37	18	6	2
L	0	48	87	162	62	71	21	7

Table 5.2.4. *Interaction analysis (Marine Corps data).*

	Grade level (high to low)							
School	1	2	3	4	5	6	7	8
B	+	+	+	$\tilde{0}$	−	−	−	0
C	$\tilde{+}$	+	0	−	$\tilde{-}$	0	+	+
E	0	+	0	0	0	0	−	−
I	0	$\tilde{+}$	+	$\tilde{-}$	0	0	0	0
A	+	0	0	0	0	−	$\tilde{-}$	0
D	0	0	0	0	0	0	0	0
G	0	0	$\tilde{+}$	0	0	0	0	+
H	0	0	0	0	0	0	0	0
F	0	−	−	+	+	0	0	0
J	0	$\tilde{-}$	$\tilde{-}$	$\tilde{+}$	$\tilde{+}$	+	+	+
K	0	0	0	0	+	+	+	0
L	−	−	−	$\tilde{+}$	$\tilde{+}$	+	+	+

combining the last three columns, (b) the schools could be divided into three groups, that is, (BCEI), (ADGH), and (FJKL). The consequent pooled results are summarized in Table 5.2.5. In particular, these three groups of schools possess respective overall pass rates of 57.1%, 49.5%, and 33.9%. A Bayesian analysis for three-way tables is proposed by Nazaret (1987), who generalizes Leonard (1975b).

Table 5.2.5. *Collapsed table (Marine Corps data).*

School	Grades 1–3	Grades 4–8	Success proportion
B	202	124	0.620
C	708	579	0.550
E	89	59	0.601
I	109	71	0.616
TOTAL	1,108	833	0.571
A	475	480	0.497
D	90	80	0.529
G	410	433	0.486
H	95	97	0.494
TOTAL	1,070	1,090	0.495
F	229	427	0.349
J	78	155	0.335
K	81	119	0.339
L	135	323	0.295
TOTAL	523	1,024	0.339

(C) *Quasi-Independence and the Remote Sensing, Occupational Mobility, and Geological Data*

For some two-way contingency tables, it is possible to achieve a model that compromises between the full-rank interaction model (5.2.4) and the independence model ($\lambda_{ij}^{AB} = 0$ for all i and j). Such quasi-independence models (Goodman, 1968) assume the form (under the Poisson sampling assumptions of Section 5.2 (B))

$$\alpha_{ij} = \begin{cases} \mu + \lambda_i^A + \lambda_j^B + \lambda_{ij}^{AB} & \text{for } (i, j) \in Q \\ \mu + \lambda_i^A + \lambda_j^B & \text{for } (i, j) \notin Q, \end{cases} \tag{5.2.14}$$

where Q contains q elements, with $q < (r-1)(s-1)$, and the constraints $\lambda_{\cdot}^A = \lambda_{\cdot}^B$ are required, without further constraints on the interactions. For some data sets, it might be useful for Q to summarize those cells in the table that are judged, via Goodman's full-rank interaction analysis, to possess nonzero interactions.

In other examples, an initial choice of Q might be dictated by the experimental design. For the remote sensing data (see Table 5.2.6), satellite observations assign $n =$ 719 objects in one of four categories by both remote sensing and visual procedures. In

Table 5.2.6. *The remote sensing data.*

		Observed category (infrared)			
		1	2	3	4
	1	12	0	3	1
		(12)	(0.261)	(3.028)	(0.711)
Observed	2	0	32	4	1
category		(0.588)	(32)	(3.573)	(0.839)
(visual)	3	1	1	18	1
		(1.017)	(0.532)	(18)	(1.450)
	4	1	0	2	21
		(0.395)	(0.207)	(2.399)	(21)

this case, the choice $Q = \{(1, 1), (2, 2), (3, 3), (4, 4)\}$ is sensible. It is straightforward (e.g., Fienberg, 1987, Section 8.2) to estimate the parameters by iterative proportionate fitting of the maximum likelihood solution. The cells in Q are always fit perfectly, leaving $\nu = (r-1)(s-1) - q$ degrees of freedom. The fitted values for the remote sensing data are given in parentheses in Table 5.2.6. These yielded a value for the chi-squared goodness-of-fit statistic of $\chi^2 = 2.80$ on 5 d.f.

From a Bayesian perspective, and following Leonard (1975b), we can imbed the quasi-independence assumption into a model of the form

$$\alpha_{ij} = \mu + \lambda_i^A + \lambda_j^B + \lambda_{ij}^{AB} + \xi_{ij} I\big[(i, j) \in Q\big]$$
$$(i = 1, \ldots, r;\ j = 1, \ldots, s), \qquad (5.2.15)$$

where $I(A)$ is the indicator function for the set A, and, given $\sigma_A^2, \sigma_B^2, \sigma_{AB}^2$, and σ_ξ^2, all effects in (5.2.15) are unconstrained and independent, with μ uniformly distributed over $(-\infty, \infty)$ and the λ_i^A, λ_j^B, and λ_{ij}^{AB} normally distributed with zero means and respective common variances σ_A^2, σ_B^2, and σ_{AB}^2. Finally the ξ_{ij} may be taken to be normally distributed with mean ξ_0 and variance σ_ξ^2. The choices $\sigma_A^2 = \sigma_B^2 = \infty$ will frequently be adequate (e.g., Laird, 1979). Whenever $\sigma_{AB}^2 = 0$, the model (5.2.14) reduces to a random effects version of the quasi-independence model (5.2.15).

Leonard (1975b) shows how to use an earlier version of the hierarchical Bayes procedures, to be discussed in Section 6.3, to estimate $\sigma_A^2, \sigma_B^2, \sigma_{AB}^2, \sigma_\xi^2$, and ξ_0 from the current data set. The 14×14 occupational mobility table (Table 5.2.7) was reported by Pearson (1904) and cross-classifies 775 fathers and sons according to their occupations. With $q = 14$, Q was taken to consist of all diagonal cells. Leonard reports the empirical estimates $\tilde{\sigma}_A^2 = 0.301$, $\tilde{\sigma}_B^2 = 1.072$, $\tilde{\sigma}_{AB}^2 = 0.000$, $\tilde{\sigma}_\xi^2 = 0.000$, and $\tilde{\xi}_0 = 1.903$. These suggest that (5.2.15) might be replaced by the very simple model

$$\alpha_{ij} = \mu + \lambda_i^A + \lambda_j^B + \xi_0 \delta_{ij} \qquad (i = 1, \ldots, 14;\ j = 1, \ldots, 14), \qquad (5.2.16)$$

Table 5.2.7. *Contingency between the occupations of fathers and sons.*

		Occupation of son													
		1	2	3	4	5	6	7	8	9	10	11	12	13	14	TOTAL
Occupation of father	1	28	0	4	0	0	0	1	3	3	0	3	1	5	2	50
	2	2	51	1	1	2	0	0	1	2	0	0	0	1	1	62
	3	6	5	7	0	9	1	3	6	4	2	1	1	2	7	54
	4	0	12	0	6	5	0	0	1	7	1	2	0	0	10	44
	5	5	5	2	1	54	0	0	6	9	4	12	3	1	13	115
	6	0	2	3	0	3	0	0	1	4	1	4	2	1	5	26
	7	17	1	4	0	14	0	6	11	4	1	3	3	17	7	88
	8	3	5	6	0	6	0	2	18	13	1	1	1	8	5	69
	9	0	1	1	0	4	0	0	1	4	0	2	1	1	4	19
	10	12	16	4	1	15	0	0	5	13	11	6	1	7	15	106
	11	0	4	2	0	1	0	0	0	3	0	20	0	5	6	41
	12	1	3	1	0	0	0	1	0	1	1	1	6	2	1	18
	13	5	0	2	0	3	0	1	8	1	2	2	3	23	1	51
	14	5	3	0	2	6	0	1	3	1	0	0	1	1	9	32
TOTAL		84	108	37	11	122	1	15	64	69	24	57	23	74	86	775

where δ_{ij} denotes the Kronecker delta function. One consequence of the model (5.2.16) is that the measures of association

$$A_{(j,j)}^{(i,i)} = \frac{\phi_{ii}\phi_{jj}}{\phi_{ij}\phi_{ji}} = e^{2\xi_0} \tag{5.2.17}$$

are constant in i and j for all $i \neq j$ and may be estimated by $\exp\{2\tilde{\xi}_0\} = 45.0$. A parallel consequence is that if a son does not follow his father's occupation, then his occupation becomes (conditionally) independent of the occupation of his father.

In Table 5.2.8, we report Leonard's Bayesian-fitted values to the 14×14 table. Dividing each fitted value by $n = 775$ would give a Bayesian estimate for the corresponding cell probability. Note that a maximum likelihood fit of the quasi-independence model, but with unequal ξ's along the diagonal, gives the value $\chi^2 = 262.36$ for the chi-squared goodness-of-fit statistic on 155 degrees of freedom. As χ^2 is less than twice the degrees of freedom, we find this model acceptable (e.g., Leonard, 1977a).

These models are of particular potential importance in geology. As a geological example, consider the data ($n = 233$) described in Table 5.2.9, which we analyzed in collaboration with Dennis Kerr. At a particular point in the earth's crust in North Central Wyoming, there are 234 geological layers described by 6 different types A, B, C, D, E, and F. The data describe the numbers of transitions, when moving from layer to layer, between the different possible types of layer. A quasi-independence model with 19 d.f., and with Q just including the six diagonal cells of the table, did not fit well ($\chi^2 = 193$). Seven iterations of a stepwise procedure were then performed, where at each step, the

Table 5.2.8. *Smoothed frequencies.*

		Occupation of son														
		1	2	3	4	5	6	7	8	9	10	11	12	13	14	TOTAL
Occupation of father	1	23.10	4.31	1.41	0.47	3.97	0.14	0.49	2.34	3.25	0.70	2.38	1.08	2.99	3.84	50.47
	2	3.84	32.30	1.58	0.52	4.43	0.16	0.55	2.61	3.62	0.79	2.65	1.20	3.33	4.29	61.87
	3	4.73	5.92	13.02	0.65	5.46	0.20	0.67	3.21	4.46	0.97	3.27	1.48	4.10	5.29	53.43
	4	4.48	5.61	1.84	4.09	5.16	0.18	0.64	3.04	4.22	0.92	3.09	1.40	3.88	5.00	43.55
	5	7.28	9.11	2.99	0.99	56.26	0.30	1.04	4.94	6.86	1.49	5.02	2.28	6.31	8.13	113.0
	5	2.94	3.68	1.21	0.40	3.39	0.81	0.42	2.00	2.77	0.60	2.03	0.92	2.55	3.29	27.01
	7	8.74	10.94	3.59	1.19	10.08	0.36	8.35	5.93	8.24	1.19	6.03	2.74	7.58	9.76	85.33
	8	5.31	6.65	2.18	0.72	6.12	0.22	0.76	24.15	5.01	1.09	3.66	1.66	4.61	5.93	68.07
	9	1.56	1.95	0.64	0.21	1.80	0.06	0.22	1.06	9.84	0.32	1.072	0.49	1.35	1.74	22.31
	10	10.17	12.72	4.17	1.39	11.72	0.42	0.49	6.90	9.58	13.94	7.01	3.19	8.81	11.35	102.86
	11	3.23	4.05	1.33	0.44	3.73	0.13	0.46	2.19	3.05	0.66	14.96	1.01	2.80	3.61	41.65
	12	1.91	2.39	0.78	0.26	2.20	0.08	0.27	1.30	1.80	0.39	1.32	4.02	1.66	2.13	20.51
	13	3.69	4.62	1.51	0.50	4.25	0.15	0.53	2.50	3.48	0.75	2.55	1.16	21.46	4.12	51.27
	14	2.23	2.79	0.91	0.30	2.57	0.09	0.32	1.51	2.10	0.46	1.54	0.70	1.93	16.67	34.12
	TOTAL	83.21	107.04	37.16	12.13	121.14	3.30	16.21	63.38	68.28	24.87	56.58	23.33	73.36	85.15	775.00

Table 5.2.9. *Observed transitions between types of geological layers.*

	A	B	C	D	E	F	TOTAL
A	–	12	12	1	0	2	27
B	3	–	17	1	0	7	28
C	8	5	–	9	0	11	33
D	8	7	7	–	0	32	54
E	1	1	1	9	–	12	24
F	3	1	0	39	24	–	67
TOTAL	23	26	37	59	24	64	233

Table 5.2.10. *Fitted frequencies (quasi-independence model with 12 d.f.).*

	A	B	C	D	E	F	TOTAL
A	–	*12.0	*12.0	1.6	0.0	1.4	27.0
B	*3.0	–	*17.0	4.2	0.0	3.8	28.0
C	*8.0	*5.0	–	10.4	0.0	9.6	33.0
D	6.0	4.5	4.0	–	0.0	39.3	53.8
E	1.5	1.1	1.0	10.6	–	9.8	24.0
F	4.5	3.4	3.1	32.2	*24.0	–	67.2
TOTAL	23.0	26.0	37.1	59.0	24.0	63.9	233.0

cell with the worst fit was added to Q at the next step. The investigators finally achieved an excellent fitting model ($\chi^2 = 19$, d.f. $= 12$) with fitted frequencies described in Table 5.2.10. The starred frequencies correspond to the extra cells added to the set Q. For example, the estimated equilibrium probabilities for the states can be calculated.

Note that our analysis is only meaningful if the extra cells included in Q are thought to possess geological meaning. Otherwise, this procedure could overfit the data in a misleading manner. In Table 5.2.11, we report an estimated transition matrix, based upon Table 5.2.10, which can be used to analyze geological layers using the theory of Markov chains. Our recommended procedure, however, would perhaps work better with a more judicious and parsimonious choice of the elements of Q. Also, the introduction of our informative priors into geological applications could be quite beneficial. Further examples of the iterative fitting procedure are reported by Kerr (1990).

Table 5.2.11. *Estimated Markov transition matrix.*

	A	B	C	D	E	F
A	–	0.444	0.444	0.059	0.000	0.052
B	0.107	–	0.607	0.150	0.000	0.136
C	0.242	0.152	–	0.315	0.000	0.219
D	0.112	0.084	0.074	–	0.000	0.730
E	0.063	0.046	0.042	0.442	–	0.448
F	0.067	0.051	0.046	0.479	0.357	–

(D) *Genetic Potential, the Ratkowsky Regression Model*

Following Bates and Watts (1988), consider nonlinear regression models of the form

$$y_i = f(x_i, \boldsymbol{\theta}) + \epsilon_i \qquad (i = 1, \ldots, n), \tag{5.2.18}$$

where the ϵ_i are independent and normally distributed with zero means and common unknown variance σ^2, f is a specified function of x_i and $\boldsymbol{\theta}$, $\boldsymbol{\theta}$ is an unknown $p \times 1$ vector of parameters, and $x_1, \ldots, x_n$ are specified constraints. Let $\boldsymbol{\theta}$ possess prior density $\pi(\boldsymbol{\theta})$, independently of $\log \sigma^2$, which is taken to be uniformly distributed over $(-\infty, \infty)$. Then the posterior density of $\boldsymbol{\theta}$ is

$$\pi_y(\boldsymbol{\theta}) \propto \pi(\boldsymbol{\theta})\left[\rho(\boldsymbol{\theta})\right]^{-\frac{1}{2}n}, \tag{5.2.19}$$

where

$$\rho(\boldsymbol{\theta}) = \sum_{i=1}^{n} \{y_i - f(x_i, \boldsymbol{\theta})\}^2. \tag{5.2.20}$$

Ratkowsky (1983) considers the regression of "loss in yield" y_i due to the value of a "density" x_i and proposes the regression function

$$f(x_i, \boldsymbol{\theta}) = \frac{\theta_1}{1 + \theta_2 x_i + \theta_3 x_i^2}, \tag{5.2.21}$$

where θ_1 is the genetic potential and θ_2 and θ_3 are linear and cubic coefficients. The vector $\boldsymbol{\theta} = (\theta_1, \theta_2, \theta_3)^T$ should not, in this case, be taken to possess an improperly uniform prior distribution, since the posterior distribution can then become improper. In the numerical example discussed below, we instead assumed uniform prior distributions for θ_1, θ_2, and θ_3 but truncated to the regions $0 \le \theta_1 \le 2{,}000$, $0 \le \theta_2 \le 1$, and $-0.001 \le \theta_3 \le 0.001$.

Ratkowsky describes his data ($n = 42$) as the "MG data set," and Leonard, Hsu, and Ritter (1994) report a full Bayesian analysis. Their Laplacian approximations to the marginal posterior densities of $\theta_1, \theta_2, \theta_3$ are algebraically explicit in terms of the least squares estimates, and virtually identical to exact results, based upon lengthy simulations using the Metropolis algorithm described by Metropolis et al. (1953) and Ritter (1992, p. 58). The Laplacian approximations are both very accurate and readily computable.

(E) *Electron Spectroscopy for Chemical Analysis*

Ritter (1992, 1994) reports that electron spectroscopy is an important tool for studying the chemistry of material surfaces and believes that adjusted counts y_i and energies x_i can often be approximately modeled by distributional assumptions leading to an approximate likelihood function of the form

$$l(\boldsymbol{\theta} \mid \mathbf{x}, \mathbf{y}) \propto \exp\left\{-\frac{1}{2}\sum_{i=1}^{n} \omega_i(y_i - \eta(x_i, \boldsymbol{\theta}))^2\right\}, \tag{5.2.22}$$

where the ω_i are specified and η is some specified function of the energy x_i and a parameter vector $\boldsymbol{\theta}$. Ritter proposes the choice

$$\eta(x, \boldsymbol{\theta}) = \eta(x, \boldsymbol{\alpha}, \boldsymbol{\beta}, \boldsymbol{\gamma}, \boldsymbol{\rho}) = \sum_{j=1}^{m} \alpha_j f(x_j, \beta_j, \gamma_j, \rho_j),$$

where $f(x_j, \beta_j, \gamma_j, \rho_j)$ denotes a complicated algebraic function, with $0 < \alpha_j < \infty$, $0 < \gamma_j < \infty$, $0 < \rho_j < 1$, and $-\infty < \beta_j < \infty$.

It is straightforward to develop a Bayesian analysis for complicated models of this type. Let $a_j = \log \alpha_j$, $b_j = \beta_j$, $g_j = \log \beta_j$, and $\sigma_j = \log \rho_j - \log(1 - \rho_j)$. Then prior information can be flexibly incorporated via a four-dimensional multivariate normal prior distribution for the a_j, b_j, g_j, and r_j. This highlights a general approach to the construction of prior distributions for nonlinear models, that is, you should seek transformations of the parameters such that it is reasonable to take the transformed parameters to possess a multivariate normal prior distribution.

Ritter (1992) considers a data set with $n = 1977$ observations, and on his page 89 he proposes particular choices of multivariate normal prior distribution based upon the scientific background, and with $m = 4$. He then follows another of our general themes by reporting Laplacian approximations to the posterior densities of the sixteen unknown parameters in the model.

Worked Example 5F: *Approximate Analysis for Nonlinear Regression*

Consider the nonlinear regression model in (5.2.18), where the prior distribution of $(\boldsymbol{\theta}, \log \sigma^2)$ is uniform. Let $\hat{\boldsymbol{\theta}}$ denote the $p \times 1$ least squares vector of $\boldsymbol{\theta}$. Suggest a generalized multivariate t-approximation to the posterior density of θ, taking the form

$$\pi_\eta^*(\theta) \propto \left[1 + \nu^{-1}(\boldsymbol{\theta} - \boldsymbol{\mu})^T \mathbf{R}(\boldsymbol{\theta} - \boldsymbol{\mu})\right]^{-\frac{1}{2}(\nu+p)},$$

for some ν, μ, and $\mathbf{R}$, which you should define. Your approximation should be valid whenever the function is twice differentiable with a unique minimum. Noting that the corresponding density for $\eta = \mathbf{a}^T\boldsymbol{\theta}$ is

$$\pi_y^*(\eta) \propto \left[1 + \nu^{-1}\left(\eta - \mathbf{a}^T\boldsymbol{\mu}\right)^T\left(\mathbf{a}^T\mathbf{R}^{-1}\mathbf{a}\right)^{-1}\left(\eta - \mathbf{a}^T\boldsymbol{\mu}\right)\right]^{-\frac{1}{2}(\nu+1)},$$

suggest a generalized t-approximation to the marginal posterior density of η for the model in (5.2.19).

Model Answer 5F:

From these assumptions, it follows from (5.2.19) that the posterior density of $\boldsymbol{\theta}$ is

$$\pi_y(\boldsymbol{\theta}) \propto \left[\rho(\boldsymbol{\theta})\right]^{-\frac{1}{2}n},$$

where

$$\rho(\boldsymbol{\theta}) = \sum_{i=1}^{n}\left\{y_i - f(x_i, \boldsymbol{\theta})\right\}^2.$$

Note that $\hat{\theta}$ minimizes $\rho(\boldsymbol{\theta})$ and that under the regularity conditions stated, the matrix

$$\mathbf{Q} = -\left[\frac{\partial^2\rho(\boldsymbol{\theta})}{\partial\left(\boldsymbol{\theta}\boldsymbol{\theta}^T\right)}\right]_{\boldsymbol{\theta}=\hat{\boldsymbol{\theta}}}$$

will be positive definite. Expanding $\rho(\boldsymbol{\theta})$ in a Taylor series, about $\boldsymbol{\theta} = \hat{\boldsymbol{\theta}}$, and neglecting cubic and higher terms yield the second-order Taylor series approximation

$$\rho^*(\boldsymbol{\theta}) = S_R^2 + \tfrac{1}{2}\left(\boldsymbol{\theta} - \hat{\boldsymbol{\theta}}\right)^T\mathbf{Q}\left(\boldsymbol{\theta} - \hat{\boldsymbol{\theta}}\right),$$

to $\rho(\boldsymbol{\theta})$, where $S_R^2 = \rho(\hat{\boldsymbol{\theta}})$. Consequently, the posterior density of $\boldsymbol{\theta}$ may be approximated by

$$\pi_y^*(\boldsymbol{\theta}) \propto \left[S_R^2 + \tfrac{1}{2}\left(\boldsymbol{\theta} - \hat{\boldsymbol{\theta}}\right)^T\mathbf{Q}\left(\boldsymbol{\theta} - \hat{\boldsymbol{\theta}}\right)\right]^{-\frac{1}{2}n}.$$

This is a generalized multivariate t-density, with $\nu = n - p$, $\boldsymbol{\mu} = \hat{\boldsymbol{\theta}}$, and $\mathbf{R} = (n-p)\,\mathbf{Q}/S_R^2$. Let $\eta = \mathbf{a}^T\boldsymbol{\theta}$. Then the marginal posterior distribution of

$$\frac{\eta - \mathbf{a}^T\hat{\boldsymbol{\theta}}}{\left(\mathbf{a}^T\mathbf{R}^{-1}\mathbf{a}\right)^{\frac{1}{2}}}$$

is approximately Gossett's t-distribution with $n - p$ degrees of freedom.

5.3 The Kalman Filter

Following Kalman (1960), Astrom (1970), Blight (1974), Leonard (1982b), Harrison and Stevens (1976), and West and Harrison (1990), consider the constant-variance two-stage Markovian model

$$y_i = \theta_i + \delta_i, \tag{5.3.1}$$

$$\theta_i = \theta_{i-1} + \epsilon_i \qquad (i = 1, 2, \ldots, m), \tag{5.3.2}$$

where the θ_i are process parameters and the δ_i and ϵ_i are independent error terms. Take the δ_i to be normally $N(0, \tau^2)$ distributed, and for $i = 2, \ldots, m$, take the ϵ_i to be $N(0, \sigma^2)$ distributed. Take ϵ_1 to be $N(0, \gamma\tau^2)$, and θ_0 and γ to be specified. Since this model involves a Markov process with superimposed random noise, it might be suitable for modeling stock prices.

Let $\alpha = \sigma^2/\tau^2$ denote the signal-to-noise ratio and let ξ denote the smaller root of the equation

$$\xi^{-1} + \xi = 2 + \alpha. \tag{5.3.3}$$

Then, whenever $\xi \leq 1$, the model in (5.3.1) and (5.3.2), for $i = 1, \ldots, m$, is equivalent to the ARIMA process, discussed by Box and Jenkins (1976, p. 8), where

$$y_i - y_{i-1} = q_i - \xi q_{i-1}, \tag{5.3.4}$$

with the q_i independent, and $N(0, \xi^{-1}\tau^2)$ distributed. This provides an alternative to spectral analysis for stationary time series. Bayesian spectral analysis is described by Bretthorst (1988).

Let $\mathbf{y}_i = (y_1, \ldots, y_i)^T$, for $i = 1, 2, \ldots, m$,

$$a_i = E\big(\theta_i \mid \mathbf{y}_i, \sigma^2, \tau^2\big), \tag{5.3.5}$$

and

$$v_i = \mathrm{var}\big(\theta_i \mid \mathbf{y}_i, \sigma^2, \tau^2\big). \tag{5.3.6}$$

When σ^2 and τ^2 are known, θ_i, given $\mathbf{y}_i$, is $N(a_i, v_i)$ distributed (as all variables possess a multivariate normal distribution), so that θ_{i+1}, given $\mathbf{y}_i$, is $N(a_i, v_i + \sigma^2)$ distributed. Regard this $N(a_i, v_i + \sigma^2)$ distribution as the prior distribution for θ_{i+1}, before y_{i+1} is observed. Then, by an easy application of the conjugate analysis, leading to the posterior density (3.4.10), the posterior distribution of θ_{i+1}, given $\mathbf{y}_{i+1}$, is $N(a_{i+1}, v_{i+1})$, where

$$a_{i+1} = \frac{\tau^{-2} y_{i+1} + \big(v_i + \sigma^2\big)^{-1} a_i}{\tau^{-2} + \big(v_i + \sigma^2\big)^{-1}} \qquad (i = 0, \ldots, m-1) \tag{5.3.7}$$

and

$$v_{i+1}^{-1} = \big(v_i + \sigma^2\big)^{-1} + \tau^{-2} \qquad (i = 1, \ldots, m-1). \tag{5.3.8}$$

A slight rearrangement of (5.3.7) and (5.3.8) yields a special case of the *Kalman filter*, that is, the updating relations

$$a_{i+1} = a_i + D_{i+1}(y_{i+1} - a_i) \qquad (i = 0, \ldots, m-1) \tag{5.3.9}$$

and

$$D_{i+1} = \frac{D_i + \alpha}{D_i + \alpha + 1} \qquad (i = 1, \ldots, m-1), \tag{5.3.10}$$

with $v_i = \tau^2 D_i$ and $D_1 = \gamma/(1+\gamma)$. In the situation where τ^2 and σ^2 are unknown, take $\nu_1\kappa_1/\tau^2$ and $\nu_2\kappa_2/\sigma^2$ to possess independent chi-squared distributions with respective degrees of freedom ν_1 and ν_2. Then the joint prior density of τ^2 and α is

$$\pi\left(\tau^2, \alpha\right) \propto \left(\tau^2\right)^{-\frac{1}{2}(\nu_1+\nu_2+2)} \alpha^{-\frac{1}{2}(\nu_2+2)} \exp\left\{-\frac{1}{2\tau^2}\nu_1\kappa_1 - \frac{1}{2\tau^2\alpha}\nu_2\kappa_2\right\}, \tag{5.3.11}$$

where $0 < \tau^2 < \infty$ and $0 < \alpha < \infty$. The density of $\mathbf{y}_m$, given α and τ^2, may be arranged in the form

$$p\left(\mathbf{y}_m \mid \tau^2, \alpha\right) \propto \left(\tau^2\right)^{-\frac{1}{2}m} \{U_1(\alpha)\}^{-\frac{1}{2}} \exp\left\{-\frac{1}{2\tau^2} U_2(\alpha)\right\}, \tag{5.3.12}$$

where

$$U_1(\alpha) = (1+\gamma) \prod_{i=1}^{m-1} (1 + \alpha + D_i), \tag{5.3.13}$$

and

$$U_2(\alpha) = (1+\gamma)^{-1}(y_1 - \theta_0)^2 + \sum_{i=1}^{m-1} (1 + \alpha + D_i)^{-1}(y_{i+1} - a_i)^2, \tag{5.3.14}$$

where the a_i and D_i depend upon α and satisfy (5.3.9) and (5.3.10). Note that the expressions in (5.3.12)–(5.3.14) can themselves be readily updated in time by multiplying (5.3.12) by the predictive density of the next observation. Consequently, the posterior density of the signal-to-noise ratio is

$$\begin{aligned} \pi(\alpha \mid \mathbf{y}_m) &\propto \int_0^\infty \pi\left(\tau^2, \alpha\right) p\left(\mathbf{y}_m \mid \tau^2, \alpha\right) d\tau^2 \qquad (0 < \alpha < \infty), \\ &\propto \left\{\mathbf{U}_1^*(\alpha)\right\}^{-\frac{1}{2}} \left\{\mathbf{U}_2^*(\alpha)\right\}^{-\frac{1}{2}\nu_T}, \end{aligned} \tag{5.3.15}$$

where

$$\mathbf{U}_1^*(\alpha) = \alpha^{(\nu_2+2)} \mathbf{U}_1(\alpha), \tag{5.3.16}$$

$$\mathbf{U}_2^*(\alpha) = \nu_1\kappa_1 + \alpha^{-1}\nu_2\kappa_2 + \mathbf{U}_2(\alpha), \tag{5.3.17}$$

and

$$\nu_T = \nu_1 + \nu_2 + m. \tag{5.3.18}$$

Now $a_m = a_m(\alpha)$, satisfying (5.3.7), denotes the posterior mean of θ_m, given $\mathbf{y}_m$, and only α. Therefore the unconditional mean a_m^* of θ_m, given $\mathbf{y}_m$, may be computed via the one-dimensional numerical integration

$$a_m^* = \int_0^\infty a_m(\alpha)\pi(\alpha \mid \mathbf{y}_m)\, d\alpha. \tag{5.3.19}$$

For $\nu_T > 2$, the unconditional posterior variance v_m^* of θ_m, given $\mathbf{y}_m$, may be computed as a similar posterior expectation, but of the quantity

$$\left(a_m - a_m^*\right)^2 + (\nu_T - 2)^{-1} U_2^*(\alpha) D_m. \tag{5.3.20}$$

Moreover, the unconditional posterior density of θ_m, given $\mathbf{y}_m$, is

$$\begin{aligned}\pi(\theta_m \mid \mathbf{y}_m) &= \int_0^\infty \pi(\theta_m, \alpha \mid \mathbf{y}_m)\, d\alpha \\ &\propto \int_0^\infty D_m^{-\frac{1}{2}}\left[U_1^*(\alpha)\right]^{-\frac{1}{2}}\left\{\mathbf{U}_2^*(\alpha) + D_m^{-1}(\theta_m - a_m)^2\right\}^{-\frac{1}{2}(\nu_T+1)} d\alpha \\ &\qquad (-\infty < \theta_m < \infty),\end{aligned}$$

and the j-step ahead predictive density of y_{m+j}, given $\mathbf{y}_m$, is

$$\begin{aligned}p(y_{m+j} \mid \mathbf{y}_m) &= \int_0^\infty p(y_{m+j}, \alpha \mid \mathbf{y}_m)\, d\alpha \\ &= \int_0^\infty D_{m,j}^{-\frac{1}{2}}\left[U_1^*(\alpha)\right]^{-\frac{1}{2}}\left\{U_2^*(\alpha) + D_{m,j}^{-1}(y_{m+j} - a_m)^2\right\}^{-\frac{1}{2}(\nu_T+1)} d\alpha,\end{aligned} \tag{5.3.21}$$

where $-\infty < y_{m+j} < \infty$ and $D_{m,j} = 1 + j\alpha + D_m$. Note that all the above procedures can be updated in time m, even when τ^2 and α are unknown, by storing all quantities that depend upon α on a grid with, say, $l = 200$ values of α, updating the values on the grid, and with them performing all appropriate numerical integrations, using Simpson's rule at each time stage. This procedure was applied to obtain the virtually exact numerical results of the next section.

The preceding fully Bayesian procedure provides an alternative to the maximum likelihood estimates for variance components. For more complicated models, variance components may also be estimated using the EM algorithm (see Sections 1.5 and 6.3).

SELF-STUDY EXERCISES

5.3.a Consider the linear growth model (a model suitable for macroeconomic prediction, often without needing further variables; see Harrison and Stevens, 1976), where

$$y_i = \theta_i + \delta_i,$$
$$\theta_i = \xi_i + \theta_{i-1} + \epsilon_i,$$

and

$$\xi_i = \xi_{i-1} + \omega_i \qquad (i = 1, \ldots, m)$$

and where the δ_i, ϵ_i, and ω_i are independent. Take the δ_i to be normally $N(0, \tau^2)$ distributed, and for $i = 2, \ldots, m$, take the ϵ_i to be $N(0, \alpha\tau^2)$ distributed and the ω_i to be $N(0, \alpha\beta\tau^2)$ distributed. Take ϵ_1 and ξ_1 to be $N(0, \gamma\tau^2)$ and $N(0, \gamma^*\alpha\tau^2)$, and θ_0, ξ_0, γ_0, and γ^* to be specified. When τ^2, α, and β are known, develop recursive relations for

$$a_i = E(\theta_i \mid \mathbf{y}_i)$$

and

$$b_i = E(\xi_i \mid \mathbf{y}_i),$$

and show how to calculate a_i and b_i. Show that two-dimensional numerical integrations would be needed to calculate these quantities, unconditionally upon τ^2, α, and β.

Hint: By combining θ_i and ξ_i into a single vector, first express this model as a special case of the general formulation of Worked Example 5G. Then apply the recursion formulas developed in Model Answer 5G.

5.3.b Consider the Markovian model in (5.3.1) and (5.3.2), but where at the first stage the y_i are independent and Poisson distributed, given their means (and variances) θ_i, and at the second stage, given θ_{i-1}, θ_i possesses, for $i = 2, \ldots, m$, Gamma distributions with parameters θ_{i-1}/α and $1/\alpha$, that is, mean θ_{i-1} and variance $\alpha\theta_{i-1}$. Suppose that θ_1 possesses a Gamma density, with parameters $\gamma\theta_0$ and where γ and θ_0 are specified.

Let $\theta^*_{m,j}$ denote our j-step ahead forecast for $\boldsymbol{\theta}_{m+j}$, given $\mathbf{y}_m$, when α is known. Generally discuss the reasonability of extending the idea expressed in Exercise 3.8.g by restricting attention to linear predictors of the form

$$\theta^*_{m,j} = \lambda_{0j} + \sum_{i=1}^{m} \lambda_{ij} y_i$$

and by choosing the λ coefficients to minimize the average risk

$$E\big[L\big(\theta^*_{m,j}, \theta_{m+j}\big)\big],$$

where the expectation should be taken with respect to both $\mathbf{y}_m$ and $\boldsymbol{\theta}_{m+j}$, and

$$L\big(\theta^*_{m,j}, \theta_{m+j}\big) = \big(\theta^*_{m,j} - \theta_{m+j}\big)^2.$$

For example, how much approximation is involved in the linearity assumption?

Note:

(a) Leonard (1982b) proves that the linear predictor, minimizing the Bayes risk, is $\theta^*_{m,j} = a_m$ for $j = 0, 1, \ldots,$ where a_m satisfies the Kalman filter recursions (5.3.9) and (5.3.10), with $a_0 = \theta_0$ and $D_1 = \gamma/(1+\gamma)$. A more general methodology is available, when α is unknown, based upon an approximation to the posterior density $\pi(\alpha \mid \mathbf{y}_m)$.

(b) The Bayesian analysis of continuous-time nonhomogeneous Poisson processes is discussed by Leonard (1978) and Clevenson and Zidek (1977). This is closely related to the theory of doubly stochastic Poisson processes, summarized with applications in electrical engineering, by Snyder (1975).

5.4 An On-Line Analysis of Chemical Process Readings

The $n = 197$ observations described in Figure 5.4.1 were reported by Box and Jenkins (1976, p. 525); 17.0 has been subtracted from each observation. Box and Jenkins (p. 239) fit the ARIMA model (5.3.4), with $\xi = 0.70$ and $\alpha = 0.13$.

We first assumed the vague prior obtained upon setting $\nu_1 = 2, \nu_2 = -2$, and $\kappa_1 = \kappa_2 = 0$ in (5.3.11). The points in Figure 5.4.2 are the resultant unconditional posterior means a_i^* of $\boldsymbol{\theta}_i$, given $\mathbf{y}_i^*$, for $i = 1, \ldots, 197$. The posterior expectation of τ^2, given $\mathbf{y}_{197}$, was equal to 0.066. The value $E(\alpha \mid \mathbf{y}_{197}) = 0.20$ suggests that the signal-to-noise ratio is low.

The posterior density $\pi(\alpha \mid \mathbf{x}_{197})$ is not reported here. However, the posterior mode at $\alpha = 0.13$ was identical to the value recommended by Box and Jenkins. Since the posterior distribution of α was skew to the right, with a very thick tail, the value $\alpha = 0.20$ may be more appropriate. The unconditional posterior density $\pi(\theta_{197} \mid \mathbf{y}_{197})$ was almost exactly a $N(0.49, 0.022)$ density. Also, the predictive densities $p(y_{198} \mid \mathbf{y}_{197}), p(y_{199} \mid \mathbf{y}_{197}), \ldots, p(y_{202} \mid \mathbf{y}_{197})$ were almost exactly normal densities with mean 0.49 and respective variances 0.101, 0.114, 0.127, 0.140, and 0.153. Therefore, our procedure permits posterior and predictive distributions, which seem much more useful than point predictions.

The analysis was repeated under an informative choice of the prior (5.3.11) with $\nu_1 = \nu_2 = 10, \kappa_1 = 0.05, \kappa_2 = 0.025$, and a prior mean of 0.63 for α. The predictions obtained were remarkably close to the results in Figure 5.4.2 and are reported by Leonard (1982b).

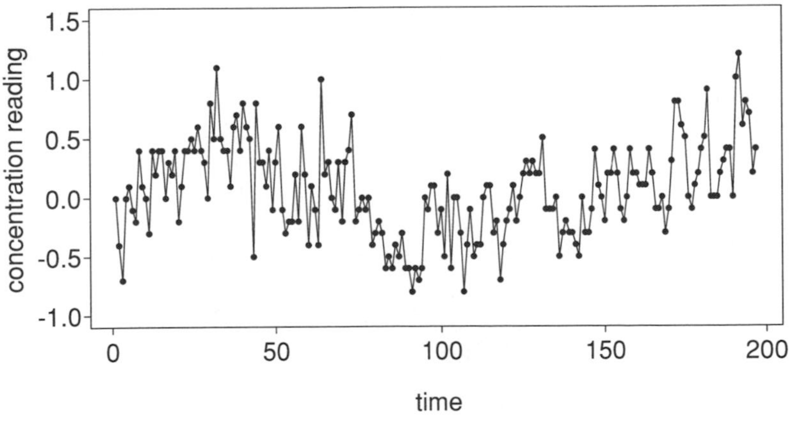

Figure 5.4.1. Chemical process readings.

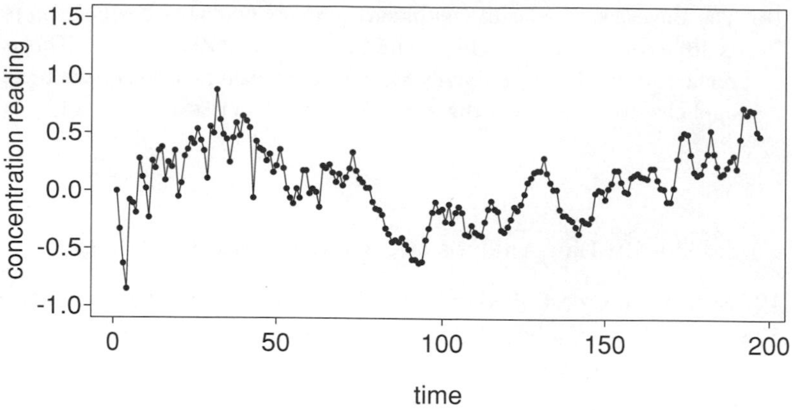

Figure 5.4.2. Forecasts for chemical process data.

5.5 An Industrial Control Chart

The fifty-two Poisson observations described in Figure 5.5.1 were reported by Hald (1952, p. 720) and represent the number of defective items in consecutive shifts of an industrial process. An approximate analysis reported by Leonard (1982b), and outlined in Exercise 5.3.b, was performed. A uniform prior was first assumed for α. The approximate unconditional posterior means $a_i^* = E(\theta_i \mid \mathbf{y}_i)$ are reported in Figure 5.5.2. The value $a_{52}^* = 0.05$ is based upon a posterior density for α, given $\mathbf{y}_{52}$, with a mode at $\alpha = 0$ and a very thick tail. There is therefore some, but not extreme, evidence that the θ_i are changing in time. The predictive variance of y_{53}, given $\mathbf{y}_{52}$, is about 1.24 times the predictive mean of 2.93, suggesting that a Pascal approximation to the predictive distribution might be appropriate.

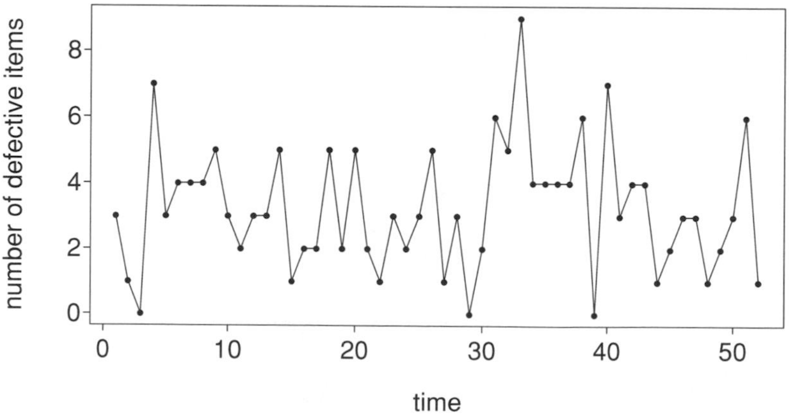

Figure 5.5.1. Number of defective items.

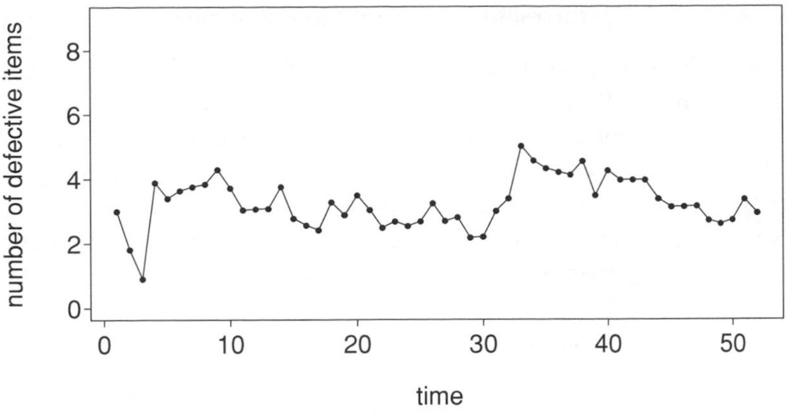

Figure 5.5.2. Forecasts of number of defective items.

The prior distribution of 5α was next assumed to be F with both degrees of freedom equal to 10. The corresponding a_i^* were quite close to the values in Figure 5.5.2 and are reported by Leonard (1982b). The posterior density of α, given $\mathbf{y}_{52}$, possessed mode 0.07, mean 0.10, and was skew to the right.

5.6 Forecasting Geographical Proportions for World Sales of Fibers

Let $\mathbf{P}_t = (P_{1t}, \ldots, P_{qt})^T$ denote the proportions of sales of fibers sold by Imperial Chemical Industries (ICI) during year t, in q different regions of the world. By considering proportions, rather than actual sales, it is possible to model the total world sales separately. Suppose that given α and $\boldsymbol{\theta}_t = (\theta_{1r}, \ldots, \theta_{qr})^T$, $\mathbf{P}_t$ possesses a Dirichlet distribution, with parameter vector $\alpha\boldsymbol{\theta}_t$, so that $E(\mathbf{P}_t \mid \alpha, \boldsymbol{\theta}_t) = \boldsymbol{\theta}_t$. Then Harrison, Leonard, and Gazard (1983) assume the "quadratic prediction" dynamic linear model, defined by the system equations

$$\boldsymbol{\theta}_t = \boldsymbol{\theta}_{t-1} + \boldsymbol{\beta}_{t-1} + \boldsymbol{\xi}_{t-1} + \boldsymbol{\epsilon}_{1t},$$

$$\boldsymbol{\beta}_t = \boldsymbol{\beta}_{t-1} + \boldsymbol{\xi}_{t-1} + \boldsymbol{\epsilon}_{2t},$$

$$\boldsymbol{\xi}_t = \boldsymbol{\xi}_{t-1} + \boldsymbol{\epsilon}_{3t} \qquad (t = 1, \ldots, m),$$

with $\boldsymbol{\xi}_t = \boldsymbol{\xi}_0, \boldsymbol{\beta}_t = \boldsymbol{\beta}_0$, and the $\boldsymbol{\epsilon}_{jt}$ independent $N(\mathbf{0}, \mathbf{C}_j)$, for $j = 1, \ldots, 3$ and $t = 1, \ldots, m$. Note that no explanatory variables (e.g., endogenous variables based upon economic considerations) are included in this model. Nevertheless, the quadratic predictions reported by Harrison, Leonard, and Gazard (1983) seemed to perform just as well as models with many more variables, primarily because of the problems of accurately modeling these extra variables. In general, a simple forecasting model can often perform better than a complicated model, and either linear or quadratic prediction often suffices.

5.7 Bayesian Forecasting in Economics

There are many splendid economic applications of Bayesian forecasting (e.g., Zellner, 1971, 1983; Highfield, 1986). Zellner and Hong (1989) and Min and Zellner (1993) forecast international growth rates using Bayesian shrinkage estimators, while Zellner, Hong, and Min (1991) forecast turning points in growth rates. Blattberg and George (1991) apply shrinkage estimation to price and promotional elasticities. Geisel (1975) analyzes simple macroeconomic models, and Hong (1989) forecasts real-output growth rates and cyclical properties of models. Lesage and Magura (1990) forecast regional payrolls.

In particular, Zellner, Hong, and Min consider a process of the form

$$y_{it} = \mathbf{x}_{it}^T \boldsymbol{\beta}_{it} + \epsilon_{it}, \tag{5.7.1}$$

$$\boldsymbol{\beta}_{it} = \boldsymbol{\beta}_{it-1} + \boldsymbol{\omega}_{it} \qquad (i = 1, 2, \ldots, N; t = 1, \ldots, T), \tag{5.7.2}$$

where the ϵ_{it} and $\boldsymbol{\omega}_{it}$ represent independent error terms, and the $\mathbf{x}_{it}$ are vectors of explanatory variables. An on-line Bayesian analysis for this model is available on lines similar to the methodology of Section 5.3.

For many data sets in economics and geology, the constant-variance assumptions in the preceding models may be relaxed to accommodate extra volatility in the time series. The concept of "stochastic volatility" has been developed in the economics literature. Recent references include Taylor (1994) and Shepherd and Pitt (1998), following an early suggestion by Leonard (1975a). Together with a stochastic model for the means of the process, it is possible to assume, say, a first-order autoregressive process for the logs of the variances. The logarithmic transformations to normality highlight one of the major themes of this volume, which is more fully developed in Chapter 6. Our probability structures for the log-variances are also currently being applied in animal sciences by Jean-Louis Foulley and Daniel Ganiola.

Worked Example 5G: *A Multivariate Kalman Filter*

Consider the hierarchical model where

$$\mathbf{y}_t = \boldsymbol{\theta}_t + \boldsymbol{\delta}_t \qquad (t = 1, 2, \ldots)$$

and

$$\boldsymbol{\theta}_t = \boldsymbol{\theta}_{t-1} + \boldsymbol{\epsilon}_t \qquad (t = 1, 2, \ldots),$$

where $\boldsymbol{\theta}_0 = \mathbf{0}$ and the $\boldsymbol{\delta}$ and $\boldsymbol{\epsilon}$ are independent multivariate normal vectors with zero mean vectors. The $\boldsymbol{\delta}_t$ possess common covariance matrix $\mathbf{G}$ and the $\boldsymbol{\epsilon}_t$ possess covariance matrix $\mathbf{C}$. The matrices $\mathbf{G}$ and $\mathbf{C}$ are known and specified. Let $\boldsymbol{\theta}_t^*$ denote the posterior mean vector of $\boldsymbol{\theta}_t$, given $\mathbf{y}_1, \mathbf{y}_2, \ldots, \mathbf{y}_t$, and let $\mathbf{D}_t$ denote the corresponding posterior covariance matrix. Find recurrence relations expressing $\mathbf{D}_t$ in terms of $\mathbf{D}_{t-1}$ and $\boldsymbol{\theta}_t^*$ in terms of $\boldsymbol{\theta}_{t-1}^*$ and $\mathbf{D}_{t-1}$. Please refer to Lemma 6.1.1 of Chapter 6 in order to complete the square in $\boldsymbol{\theta}_t$. You will also need to show that Bayes' theorem reduces to a relation to the form

$$\pi(\boldsymbol{\theta}_t \mid \mathbf{y}_1, \mathbf{y}_2, \ldots, \mathbf{y}_t) \propto p(\mathbf{y}_t \mid \boldsymbol{\theta}_t)\pi(\boldsymbol{\theta}_t \mid \mathbf{y}_1, \mathbf{y}_2, \ldots, \mathbf{y}_{t-1}),$$

in this special case. This question is good practice for some of the methodology in Chapter 6. Note that a very general formulation, including the models in Exercise 5.3.a and Section 5.7, would prefix the first terms on the right-hand sides of our hierarchical model by matrices $\mathbf{A}_i$ and $\mathbf{B}_t$.

Model Answer 5G:

Note that with $\mathbf{Y}_{t-1} = (\mathbf{y}_1, \mathbf{y}_2, \ldots, \mathbf{y}_{t-1})^T$,

$$\theta_{t-1} \mid \mathbf{Y}_{t-1} \sim N(\boldsymbol{\theta}^*_{t-1}, \mathbf{D}_{t-1}),$$

a multivariate normal distribution, with mean vector $\boldsymbol{\theta}^*_{t-1}$ and covariance matrix $\mathbf{D}_{t-1}$.

Since $\boldsymbol{\theta}_t = \boldsymbol{\theta}_{t-1} + \boldsymbol{\epsilon}_t$, where $\boldsymbol{\epsilon}_t \sim N(\mathbf{0}, \mathbf{C})$,

$$\boldsymbol{\theta}_t \mid \mathbf{Y}_{t-1} \sim N(\boldsymbol{\theta}^*_{t-1}, \mathbf{D}_{t-1} + \mathbf{C}).$$

By Bayes' theorem, conditioning on $\mathbf{Y}_{t-1}$ in each density,

$$\begin{aligned}\pi(\boldsymbol{\theta}_t \mid \mathbf{Y}_t) &= \pi(\boldsymbol{\theta}_t \mid \mathbf{y}_t, \mathbf{Y}_{t-1})\\ &\propto p(\mathbf{y}_t \mid \boldsymbol{\theta}_t, \mathbf{Y}_{t-1})\pi(\boldsymbol{\theta}_t \mid \mathbf{Y}_{t-1})\\ &\propto p(\mathbf{y}_t \mid \boldsymbol{\theta}_t)\pi(\boldsymbol{\theta}_t \mid \mathbf{Y}_{t-1}),\end{aligned}$$

since, when $\boldsymbol{\theta}_t$ is specified, the distribution of $\mathbf{Y}_t$ is unaffected by $\mathbf{Y}_{t-1}$.

As $\mathbf{Y}_t \mid \boldsymbol{\theta}_t \sim N(\boldsymbol{\theta}_t, \mathbf{G})$,

$$\begin{aligned}\pi(\boldsymbol{\theta}_t \mid \mathbf{Y}_t) \propto \exp\big\{ &-\tfrac{1}{2}(\mathbf{y}_t - \boldsymbol{\theta}_t)^T \mathbf{G}^{-1}(\mathbf{y}_t - \boldsymbol{\theta}_t)\\ &-\tfrac{1}{2}(\boldsymbol{\theta}_t - \boldsymbol{\theta}^*_{t-1})^T(\mathbf{D}_{t-1} + \mathbf{C})^{-1}(\theta_t - \boldsymbol{\theta}^*_{t-1})\big\}.\end{aligned}$$

Now complete the square in $\boldsymbol{\theta}_t$, using the hint, with $\mathbf{A} = \mathbf{G}^{-1}$, $\mathbf{a} = \mathbf{y}_t$, $\mathbf{B} = (\mathbf{D}_{t-1} + \mathbf{C})^{-1}$, and $\mathbf{b} = \boldsymbol{\theta}_{t-1}$. The second contribution on the right side of the completed square will not be needed, since it does not depend upon $\boldsymbol{\theta}_t$, and gets absorbed in the proportionality constant of $\pi(\boldsymbol{\theta}_t \mid \mathbf{Y}_t)$. The first contribution yields

$$\pi(\boldsymbol{\theta}_t \mid \mathbf{Y}_t) \propto \exp\big\{ -\tfrac{1}{2}(\boldsymbol{\theta}_t - \boldsymbol{\theta}^*_t)^T \mathbf{D}_t^{-1}(\boldsymbol{\theta}_t - \boldsymbol{\theta}^*_t)\big\},$$

where $\boldsymbol{\theta}^*_t = (\mathbf{A} + \mathbf{B})^{-1}(\mathbf{A}\mathbf{a} + \mathbf{B}\mathbf{b})$ and $\mathbf{D}_t^{-1} = \mathbf{A} + \mathbf{B}$, so that

$$\boldsymbol{\theta}^*_t = \{G^{-1} + (\mathbf{D}_{t-1} + \mathbf{C})^{-1}\}^{-1}\{G^{-1}\mathbf{y}_t + (\mathbf{D}_{t-1} + \mathbf{C})^{-1}\boldsymbol{\theta}^*_{t-1}\} \qquad (t = 1, 2, \ldots)$$

and

$$\mathbf{D}_t^{-1} = G^{-1} + (\mathbf{D}_{t-1} + \mathbf{C})^{-1} \qquad (t = 1, 2, \ldots).$$

Hence, our recurrence relations are well defined. Since $\boldsymbol{\theta}^*_0 = \mathbf{0}$ and $\mathbf{D}_0 = \mathbf{0}$, all elements in the sequence can be calculated recursively.

6

Prior Structures, Posterior Smoothing, and Bayes–Stein Estimation

6.0 Preliminaries and Overview

Stein (1956) showed that the UMVU estimators for several normal means are inadmissible, with respect to squared error loss, in dimensions higher than two. Lindley (1962) proposed shrinkage estimators that not only smooth the UMVU estimators in a sensible manner, but possess even better mean squared error properties by shrinking each estimate towards the grand mean. This is a key issue that influences Bayesian and Fisherian thinking to this day.

For many problems with several parameters, it is possible to find shrinkage estimators with similarly superior frequency properties. However, it is also important for the shrinkage estimators to smooth, say, the maximum likelihood estimators in a statistically sensible manner. In this chapter, we exploit the duality between the "prior structure," that is, the functional form of the prior, and the posterior smoothing. Unknown prior parameters appearing in the prior structure can be estimated from the current data set by empirical or hierarchical Bayesian techniques. We hence obtain, in many nonlinear situations, sensible data-based smoothing estimators with potentially superior frequency properties. The latter may be evaluated either theoretically or via computer simulation in any particular case.

In Section 6.1, we continue our unifying theme of seeking normalizing transformations for the parameters, which can suggest both a multivariate normal prior and an approximate multivariate normal likelihood. Related approximations also justify BIC (see Schwarz, 1978, and equation (1.1.6)) in a manner much more general than Schwarz's original proof and also lead (Exercise 6.1.a) to an interesting large sample approximation to the model-checking criterion developed by Box (1980). The consequent "unifying approach to multiparameter problems" may also be used to analyze nonlinear random effects models with many applications, such as for animal genetics, psychometrics, the ozone layer, and so on.

In Exercise 6.1.b, it is demonstrated that as well as smoothing the estimates for several parameters, it is possible to semi-parametrically smooth the estimate for an entire (logistic) regression function. This parallels density-smoothing techniques; the Bayesian methodology can also be interpreted in terms of the theory of splines, using a unifying duality first observed by Kimeldorf and Wahba (1970). The mathematics is related to quantum theory, to Dirichlet boundary-value problems in physics, and to the theory of nonlinear Fredholm integral equations (Silver, Martz, and Wallstrom, 1993, and Leonard, 1978, discuss these dualities).

In Exercise 6.1.c, a related method is discussed for smoothing relative frequency histograms, Exercises 6.1.d–6.1.f are concerned with a general conjugate analysis for the linear statistical model, and extra reading is suggested regarding the Bayesian smoothing of normal regression functions. It is also suggested that researchers either work out special cases of the linear model themselves or refer to Box and Tiao (1973).

In Section 6.2, approximations are provided for posterior densities under multivariate normal priors, and the results are applied to a simple special case in Exercises 6.2.a and 6.2.b. A general theory of prior structures and the estimation of hyperparameters is described in Section 6.3 (A). The theory is relevant under a range of philosophies, for example, empirical and hierarchical Bayes; but whichever philosophy is employed, appealing smoothing estimators can be computed for the first-stage prior parameters. In Section 6.3 (B), it is shown that these ideas are also relevant to the development of sampling models, which are appropriate for nonrandomized data, for example, taking account of overdispersion and serial or spatial correlation, or nonlinear relationships. Sampling models generalizing the multinomial and multivariate normal are described.

Self-Study Exercises 6.3.a–6.3.i develop and investigate the general theory in a variety of special cases, such as several exchangeable parameters, binomial proportions, normal or Poisson means, autoregressive models for nonnormal data, M-group regression, and the analysis of several multinomial distributions. Exercise 6.3.j describes Stein's mean squared error results for several normal means via integration by parts. This motivates the consideration of many of the shrinkage estimators in this chapter.

In Section 6.4, Monte Carlo and importance sampling methods are described, together with related convergence theorems. These have already been applied to several examples in Chapter 5. In Section 6.5, we describe a number of exciting examples that emphasize the real practical advantages of the Bayesian approach. Frequency properties are emphasized in Sections 6.5 (B), (G), and (I). MCMC procedures are considered in Sections 6.6–6.9.

Further semi-parametric procedures are developed in Section 6.7, via Bayesian inference for sampling models that can be expressed as mixtures. In particular, the Lavine–West Gibbs sampling approach is described, followed by a practical case study in Section 6.8. The rejection-sampling method (Zeger and Karim, 1991; Carlin and Gelfand, 1991) in Section 6.9, for generalized linear models with random effects, takes us towards the frontiers of Bayesian research. Some self-study exercises in this chapter describe challenging research-level problems. Many important research results are preserved in these exercises.

6.1 Multivariate Normal Priors for the Transformed Parameters

In Section 1.5 (B), it is demonstrated that in many situations, a preliminary transformation of the parameters can ensure that the (twice-differentiable) likelihood

$$l(\boldsymbol{\gamma} \mid \mathbf{y}) = p(\mathbf{y} \mid \boldsymbol{\gamma}) \qquad (\boldsymbol{\gamma} \in R^p) \tag{6.1.1}$$

of a $p \times 1$ vector of parameters $\boldsymbol{\gamma} = (\gamma_1, \ldots, \gamma_p)^T$, given an $n \times 1$ observation vector $\mathbf{y}$, can be approximated, when the maximum likelihood vector $\hat{\boldsymbol{\gamma}}$ is finite, by

$$l^*(\boldsymbol{\gamma} \mid \mathbf{y}) = l(\hat{\boldsymbol{\gamma}} \mid \mathbf{y}) \exp\left\{-\tfrac{1}{2}(\boldsymbol{\gamma} - \hat{\boldsymbol{\gamma}})^T \mathbf{R} (\boldsymbol{\gamma} - \hat{\boldsymbol{\gamma}})\right\} \qquad (\boldsymbol{\gamma} \in R^p), \tag{6.1.2}$$

where $\mathbf{R}$ is the likelihood information matrix. Under quite broad regularity conditions, it is true (see, Johnson, 1967, 1970; Walker, 1969; Hoadley, 1971) that whenever $\mathbf{R}^{\frac{1}{2}}(\hat{\boldsymbol{\gamma}} - \boldsymbol{\gamma})$ converges in (sampling) distribution to a standard $N(\mathbf{0}, \mathbf{I}_p)$ distribution, as $n \to \infty$, then $\mathbf{R}^{\frac{1}{2}}(\boldsymbol{\gamma} - \hat{\boldsymbol{\gamma}})$ converges in (posterior) distribution to a $N(\mathbf{0}, \mathbf{I}_p)$ distribution, in situations where the observed vector $\hat{\boldsymbol{\gamma}}$ has finite elements, and for any prior density $\pi(\boldsymbol{\gamma})$, with $\pi(\boldsymbol{\gamma}) > 0$ for all $\boldsymbol{\gamma} \in R^p$. It is frequently also true that when n is large, the ratio of (6.1.2) to the exact likelihood behaves like $1 + Kn^{-\frac{1}{2}}$ where K is constant in n.

In addition to an approximate prior-to-posterior analysis, (6.1.2) permits the derivation of an important result for model comparison. Note that Bayes' theorem, applied backwards, tells us that the prior predictive density $p(\mathbf{y}) = \int p(\mathbf{y} \mid \boldsymbol{\gamma})\pi(\boldsymbol{\gamma})d\boldsymbol{\gamma}$ satisfies

$$p(\mathbf{y}) = \frac{\pi(\boldsymbol{\gamma}) l(\boldsymbol{\gamma} \mid \mathbf{y})}{\pi(\boldsymbol{\gamma} \mid \mathbf{y})} \qquad (\boldsymbol{\gamma} \in R^p), \tag{6.1.3}$$

where $\pi(\boldsymbol{\gamma} \mid \mathbf{y})$ denotes the corresponding posterior density. Under the conditions of the preceding paragraph, constrain attention to situations where $|\mathbf{R}|$ behaves like Cn, as $n \to \infty$, with C constant in n. Then, setting $\boldsymbol{\gamma} = \hat{\boldsymbol{\gamma}}$ in (6.1.3) gives us the large sample approximation

$$p(\mathbf{y}) \approx \frac{\pi(\boldsymbol{\gamma}) l(\boldsymbol{\gamma} \mid \mathbf{y})}{(2\pi)^{-\frac{p}{2}} |\mathbf{R}|^{\frac{1}{2}} \exp\left\{-\tfrac{1}{2}(\boldsymbol{\gamma} - \hat{\boldsymbol{\gamma}})^T \mathbf{R} (\boldsymbol{\gamma} - \hat{\boldsymbol{\gamma}})\right\}} \qquad (\boldsymbol{\gamma} \in R^p) \tag{6.1.4}$$

$$\approx \frac{\pi(\hat{\boldsymbol{\gamma}}) l(\hat{\boldsymbol{\gamma}} \mid \mathbf{y})}{(2\pi)^{-\frac{p}{2}} |\mathbf{R}|^{\frac{1}{2}}} \tag{6.1.5}$$

$$\propto l(\hat{\boldsymbol{\gamma}} \mid \mathbf{y}) \left(\frac{n}{2\pi}\right)^{-\frac{p}{2}} \qquad (n \to \infty). \tag{6.1.6}$$

This provides a justification for Schwarz's BIC in (1.1.6) and for the approximate posterior probability (2.3.3). The derivation is more general than the exponential family justification due to Schwarz (1978). Note that constant multiplicative components of $|\mathbf{R}|^{\frac{1}{2}}$ should not be included in (6.1.6), since application of the approximation to a $p \times 1$ linear transformation of the form $\boldsymbol{\beta} = \mathbf{A}\boldsymbol{\gamma}$, where $\mathbf{A}$ is arbitrary, can then give different results, depending upon $\mathbf{A}$.

Returning to our prior-to-posterior analysis for finite n, suppose that the prior distribution of $\boldsymbol{\gamma}$, given $\boldsymbol{\mu}$ and $\mathbf{C}$, is $N(\boldsymbol{\mu}, \mathbf{C})$. This specification is frequently more flexible than an exactly conjugate prior, since the nondiagonal elements of $\mathbf{C}$ permit prior dependencies between the elements of $\boldsymbol{\gamma}$. Also, for many models, such as the linear logistic regression model with likelihood in (1.3.18), a convenient conjugate prior distribution does not obviously exist.

Lemma 6.1.1 (Completing the square): *For any $p \times p$ positive definite matrix* $\mathbf{A}$, *$p \times p$ positive semidefinite matrix* $\mathbf{B}$, *and $p \times p$ vectors* $\mathbf{a}$ *and* $\mathbf{b}$,

$$(\boldsymbol{\gamma} - \mathbf{a})^T \mathbf{A}(\boldsymbol{\gamma} - \mathbf{a}) + (\boldsymbol{\gamma} - \mathbf{b})^T \mathbf{B}(\boldsymbol{\gamma} - \mathbf{b}) = (\boldsymbol{\gamma} - \boldsymbol{\gamma}^*)^T (\mathbf{A} + \mathbf{B})(\boldsymbol{\gamma} - \boldsymbol{\gamma}^*) + (\mathbf{a} - \mathbf{b})^T \mathbf{H}(\mathbf{a} - \mathbf{b}), \tag{6.1.7}$$

where

$$\boldsymbol{\gamma}^* = (\mathbf{A} + \mathbf{B})^{-1}(\mathbf{A}\mathbf{a} + \mathbf{B}\mathbf{b}) \tag{6.1.8}$$

and

$$\mathbf{H} = \mathbf{A}(\mathbf{A} + \mathbf{B})^{-1}\mathbf{B}. \tag{6.1.9}$$

If, furthermore, $\mathbf{B}$ is positive definite, then

$$\mathbf{H} = \left(\mathbf{A}^{-1} + \mathbf{B}^{-1}\right)^{-1}. \tag{6.1.10}$$

Proof. *Outline:* The vector $\boldsymbol{\gamma}^*$ in (6.1.8) minimizes the left-hand side of (6.1.7), since the first derivative, with respect to $\boldsymbol{\gamma}$, of this expression is

$$2\mathbf{A}(\boldsymbol{\gamma} - \mathbf{a}) + 2\mathbf{B}(\boldsymbol{\gamma} - \mathbf{b}).$$

Since the quadratic term $\boldsymbol{\gamma}^T(\mathbf{A} + \mathbf{B})\boldsymbol{\gamma}$ is common to both the left-hand side and the first term on the right-hand side, it follows that both sides contain linear and quadratic terms in $\boldsymbol{\gamma}$. Finally, both left- and right-hand sides, with $\boldsymbol{\gamma}$ replaced by $\boldsymbol{\gamma}^*$, reduce to $(\mathbf{a} - \mathbf{b})^T \mathbf{H}(\mathbf{a} - \mathbf{b})$. Consequently, the constant terms on either side are also identical. ■

An Approximation to the Posterior Density

It is an immediate consequence of Bayes' theorem and Lemma 6.1.1 that under the likelihood approximation (6.1.2) and an $N(\boldsymbol{\mu}, \mathbf{C})$, the posterior distribution of $\boldsymbol{\gamma}$ is $N(\boldsymbol{\gamma}^*, \mathbf{D}^*)$, where

$$\boldsymbol{\gamma}^* = \left(\mathbf{R} + \mathbf{C}^{-1}\right)^{-1}\left(\mathbf{R}\hat{\boldsymbol{\gamma}} + \mathbf{C}^{-1}\boldsymbol{\mu}\right) \tag{6.1.11}$$

and

$$(\mathbf{D}^*)^{-1} = \mathbf{R}^* = \mathbf{R} + \mathbf{C}^{-1}. \tag{6.1.12}$$

Furthermore, if $\boldsymbol{\gamma}$ has an improper prior density proportional to

$$\exp\left\{-\tfrac{1}{2}(\boldsymbol{\gamma} - \boldsymbol{\mu})^T \mathbf{R}_0(\boldsymbol{\gamma} - \boldsymbol{\mu})\right\},$$

where the precision matrix $\mathbf{R}_0$ is singular, then just replace $\mathbf{C}^{-1}$ by $\mathbf{R}_0$, in (6.1.11) and (6.1.12). The geometry of matrix-weighted average forms, like (6.1.11), is discussed by

Chamberlain and Leamer (1976). The elements of $\boldsymbol{\gamma}^*$ do not necessarily fall between the corresponding elements of $\hat{\boldsymbol{\gamma}}$ and $\boldsymbol{\mu}$. Positive values for the off-diagonal elements of $\mathbf{C}$ can lead to substantial smoothing of the maximum likelihood vector $\hat{\boldsymbol{\gamma}}$. Note that approximate marginalization is very straightforward for linear transformations $\eta = \mathbf{a}^T\boldsymbol{\gamma}$, since these possess approximate $N(\mathbf{a}^T\boldsymbol{\gamma}^*, \mathbf{a}^T\mathbf{D}^*\mathbf{a})$ distributions. These, however, can be replaced by more accurate, nonnormal approximations, which are still algebraically explicit, in terms of $\hat{\boldsymbol{\gamma}}$ and $\mathbf{R}$. See the Leonard, Hsu, and Tsui procedure outlined at the end of Section 5.1 (C). For applications in multivariate analysis, see Press (1982).

Our two-stage distributional assumptions for $\mathbf{y}$ and $\boldsymbol{\gamma}$ can alternatively be interpreted as together comprising a two-stage "random effects" sampling model for $\mathbf{y}$. Such models are particularly applicable to human and animal genetics. For example, many papers contained in the volume by Gianola and Hammond (1990) apply random effects or fixed effects models to the genetic improvement of livestock. One example is provided by the Foulley–Gianola–Im model (1.8.1). These ideas are also applicable to psychometrics, for example, the generalization of Birnbaum's model introduced in Exercise 1.1.i, and also to doubly stochastic Poisson processes (Arjas and Gasbarra, 1994; Dayan, 1987; Heikkinen and Arjas, 1997; and Arjas and Heikkinen, 1997). Alternative random effects models in survival analysis are described by Susarla and Van Ryzin (1976). They are applied to multivariate binary models for interviewer variability by Anderson and Aitkin (1985) and to similar models with serial correlations by Stiratelli, Laird, and Ware (1984).

Worked Example 6A: *A Likelihood Approximation*

A single parameter γ possesses likelihood $l(\gamma \mid \mathbf{y})$ that is twice differentiable for all $\gamma \in (-\infty, \infty)$ and an $N(\mu, \sigma^2)$ prior. Find algebraically explicit approximations to the posterior mean and variance of γ and show how to make posterior probability statements regarding γ.

Model Answer 6A:

In the scalar case, the likelihood approximation (6.1.2) reduces to

$$l^*(\gamma, \mathbf{y}) = l(\hat{\gamma} \mid \mathbf{y}) \exp\left\{\frac{1}{2v}(\gamma - \hat{\gamma})^2\right\},$$

where $\hat{\gamma}$ is the maximum likelihood estimate of γ, and

$$v^{-1} = \left[-\frac{\partial^2}{\partial\gamma^2} \log l(\gamma \mid \mathbf{y})\right]_{\gamma=\hat{\gamma}}.$$

Under our $N(\mu, \sigma^2)$ prior, the posterior density for γ is therefore approximated by

$$\pi^*(\gamma \mid \mathbf{y}) \propto \exp\left\{-\frac{1}{2v}(\gamma - \hat{\gamma})^2 - \frac{1}{2\sigma^2}(\gamma - \mu)^2\right\}.$$

Since

$$v^{-1}(\gamma-\hat{\gamma})^2+\sigma^{-2}(\gamma-\mu)^2=\left(v^{-1}+\sigma^{-2}\right)(\gamma-\gamma^*)^2 + \left(v+\sigma^2\right)^{-1}(\hat{\gamma}-\mu)^2,$$

where $\gamma^*=(v^{-1}+\sigma^{-2})^{-1}(v^{-1}\hat{\gamma}+\sigma^{-2}\mu)$, we have

$$\pi^*(\gamma\mid\mathbf{y})\propto\exp\left\{-\tfrac{1}{2}\left(v^{-1}+\sigma^{-2}\right)(\gamma-\gamma^*)^2\right\}.$$

Hence the posterior distribution of γ is approximately $N(\gamma^*,(v^{-1}+\sigma^{-2})^{-1})$, and the posterior c.d.f. is approximately $\Phi\{(v^{-1}+\sigma^{-2})^{\frac{1}{2}}(\gamma-\gamma^*)\}$.

Worked Example 6B: *Posterior Analysis under Prior Exchangeability*

Consider the approximation (6.1.2) to the likelihood of $\boldsymbol{\gamma}$. Assume the exchangeable prior distribution, where, given ξ, the elements of $\boldsymbol{\gamma}$ are a random sample from a normal $N(\xi,\sigma^2)$ distribution. However, ξ has a normal $N(\mu,\tau^2)$ distribution. Assume that μ, σ^2, and τ^2 are specified. By referring to the corresponding approximation to the posterior distribution of $\boldsymbol{\gamma}$ and ξ,

(a) find the posterior density of ξ;
(b) find an approximation to the posterior distribution of $\boldsymbol{\gamma}$, unconditional upon ξ, by showing how (6.1.11) and (6.1.12) reduce in this case;
(c) state your choices of the prior mean vector $\boldsymbol{\mu}$ and covariance matrix $\mathbf{C}$ of $\boldsymbol{\gamma}$.

Model Answer 6B:

(a) Under the likelihood approximation (6.1.2), the joint posterior density of $\boldsymbol{\gamma}$ and ξ is

$$\pi^*(\boldsymbol{\gamma},\xi)\propto\exp\left\{-\frac{1}{2}(\boldsymbol{\gamma}-\hat{\boldsymbol{\gamma}})^T)\mathbf{R}(\boldsymbol{\gamma}-\hat{\boldsymbol{\gamma}}) - \frac{1}{2\sigma^2}(\boldsymbol{\gamma}-\xi\mathbf{e}_p)^T\mathbf{R}(\boldsymbol{\gamma}-\xi\mathbf{e}_p)-\frac{1}{2\tau^2}(\xi-\mu)^2\right\},$$

where $\mathbf{e}_p$ denotes the $p\times 1$ unit vector. Completing the square, for the first two terms, gives

$$\pi^*(\boldsymbol{\gamma},\xi\mid\mathbf{y})\propto\exp\left\{-\frac{1}{2}(\boldsymbol{\gamma}-\boldsymbol{\gamma}_\xi)\left(\mathbf{R}+\sigma^{-2}\mathbf{I}_p\right)(\boldsymbol{\gamma}-\boldsymbol{\gamma}_\xi) -\frac{1}{2}(\hat{\boldsymbol{\gamma}}-\xi\mathbf{e}_p)^T\left(\mathbf{R}^{-1}+\sigma^2\mathbf{I}_p\right)^{-1}(\hat{\boldsymbol{\gamma}}-\xi\mathbf{e}_p) -\frac{1}{2\tau^2}(\xi-\mu)^2\right\},$$

where

$$\boldsymbol{\gamma}_\xi = \left(\mathbf{R} + \sigma^2 \mathbf{I}_p\right)^{-1}\left(\mathbf{R}\hat{\boldsymbol{\gamma}} + \sigma^{-2}\xi \mathbf{e}_p\right),$$

and $\mathbf{I}_p$ denotes the $p \times p$ identity matrix. Therefore, conditional on ξ, the posterior distribution of $\boldsymbol{\gamma}$ is approximately $N(\boldsymbol{\gamma}_\xi, \{\mathbf{R} + \sigma^{-2}\mathbf{I}_p\}^{-1})$. By integrating out $\boldsymbol{\gamma}$, we find that the posterior density of ξ is approximated by

$$\pi^*(\xi \mid \mathbf{y}) \propto \exp\Big\{ -\tfrac{1}{2}(\hat{\boldsymbol{\gamma}} - \xi \mathbf{e}_p)^T\left(\mathbf{R}^{-1} + \sigma^{-2}\mathbf{I}_p\right)(\hat{\boldsymbol{\gamma}} - \xi \mathbf{e}_p) \\ - \tfrac{1}{2}\tau^{-2}(\xi - \mu)^2\Big\}.$$

This density is normal with variance

$$v = \left(u + \tau^{-2}\right)^{-1},$$

with $u = \mathbf{e}_p^T(\mathbf{R}^{-1} + \sigma^2 \mathbf{I}_p)^{-1}\mathbf{e}_p$, since $-\frac{1}{2}v^{-1}$ is the coefficient of ξ^2 in the expression within the exponential. The mean of ξ is

$$\xi^* = \left\{u + \tau^{-2}\right\}^{-1}\left\{\mathbf{e}_p^T\left(\mathbf{R}^{-1} + \sigma^2\mathbf{I}_p\right)^{-1}\hat{\boldsymbol{\gamma}} + \tau^{-2}\mu\right\},$$

since this is also the value of ξ maximizing the expression within the exponential. Consequently, the posterior distribution of ξ is approximately $N(\xi^*, v)$.

(b) Since $\boldsymbol{\gamma} \mid \mathbf{y}, \xi \sim N(\boldsymbol{\gamma}_\xi, \tilde{\mathbf{D}})$, where $\tilde{\mathbf{D}} = (\mathbf{R} + \sigma^{-2}\mathbf{I}_p)^{-1}$, we have $\boldsymbol{\gamma} \mid \mathbf{y} \sim N(\boldsymbol{\gamma}^*, \mathbf{D}^*)$, where

$$\boldsymbol{\gamma}^* = \underset{\xi \mid \mathbf{y}}{E}(\boldsymbol{\gamma}_\xi) = \left(\mathbf{R} + \sigma^{-2}\mathbf{I}_p\right)^{-1}\left(\mathbf{R}\hat{\boldsymbol{\gamma}} + \sigma^{-2}\xi^*\mathbf{e}_p\right)$$

and

$$\mathbf{D}^* = \tilde{\mathbf{D}} + \underset{\xi \mid \mathbf{y}}{\mathbf{cov}}(\boldsymbol{\gamma}_\xi) = \tilde{\mathbf{D}} + \sigma^{-4}v\tilde{\mathbf{D}}\mathbf{e}_p\mathbf{e}_p^T\tilde{\mathbf{D}}.$$

(c) We have thus derived a special case of (6.1.11) and (6.1.12), when $\boldsymbol{\mu} = \mu\mathbf{e}_p$ and $\mathbf{C} = \sigma^2\mathbf{I}_p + \tau^2\mathbf{J}_p$, with $\mathbf{J}_p$ denoting the $p \times p$ unit matrix.

SELF-STUDY EXERCISES

6.1.a *Box's Model-Checking Criterion:* Consider the significance probability

$$\alpha_p = \text{prob}\big\{\log p(\mathbf{Y}) \le \log p(\mathbf{y})\big\}, \tag{6.1.13}$$

where $p(\mathbf{y})$ is the prior predictive density or p.m.f. in (6.1.3), and the probability similarly relates to the prior predictive distribution of $\mathbf{Y}$. Under reasonable broad regularity conditions, use Schwarz's criterion in (6.1.6) to show that α_p is approximated, for large n, by

$$\alpha_p^* = \text{prob}\big\{L(\mathbf{Y}) \le L(\mathbf{y})\big\}, \tag{6.1.14}$$

where $L(\mathbf{y})$ evaluates the log-likelihood $\log l(\boldsymbol{\gamma} \mid \mathbf{y})$ at the maximum likelihood estimate $\hat{\boldsymbol{\gamma}} = \hat{\boldsymbol{\gamma}}(\mathbf{y})$. Hence, show that in situations where $n^{-1}L(\mathbf{Y})$ is not identical to its expectation, but nevertheless converges in (sampling) distribution to the expectation

$$\zeta(\boldsymbol{\gamma}) = \int_{R^p} l(\boldsymbol{\gamma} \mid \mathbf{y}) \log l(\boldsymbol{\gamma} \mid \mathbf{y})\, d\mathbf{y}$$

with probability one as $n \to \infty$, α_p approaches

$$\alpha_p^* = \text{prior prob}\big(L(\mathbf{y}) \geq \zeta(\boldsymbol{\gamma})\big). \tag{6.1.15}$$

Compare this criterion with more standard fixed-size goodness-of-fit tests that would reject the null model with sampling probability one, as $n \to \infty$.

6.1.b *Smoothing a Logistic Regression Function* (A semi-parametric procedure):

(a) Consider the situation described in Exercise 1.3.c, but where there is a single explanatory variable x_i for each binomial frequency Y_i, and x_i confined to a bounded interval (a, b). Temporarily replace (1.3.17) by

$$\alpha_i = \log\theta_i - \log(1-\theta_i) = g(x_i), \tag{6.1.16}$$

where g is a totally unknown function. Show, by replacing the θ_i in the likelihood of the θ_i by the corresponding $\theta_i = e^{g(x_i)}/(1+e^{g(x_i)})$, that the log-likelihood (functional) of g is

$$L(g) = \sum_{i=1}^{m} y_i g(x_i) - \sum_{i=1}^{m} n_i \log\left\{1 + e^{g(x_i)}\right\}, \tag{6.1.17}$$

and that any function satisfying $g(x_i) = \log\{y_i/(n_i - x_i)\}$ for $i = 1, \ldots, m$ is a maximum likelihood estimate.

(b) Silverman (1978) and Leonard (1978) recommend smoothly estimating g by instead maximizing a penalized log-likelihood functional, such as

$$L^*(g) = L(g) - \frac{1}{2}\lambda \int_a^b \left[g^{(2)}(t)\right]^2 dt, \tag{6.1.18}$$

where $g^{(2)}(t) = \partial g^2(t)/\partial t^2$. As an approximate solution (Leonard, 1982c), apply an idea, expressed by Kimeldorf and Wahba (1971) in a simpler context, by constraining attention to the class of g's satisfying

$$g(t) = \boldsymbol{\phi}^T(t)\boldsymbol{\gamma} \qquad \text{for } t \in (a, b), \tag{6.1.19}$$

where $\boldsymbol{\phi}(t) = [\phi_1(t), \ldots, \phi_p(t)]^T$ is a vector of specified functions, and $\boldsymbol{\gamma}$ is a $p \times 1$ vector of unknown parameters. Show that under the restriction (6.1.19), maximization of (6.1.18), with respect to g, is equivalent to maximization, with respect to $\boldsymbol{\gamma}$, of

$$\log l(\boldsymbol{\gamma} \mid \mathbf{y}) - \tfrac{1}{2}\lambda\boldsymbol{\gamma}^T\mathbf{R}_0\boldsymbol{\gamma}, \tag{6.1.20}$$

where

$$\mathbf{R}_0 = \int_a^b \boldsymbol{\phi}^{(2)}(t)\left[\boldsymbol{\phi}^{(2)}(t)\right]^T dt, \tag{6.1.21}$$

and describe the likelihood contribution to (6.1.20).

(c) If instead $\boldsymbol{\gamma}$ possesses a $N(\mathbf{0}, \lambda^{-1}\mathbf{R}_0^{-1})$ prior, show that the posterior density of $\boldsymbol{\gamma}$ is

$$\pi(\boldsymbol{\gamma} \mid \mathbf{y}) \propto l(\boldsymbol{\gamma} \mid \mathbf{y}) \exp\left\{ -\tfrac{1}{2}\lambda\boldsymbol{\gamma}^T\mathbf{R}_0\boldsymbol{\gamma}\right\}. \tag{6.1.22}$$

Given the close resemblance of $\log \pi(\boldsymbol{\gamma} \mid \mathbf{y})$ and (6.1.20), do you think that this choice of prior is reasonable?

(d) Under the likelihood approximation (6.1.2), and prior in part (c), show that for each $t \in (a, b)$, the posterior density of $g(t)$ is normal, with mean

$$\boldsymbol{\phi}^T(t)(\mathbf{R} + \lambda\mathbf{R}_0)^{-1}\mathbf{R}\hat{\boldsymbol{\gamma}} \tag{6.1.23}$$

and variance

$$\boldsymbol{\phi}^T(t)(\mathbf{R} + \lambda\mathbf{R}_0)^{-1}\boldsymbol{\phi}(t). \tag{6.1.24}$$

Do you think that (6.1.23) provides a reasonable way of smoothing a logistic regression function? Would you prefer a different prior mean vector?

Note: For further references relating to the linear approximation device (6.1.19), a practical example on heart disease data, the estimation of λ by cross-validation, and analogous Bayesian smoothing procedures for normal regression functions, see O'Sullivan, Yandell, and Raynor (1986), Yandell and Green (1986), Stone (1974), Geisser (1974), Blight and Ott (1975), O'Hagan (1978), and Wahba (1978, 1985a). For a fascinating duality between smoothing splines and the Bayesian posterior mean value function, see Kimeldorf and Wahba (1970). For an excellent account of the literature of smoothing splines, see Wahba (1990). This is of course a subset of the Bayesian approach and is therefore well justified by the frequency properties of Section 3.10.

In this spirit, Hsu and Leonard (1997) assume that $g(\mathbf{t})$ has a Gaussian prior for $\mathbf{t}$ in an arbitrary space τ, with $g(\mathbf{t})$ possessing a mean value function of the form $\mu(\mathbf{t}) = \boldsymbol{\phi}_0^T(\mathbf{t})\boldsymbol{\beta}$, representing a hypothesized model for g, and $K(\mathbf{s}, \mathbf{t}) = \text{cov}\,[g(\mathbf{s}), g(\mathbf{t})]$ assuming an arbitrary form, for example, $K(\mathbf{s}, \mathbf{t}) = \sigma^2 \exp\{-\xi(\mathbf{s}-\mathbf{t})^T(\mathbf{s}-\mathbf{t})\}$. (See the normal theory methods by O'Hagan, 1978.) Morris, Mitchell, and Ylivisaker (1993) confirm that this formulation contributes also well to the choice of experimental design. The computations for exact Bayesian posterior inferences do not increase in complexity with the dimension of $\mathbf{t}$.

Interdisciplinary note: A large amount of the function space mathematics underlying criteria like (6.1.18), for example, for smoothing probability densities rather than regression functions (see Leonard, 1978; Lenk, 1988, 1991; and Section 6.5 (J)), is surprisingly close to the related mathematics of quantum theory (see Silver, Martz, and Wallstrom, 1993). Similar methods have been applied to good effect in meteorology; for example, Wahba and Wendelberger (1980) use similar criteria to smooth weather maps, and Wahba (1985b) develops an interesting approach for satellite-computed tomography.

6.1.c *The Semi-Parametric Smoothing of Multinomial Cell Probabilities* (Leonard, 1973a and 1973b):

(a) Let $y_1, \ldots, y_p$ possess a multinomial distribution with all probabilities $\theta_1, \ldots, \theta_p$ falling in the unit simplex S_U, and sample size $n = \sum y_i$. Consider the logits $\gamma_1, \ldots, \gamma_p$ satisfying

$$\theta_j = \frac{e^{\gamma_j}}{\sum_{g=1}^{p} e^{\gamma_g}} \qquad (j = 1, \ldots, p). \tag{6.1.25}$$

Show that the likelihood of $\boldsymbol{\gamma} = (\gamma_1, \ldots, \gamma_p)^T$ is

$$l(\boldsymbol{\gamma} \mid \mathbf{y}) \propto \exp\left\{\sum_{j=1}^{p} \gamma_j y_j - n \log \sum_{j=1}^{p} e^{\gamma_j}\right\}. \tag{6.1.26}$$

Setting $\hat{\boldsymbol{\gamma}} = (\hat{\gamma}_1, \ldots, \hat{\gamma}_p)^T = (\log y_1, \ldots, \log y_p)^T$ shows that the likelihood approximation (6.1.2) depends upon the choice

$$\mathbf{R} = \operatorname{diag}(y_1, y_2, \ldots, y_p) - n^{-1}\mathbf{y}\mathbf{y}^T \tag{6.1.27}$$

for the likelihood information matrix $\mathbf{R}$, where $\mathbf{y} = (y_1, \ldots, y_p)^T$. Note that $\mathbf{R}$ is a singular matrix.

(b) Show that

$$l^*(\boldsymbol{\gamma} \mid \mathbf{y}) = l(\hat{\boldsymbol{\gamma}} \mid \mathbf{y}) \exp\left\{-\frac{1}{2}\sum_{j=1}^{p} y_j(\gamma_j - \hat{\gamma}_j - d)^2\right\}, \tag{6.1.28}$$

where

$$d = n^{-1}\sum_{j=1}^{p} y_j(\gamma_j - \hat{\gamma}_j). \tag{6.1.29}$$

(c) As an alternative to the more restrictive Dirichlet assumptions of Exercise 5.1.f, suppose that $\boldsymbol{\gamma}$ possesses a normal $N(\boldsymbol{\mu}, \mathbf{C})$ prior distribution, where $\mathbf{C}$ possesses the (j, k)th element

$$c_{jk} = \sigma^2 \rho^{|j-k|} \qquad \left(0 < \sigma^2 < \infty, -1 < \rho < +1\right). \tag{6.1.30}$$

Note that

$$\mathbf{C}^{-1} = \tau \begin{bmatrix} 1 & -\rho & & & & \bigcirc \\ -\rho & 1+\rho^2 & -\rho & \ddots & & \\ & \ddots & \ddots & & & \\ & -\rho & \ddots & \ddots & & \\ & & \ddots & \ddots & -\rho & \\ & & & \ddots & \ddots & \\ & & & -\rho & 1+\rho^2 & -\rho \\ \bigcirc & & & & & \\ & & & & \rho & 1 \end{bmatrix}, \tag{6.1.31}$$

$$\tag{6.1.32}$$

where $\tau = \sigma^{-2}(1 - \rho^2)^{-1}$. Express the prior density in the form

$$\pi(\boldsymbol{\gamma}) \propto \exp\left[-\tfrac{1}{2}\tau Q_\rho\{(\boldsymbol{\gamma} - \boldsymbol{\mu})\}\right], \tag{6.1.33}$$

where, for any $p \times 1$ vector $\boldsymbol{\lambda} = (\lambda_1, \ldots, \lambda_p)^T$,

$$Q_\rho\{\boldsymbol{\lambda}\} = \left(1 - \rho^2\right)\lambda_1^2 + \sum_{j=2}^{p}(\lambda_j - \rho\lambda_{j-1})^2. \tag{6.1.34}$$

By combining (6.1.33) with the likelihood approximation (6.1.28), show that the posterior modes γ_j^* of the γ_j, or the posterior means, approximately satisfy a particular set of linear equations. Interpret how these equations smooth the maximum likelihood estimates $\hat{\gamma}_j$. Suggest how the μ_j may be based upon a hypothesized sampling model.

6.1.d *The Linear Statistical Model*: Assume that an $n \times 1$ vector $\mathbf{y}$ possesses a multivariate normal $N(\mathbf{X}\boldsymbol{\beta}, \phi\mathbf{I}_n)$ distribution, where $\mathbf{X}$ is a specified $n \times p$ matrix, and $\mathbf{R} = \mathbf{X}^T\mathbf{X}$ is nonsingular. Let ϕ and $\boldsymbol{\beta}$ possess a conjugate distribution, which we refer to as the $C(\nu, \lambda, \mu, \mathbf{R}_0)$ distribution, that is, conditionally on ϕ, $\boldsymbol{\beta}$ is $N(\mu, \phi^{-1}, \mathbf{R}_0^{-1})$, and $\nu\lambda/\phi$ unconditionally possesses a chi-squared distribution with ν degrees of freedom.

(a) By reference to Exercise 5.1.c and Lemma 6.1.1, show that the posterior distribution of ϕ and $\boldsymbol{\beta}$ is $C(\nu^*, \lambda^*, \boldsymbol{\mu}^*, \mathbf{R}^*)$, where $\nu^* = \nu + n$,

$$\begin{aligned} \boldsymbol{\mu}^* &= (\mathbf{R}_0 + \mathbf{R})^{-1}\left(\mathbf{R}_0\boldsymbol{\mu} + \mathbf{R}\hat{\boldsymbol{\beta}}\right), \\ \mathbf{R}^* &= \mathbf{R}_0 + \mathbf{R}, \\ \text{and} \quad \lambda^* &= (\nu + n)^{-1}\left[\nu\lambda + S_R^2 + \left(\hat{\boldsymbol{\beta}} - \boldsymbol{\mu}\right)^T\left\{\mathbf{R}_0^{-1} + \mathbf{R}^{-1}\right\}^{-1}\left(\hat{\boldsymbol{\beta}} - \boldsymbol{\mu}\right)\right], \end{aligned}$$

where $\hat{\boldsymbol{\beta}}$ and S_R^2 are the least squares vector and residual sum of squares. Note that

$$(\mathbf{y} - \mathbf{X}\boldsymbol{\beta})^T(\mathbf{y} - \mathbf{X}\boldsymbol{\beta}) = S_R^2 + n\left(\boldsymbol{\beta} - \hat{\boldsymbol{\beta}}\right)^T\mathbf{R}\left(\boldsymbol{\beta} - \hat{\boldsymbol{\beta}}\right).$$

(b) Hence describe the unconditional posterior mean vector of $\boldsymbol{\beta}$ and mean of ϕ^{-1}.

(c) Find a Bayesian justification for "ridge estimators" of the form $\mu^* = (\kappa\mathbf{I}_p + \mathbf{R})^{-1}\mathbf{R}\hat{\boldsymbol{\beta}}$? Can these estimators be judged by a sensible choice of prior? (See Draper and Van Nostrand, 1979, for further reading. Other methods for smoothing regression functions are described by Blight and Ott, 1975, Wahba (1978), and O'Hagan, 1978.)

Note: Shiau (1985) applies these Bayesian results to the theory of smoothing splines, and with some clever extensions shows how to handle regression functions with discontinuities. Shiau, Wahba, and Johnson (1987) apply related ideas to atmospheric technology. More recently, van der Linde (1995, 1996, 1997) discusses smoothing spline results that parallel the Bayesian smoothing, created by our weighted average form for μ^*.

6.1.e In Exercise 6.1.d, the prior density of $\boldsymbol{\beta}$ is $t_\theta(\nu, \boldsymbol{\mu}, \lambda^{-1}\mathbf{R}_0)$, that is, generalized multivariate t, with ν degrees of freedom, and density

$$\pi(\boldsymbol{\beta}) \propto \left[1 + \nu^{-1}\lambda^{-1}(\boldsymbol{\beta} - \boldsymbol{\mu})^T\mathbf{R}_0(\boldsymbol{\beta} - \boldsymbol{\mu})\right]^{-\frac{1}{2}(\nu+p)}, \tag{6.1.35}$$

and the prior density of $\eta = \mathbf{a}^T\boldsymbol{\theta}$ is $t_\eta(\nu, \mathbf{a}^T\boldsymbol{\mu}, \lambda^{-1}\{\mathbf{a}^T\mathbf{R}_0^{-1}\mathbf{a}\}^{-1})$.

Use this result to find the posterior density of η. Show how to obtain a 95% Bayesian interval for η. Show that when $\nu = -p$, $\lambda = 0$, and $\mathbf{R}_0$ is the $p \times p$ matrix of zeros, this gives a 95% confidence interval for η.

Give three examples (e.g., linear regression, analysis of variance, split plot or factorial designs) showing how the results in Exercises 6.1.d and 6.1.e reduce for special cases of the linear statistical model.

6.1.f Investigate whether or not the term $S^2(\boldsymbol{\mu}) = S_R^2 + (\hat{\boldsymbol{\beta}} - \boldsymbol{\mu})^T[\mathbf{R}_0^{-1} + \mathbf{R}^{-1}]^{-1}(\hat{\boldsymbol{\beta}} - \boldsymbol{\mu})$ occuring in the expression for λ^* in Exercise 6.1.d compromises between S_R^2 and $S_T^2 = (\mathbf{y} - \mathbf{X}\boldsymbol{\mu})^T(\mathbf{y} - \mathbf{X}\boldsymbol{\mu})$. Is it necessarily true that $S_R^2 \leq S^2(\boldsymbol{\mu}) \leq S_T^2$? Give conditions on $\mathbf{R}_0$ and $\mathbf{R}$ such that this is true. Is there any connection with Chamberlain and Leamer (1976)?

6.1.g If the prior mean vector $\boldsymbol{\mu}$ and covariance matrix $\mathbf{C}$ of a vector $\boldsymbol{\gamma}$ of unknown parameters are completely specified, but there is no further prior information about $\boldsymbol{\gamma}$, show that Boltzmann's theorem (Theorem 3.4.1) justifies an $N(\boldsymbol{\mu}, \mathbf{C})$ prior distribution for $\boldsymbol{\gamma}$. If a common prior mean μ and variance σ^2 is specified for each element of $\boldsymbol{\gamma}$, but there is no further information about $\boldsymbol{\gamma}$, show that Boltzmann's theorem justifies an exchangeable prior distribution, where the elements of $\boldsymbol{\gamma}$ are independent $N(\mu, \sigma^2)$.

6.2 Posterior Mode Vectors and Laplacian Approximations

The approximation (6.1.11) to the posterior mean vector fails whenever $\hat{\boldsymbol{\gamma}}$ is not finite. In any case, a superior multivariate normal approximation to the posterior distribution may be computed by reference to the exact posterior mode vector $\tilde{\boldsymbol{\gamma}}$ (e.g., Ladalla, 1976; Johnson and Ladalla, 1979).

Expanding $\log \pi(\boldsymbol{\gamma} \mid \mathbf{y})$ in a Taylor series about $\boldsymbol{\gamma} = \tilde{\boldsymbol{\gamma}}$ (in fashion similar to the expansion of $\log l(\boldsymbol{\gamma} \mid \mathbf{y})$ described in Section 1.5 (B)), neglecting cubic and higher terms in the expansion, and taking exponentials yield the second-order approximation

$$\pi^*(\boldsymbol{\gamma} \mid \mathbf{y}) = \pi(\tilde{\boldsymbol{\gamma}} \mid \mathbf{y}) \exp\left\{ -\tfrac{1}{2}(\boldsymbol{\gamma} - \tilde{\boldsymbol{\gamma}})^T \tilde{\mathbf{R}} (\boldsymbol{\gamma} - \tilde{\boldsymbol{\gamma}}) \right\} \qquad (\boldsymbol{\gamma} \in R^p), \tag{6.2.1}$$

where $\tilde{\mathbf{R}} = -\partial^2 \log \pi(\boldsymbol{\gamma} \mid \mathbf{y}) / \partial(\boldsymbol{\gamma}\boldsymbol{\gamma}^T)$ denotes the posterior information matrix.

Johnson and Ladalla recommend further terms, based upon an Edgeworth expansion. However, (6.2.1) can itself be remarkably accurate if a preliminary transformation for the parameters is employed. It follows that the marginal posterior density of $\eta = \mathbf{a}^T \gamma$ is approximately $N(\mathbf{a}^T \tilde{\boldsymbol{\gamma}}, (\mathbf{a}^T \tilde{\mathbf{R}}^{-1} \mathbf{a})^{-1})$. In the situation described in Section 6.1, where a priori $\boldsymbol{\gamma}$ is $N(\boldsymbol{\mu}, \mathbf{C})$, given $\boldsymbol{\mu}$ and $\mathbf{C}$, the posterior mode vector $\tilde{\boldsymbol{\gamma}}$ satisfies the typically nonlinear equation

$$\frac{\partial \log l(\tilde{\boldsymbol{\gamma}} \mid \mathbf{y})}{\partial \tilde{\boldsymbol{\gamma}}} = \mathbf{C}^{-1}(\tilde{\boldsymbol{\gamma}} - \boldsymbol{\mu}), \tag{6.2.2}$$

and the posterior information matrix $\tilde{\mathbf{R}}$ is

$$\tilde{\mathbf{R}} = \mathbf{R}(\tilde{\boldsymbol{\gamma}}) + \mathbf{C}^{-1}, \tag{6.2.3}$$

where $\mathbf{R}(\boldsymbol{\gamma}) = -\partial^2 \log l(\boldsymbol{\gamma} \mid \mathbf{y}) / \partial(\boldsymbol{\gamma}\boldsymbol{\gamma}^T)$. Equation (6.2.2) can be solved by a simple modification of the Newton–Raphson procedure discussed in Section 1.5 (B) for maximum likelihood estimation. The matrix (6.2.3) also serves as Hessian in the Newton–Raphson iterations, and the $\mathbf{C}^{-1}$ contribution can greatly speed up the iterations. The procedure can frequently converge even when the maximum likelihood estimate $\hat{\boldsymbol{\gamma}}$ is not finite.

The related Laplacian approximation (5.1.5) can give accurate approximations to the marginal posterior distributions of each γ_j and does not constrain the approximations to be normal. Let $\boldsymbol{\gamma}^T = (\eta, \boldsymbol{\xi}^T)$, where $\eta = \gamma_1$ and $\boldsymbol{\xi} = (\gamma_2, \ldots, \gamma_p)^T$. This can be justified by noting that in similar fashion to (6.1.3), the marginal posterior density of η is

$$\pi(\eta \mid \mathbf{y}) = \frac{\pi(\eta, \boldsymbol{\xi} \mid \mathbf{y})}{\pi(\boldsymbol{\xi} \mid \eta, \mathbf{y})} \qquad (\boldsymbol{\xi} \in R^{p-1}). \tag{6.2.4}$$

However, the conditional posterior distribution of $\boldsymbol{\xi}$, given η, may by analogy with (6.2.1) be approximated by an $N(\boldsymbol{\xi}_\eta, \mathbf{R}_\eta^{-1})$ distribution, where $\boldsymbol{\xi}_\eta$ conditionally maximizes $\pi(\boldsymbol{\xi} \mid \eta, \mathbf{y})$ with respect to $\boldsymbol{\xi}$, given η, and hence conditionally maximizes $\pi(\eta, \boldsymbol{\xi} \mid \mathbf{y}) = \pi(\boldsymbol{\xi} \mid \eta, \mathbf{y})\pi(\eta \mid \mathbf{y})$ with respect to $\boldsymbol{\xi}$, given η, and

$$\mathbf{R}_\eta = -\partial^2 \log \pi(\eta, \boldsymbol{\xi}_\eta \mid \mathbf{y}) / \partial\left(\boldsymbol{\xi}_\eta \boldsymbol{\xi}_\eta^T\right).$$

Therefore (6.2.4) gives, approximately,

$$\pi(\eta \mid \mathbf{y}) \sim \pi(\eta, \boldsymbol{\xi} \mid \mathbf{y})(2\pi)^{\frac{1}{2}p} |\mathbf{R}_\eta|^{-\frac{1}{2}} \exp\left\{\tfrac{1}{2}(\xi - \xi_\eta)^T \mathbf{R}_\eta (\xi - \xi_\eta)\right\}. \tag{6.2.5}$$

Setting $\boldsymbol{\xi} = \boldsymbol{\xi}_\eta$ gives our Laplacian approximation (5.1.5),

$$\pi^*(\eta \mid \mathbf{y}) \propto \pi(\eta, \boldsymbol{\xi}_\eta \mid \mathbf{y}) |\mathbf{R}_\eta|^{-\frac{1}{2}}, \tag{6.2.6}$$

which is exact whenever the conditional density $\pi(\boldsymbol{\xi} \mid \eta, \mathbf{y})$ is either multivariate normal or assumes the form of the multivariate t-density (6.1.34).

Worked Example 6C: Laplacian Methods for Log-Contrasts of Exponential Parameters

Observations $y_1, y_2, \ldots, y_n$ are independent, given $\lambda_1, \lambda_2, \ldots, \lambda_n$, and y_i possesses an exponential distribution with density

$$f(y_i \mid \lambda_i) = \lambda_i \exp\{-\lambda_i y_i\} \qquad (0 < y_i < \infty).$$

In the prior assessment, the λ_i are Gamma distributed with common mean α/β and variance α/β^2.

(a) Find the posterior density of $\gamma_i = \log \lambda_i$ and the posterior mode of γ_i.
(b) Let $\eta = \sum_{i=1}^n a_i \gamma_i$, where the a_i are specified constants that sum to zero. Show how the conditional posterior modes of the γ_i, given that $\sum_{i=1}^n a_i \gamma_i = \eta$, can be calculated explicitly in terms of a Lagrange multiplier ξ, and show how to calculate ξ.
(c) Use the Tierney–Kass–Kadane suggestion in equation (5.1.7) to derive, in terms of ξ, an algebraically explicit Laplacian approximation to the marginal posterior density of η.

(d) Indicate how your results might be modified if, instead, the γ_i are independent $N(\mu, \sigma^2)$ distributed in the prior assessment.

Model Answer 6C:

(a) The posterior density of λ_i is

$$\begin{aligned}\pi(\lambda_i \mid y_i) &\propto \lambda_i \exp\{-y_i\lambda_i\}\lambda_i^{\alpha-1}\exp\{-\beta\lambda_i\}\\ &\propto \lambda_i^{\alpha}\exp\big\{-(y_i+\beta)\lambda_i\big\} \qquad (0<\lambda_i<\infty),\end{aligned}$$

so that λ_i possesses a Gamma $(\alpha+1,\, y_i+\beta)$ distribution. Consequently, the posterior density of $\gamma_i = \log\lambda_i$ is

$$\pi(\gamma_i \mid y_i) \propto \exp\big\{(\alpha+1)\gamma_i - (y_i+\beta)e^{\gamma_i}\big\}.$$

Differentiating the term within the exponential, with respect to γ_i, we find that the posterior mode of γ_i is $\tilde{\gamma}_i = \log[(\alpha+1)/(y_i+\beta)]$.

(b) We need to maximize the expression

$$\sum_{i=1}^{n}\log\pi(\gamma_i \mid y_i) = \text{ constant } + (\alpha+1)\sum_{i=1}^{n}\gamma_i - \sum_{i=1}^{n}(y_i+\beta)e^{\gamma_i},$$

subject to $\sum_{i=1}^{n} a_i\gamma_i = \eta$. The conditional maxima γ_i^* satisfy

$$(\alpha+1) - (y_i+\beta)e^{\gamma_i^*} + a_i\xi = 0,$$

where ξ is our Lagrange multiplier. Therefore

$$\gamma_i^* = \log\left(\frac{1+\alpha+a_i\xi}{y_i+\beta}\right),$$

where

$$\eta = \sum_{i=1}^{n} a_i\gamma_i^* = \sum_{i=1}^{n}\log\left(\frac{1+\alpha+a_i\xi}{y_i+\beta}\right).$$

Hence η can be plotted as a function of ξ. However, these functions would need to be inverted computationally in order to obtain ξ in terms of η.

(c) The $\mathbf{R}_\eta$ matrix in (5.1.8) reduces to the diagonal matrix, whose ith diagonal element is $(y_i+\beta)e^{\gamma_i^*}$. Hence

$$\begin{aligned}\pi_y^*(\eta) &\propto \frac{\exp\big\{(\alpha+1)\sum_{i=1}^{n}\gamma_i^* - \sum_{i=1}^{n}(y_i+\beta)e^{\gamma_i^*}\big\}}{\prod_{i=1}^{n} e^{\frac{1}{2}\gamma_i^*}\big\{\sum_{i=1}^{n} a_i^2(y_i+\beta)^{-1}e^{-\gamma_i^*}\big\}^{\frac{1}{2}}}\\ &\propto \frac{\exp\big[\sum_{i=1}^{n}\big\{(\alpha+\frac{1}{2})\gamma_i^* - (y_i+\beta)e^{\gamma_i^*}\big\}\big]}{\big\{\sum_{i=1}^{n} a_i^2(y_i+\beta)^{-1}e^{-\gamma_i^*}\big\}^{\frac{1}{2}}}\\ &\propto \frac{\prod_{i=1}^{n}(1+\alpha+a_i\xi)^{\alpha+\frac{1}{2}}}{\big[\sum_{i=1}^{n}\big\{a_i^2/(1+\alpha+a_i\xi)\big\}\big]^{\frac{1}{2}}}.\end{aligned}$$

(d) Under this informative prior, we would instead need to maximize

$$\sum_{i=1}^{n} \log \pi(\gamma_i \mid y_i) = \text{constant} + \sum_{i=1}^{n} \gamma_i - \sum_{i=1}^{n} y_i e^{\gamma_i} - \frac{1}{2\sigma^2} \sum_{i=1}^{n} (\gamma_i - \mu)^2$$

subject to $\sum_{i=1}^{n} a_i \gamma_i = \eta$. This could be achieved computationally, using standard maximization procedures.

SELF-STUDY EXERCISES

6.2.a Consider a cell frequency y, which possesses a binomial distribution with probability θ and sample size n, and take $\gamma = \log \theta - \log(1 - \theta)$ to possess a normal prior distribution, with mean μ and variance σ^2. Use the methodology of Section 6.1 to show that the posterior distribution of γ is approximately $N(\gamma^*, (nv^{-1} + \sigma^{-2})^{-1})$, where $\gamma^* = (nv^{-1}\hat{\gamma} + \sigma^{-2}\mu)/(nv^{-1} + \sigma^{-2})$, $\hat{\gamma} = \log(p/(1-p))$, and $v = 1/p(1-p)$, with $p = y/n$. (These approximations are reasonable if $y \geq 5$ and $n - y \geq 5$.) Show that the exact posterior mode $\tilde{\gamma}$ of γ satisfies the equation

$$\tilde{\theta} = p - \frac{\tilde{\gamma} - \mu}{\sigma^2}, \tag{6.2.7}$$

where $\tilde{\theta} = e^{\tilde{\gamma}}/(1 + e^{\tilde{\gamma}})$. Show that γ^* may be alternatively derived as an approximation to $\tilde{\gamma}$ by noting that

$$\tilde{\gamma} = \hat{\gamma} + \log\left\{1 - \frac{\tilde{\gamma} - \mu}{np\sigma^2}\right\} - \log\left\{1 + \frac{\tilde{\gamma} - \mu}{n(1-p)\sigma^2}\right\} \tag{6.2.8}$$

and by using, for small z, the approximation

$$\log(1 + z) = z + \sum_{r=2}^{\infty} r^{-1}(-1)^{r+1} z^r \simeq z. \tag{6.2.9}$$

6.2.b In Exercise 6.2.a, show that

$$\left| \frac{\tilde{\gamma} - \gamma^*}{\hat{\gamma} - \gamma^*} \right| \leq n^{-1} v^{-1} \sum_{r=2}^{\infty} |K_r| |\hat{\alpha} - \mu|^{r-1}, \tag{6.2.10}$$

where

$$K_r = r^{-1} n^{-r} \sigma^{-2r} \{p^{-r} + (-1)^{r+1}(1-p)^{-r}\}. \tag{6.2.11}$$

Hence, show that

$$\left| \frac{\tilde{\gamma} - \gamma^*}{\hat{\gamma} - \gamma^*} \right| = n^{-1}\sigma^{-2} \frac{|p - \frac{1}{2}|}{p(1-p)} |\hat{\gamma} - \mu| + 0(n^{-2}) \qquad (n \to \infty), \tag{6.2.12}$$

where $0(n^{-2})$ behaves like some constant multiple of n^{-2}, as $n \to \infty$. Hence show that γ^* will lie much closer to $\tilde{\gamma}$ than to $\hat{\gamma}$, as n gets large.

6.2.c In Exercise 6.1.c, show that the posterior modes $\tilde{\gamma}_1, \ldots, \tilde{\gamma}_p$ satisfy

$$n\tilde{\theta}_j = y_j - \tau\big[(1+\rho^2)\tilde{\lambda}_j - \rho\tilde{\lambda}_{j-1} - \rho\tilde{\lambda}_{j+1}\big] \qquad (i = 2, \ldots, p-1), \tag{6.2.13}$$

where $\tilde{\theta}_j = e^{\tilde{\gamma}_j} / \sum_{g=1}^{p} e^{\tilde{\gamma}_g}$ and $\tilde{\lambda}_j = \tilde{\gamma}_j - \mu_j$. Obtain two further equations, involving $\tilde{\gamma}_1$ and $\tilde{\gamma}_p$.

6.2.d Consider the Dirichlet prior and posterior for multinomial cell probabilities $\theta_1, \ldots, \theta_p$, as described in Exercise 5.1.f. The posterior density of the logits, satisfying (6.1.25), under this Dirichlet prior is

$$\pi(\boldsymbol{\gamma} \mid \mathbf{y}) \propto \exp\left\{\sum_{j=1}^{p} \gamma_j(\alpha\xi_j + y_j) - (\alpha + n)\log\sum_{j=1}^{p} e^{\gamma_j}\right\} \qquad (\boldsymbol{\gamma} \in R^p). \tag{6.2.14}$$

Use the results of Exercise 6.1.c to show that this density may be approximated by

$$\pi^*(\boldsymbol{\gamma} \mid \mathbf{y}) \propto \exp\left\{-\frac{1}{2}\sum_{j=1}^{p}(\alpha\xi_j + y_j)(\gamma_j - \hat{\gamma}_j - d)^2\right\} \qquad (\boldsymbol{\gamma} \in R^p), \tag{6.2.15}$$

where $\tilde{\gamma}_j = \log(\alpha\xi_j + y_j)$ and $d = (\alpha + n)^{-1}\sum_{j=1}^{p}(\alpha\xi_j + y_j)(\gamma_j - \tilde{\gamma}_j)$. Show that this approximation posterior distribution may be represented by the following:

Stage I: Given κ and y_i, $\gamma_j \sim N(\kappa + \tilde{\gamma}_j, (\alpha\xi_j + y_j)^{-1})$, with $\gamma_1, \ldots, \gamma_p$ independent.

Stage II: κ is uniformly distributed over $(-\infty, \infty)$.

6.2.e Let

$$\begin{aligned}\eta = \mathbf{a}^T\boldsymbol{\gamma} &= a_1\gamma_1 + a_2\gamma_2 + \cdots + a_p\gamma_p \\ &= a_1\log\theta_1 + a_2\log\theta_2 + \cdots + a_p\log\theta_p\end{aligned}$$

be a contrast in the γ_j, and a log-contrast in the θ_j, with $\sum_{j=1}^{p} a_j = 0$. Use the construction at the end of Exercise 6.2.d to show that the posterior distribution of η is approximately $N(\sum_j a_j \log(\alpha\xi_j + y_j), \sum_j a_j^2(\alpha\xi_j + y_j)^{-1})$. If $\boldsymbol{\beta} = \mathbf{A}\boldsymbol{\alpha}$ is a vector of $p-1$ linear independent log-contrasts, show that the posterior distribution of $\boldsymbol{\beta}$ is approximately $N(\mathbf{A}\tilde{\boldsymbol{\gamma}}, \mathbf{A}\mathbf{V}\mathbf{A}^T)$, where $\tilde{\boldsymbol{\gamma}} = (\tilde{\gamma}_1, \ldots, \tilde{\gamma}_p)^T$ and $\mathbf{V} = \mathrm{diag}[(\alpha\xi_1 + y_1)^{-1}, \ldots, (\alpha\xi_p + y_p)^{-1}]$. Summarize your results, when $\alpha\xi_j = \frac{1}{2}$, $\xi_j = 1/p$, for $j = 1, \ldots, p$, and $\alpha = \frac{1}{2}p$. See also Lindley (1964) and Aitchison (1986).

6.2.f *Goodman's Full-Rank Interaction Model* (Goodman, 1964): Consider the cell frequencies $\{y_{ij}; i = 1, \ldots, r; j = 1, \ldots, s\}$ for an $r \times s$ contingency table, which satisfy $\sum y_{ij} = n$. Take the cell frequencies to possess a multinomial distribution, with corresponding cell probabilities $\{\theta_{ij}; i = 1, \ldots, r; j = 1, \ldots, r\}$ satisfying $\sum \theta_{ij} = 1$, so that the model has $rs - 1$ distinct parameters. Consider multivariate logits $\{\gamma_{ij}; i = 1, \ldots, r; j = 1, \ldots, s\}$ satisfying

$$\theta_{ij} = \frac{e^{\gamma_{ij}}}{\sum_{k=1}^{r}\sum_{g=1}^{s} e^{\gamma_{kg}}} \qquad (i = 1, \ldots, r; j = 1, \ldots, s), \tag{6.2.16}$$

and assume the full-rank interaction model

$$\gamma_{ij} = \lambda_i^A + \lambda_j^B + \lambda_{ij}^{AB} \qquad (i = 1, \ldots, r;\ j = 1, \ldots, s), \tag{6.2.17}$$

where λ_i^A is the ith *row effect*, λ_j^B is the jth *column effect*, and λ_{ij}^{AB} is the (i, j)th *interaction effect*. Under the constraints $\lambda_.^A = \lambda_.^B = \lambda_{i.}^{AB} \equiv \lambda_{.j}^{AB} \equiv 0$, there still are $rs - 1$ distinct parameters in the model, including $(r - 1)(s - 1)$ distinct interaction interaction effects. Show that subject to these constraints,

$$\lambda_i^A = \gamma_{i.} - \gamma_{..}\,,$$

$$\lambda_j^B = \gamma_{.j} - \gamma_{..}\,,$$

and

$$\lambda_{ij}^{AB} = \gamma_{ij} - \gamma_{i.} - \gamma_{.j} + \gamma_{..}\,. \tag{6.2.18}$$

Assume that the prior density of the θ_{ij} is proportional to $\prod_{ij} \theta_{ij}^{-\frac{1}{2}}$ on the unit simplex. Use the results of Exercise 6.2.e to show that the posterior distribution of λ_{ij}^{AB} is approximately $N(\tilde{\gamma}_{ij} - \tilde{\gamma}_{i.} - \tilde{\gamma}_{.j} + \tilde{\gamma}_{..}, \epsilon_{ij}^{AB})$, where $\tilde{\gamma}_{ij} = \log(y_{ij} + \frac{1}{2})$, and

$$\begin{aligned}\epsilon_{ij}^{AB} &= (1 - 2r^{-1})(1 - 2s^{-1})u_{ij} + s^{-1}(1 - r^{-1})u_{i.} \\ &\quad + r^{-1}(1 - s^{-1})u_{.j} + s^{-1}r^{-1}u_{..}\end{aligned} \tag{6.2.19}$$

with $u_{ij} = (y_{ij} + \frac{1}{2})^{-1}$.

6.2.g In Exercise 6.2.f, use (6.2.14) to show that the exact posterior density of the γ_{ij} is

$$\pi(\boldsymbol{\gamma} \mid \mathbf{y}) \propto \frac{\exp\left\{\sum_{i=1}^r \sum_{j=1}^s \gamma_{ij}\left(y_{ij} + \frac{1}{2}\right)\right\}}{\left(\sum_{i=1}^r \sum_{j=1}^s e^{\gamma_{ij}}\right)^{(2n+p)/2}}, \tag{6.2.20}$$

but with a simple constraint (e.g., $\gamma_{..} = 0$) to maintain the correct dimensionality of the parameter space. Hence, use the representation (5.1.7) to obtain and develop full technical details of the Laplacian approximation to the marginal posterior density of $\eta = \gamma_{ij} - \gamma_{i.} - \gamma_{.j} + \gamma_{..}$.

Note: Leonard (1975b) and Laird (1979) consider the alternative formulation where, rather than imposing constraints, the λ_i^A in (6.2.17) are taken to be independent $N(0, \sigma_A^2)$, the λ_j^B to be independent $N(0, \sigma_B^2)$, and the λ_{ij}^{AB} to be independent $N(0, \sigma_{AB}^2)$. Laird lets $\sigma_A^2 \to \infty$ and $\sigma_B^2 \to \infty$. Under the Dirichlet prior, Altham (1969) describes an exact analysis for a 2×2 table, in which $\lambda_{11}^{AB} = -\lambda_{12}^{AB} = -\lambda_{21}^{AB} = \lambda_{22}^{AB} = \frac{1}{4}(\gamma_{11} + \gamma_{22} - \gamma_{12} - \gamma_{21}) = \frac{1}{4}\log\phi$, where $\phi = \theta_{22}\theta_{11}/\theta_{12}\theta_{21}$ denotes the measure of association. Under Jeffreys' prior, the posterior probability that $\phi < 1$ compromises between two significance probabilities (with weak and strong inequality) for Fisher's exact test (see Leonard and Hsu, 1994).

6.2.h Show that Akaike's information criterion for Goodman's full-rank interaction model in 6.2.f is

$$\mathrm{AIC} = n \sum_{i=1}^{r} \sum_{j=1}^{s} p_{ij} \log p_{ij} - K - rs + 1, \tag{6.2.21}$$

where $K = \log[n!/\prod_{i=1}^{r}\prod_{j=1}^{s} y_{ij}!]$ and $p_{ij} = \log y_{ij}$. Show that when all the interaction effects λ_{ij}^{AB} in (6.2.17) are zero, the θ_{ij} in (6.2.16) satisfy

$$\theta_{ij} = \phi_i^A \phi_j^B \qquad (i = 1, \ldots, r;\ j = 1, \ldots, s), \tag{6.2.22}$$

where $\phi_i^A = e^{\lambda_l}/\sum_{k=1}^{r} e^{\lambda_k^A}$, and $\phi_j^B = e^{\lambda_j^B}/\sum_{g=1}^{s} e^{\lambda_k^B}$. This property is associated only with the multivariate logit transformation. Equation (6.2.22) represents the null hypothesis H_0: the rows and columns of the contingency table are independent. Show that under H_0,

$$\text{AIC} = n \sum_{i=1}^{r}\sum_{j=1}^{s} p_{ij}(\log p_{i.} + \log p_{.j}) - K - r - s + 2, \tag{6.2.23}$$

where the star notation denotes summation with respect to that subscript. Show that AIC prefers the full-rank interaction model to H_0 whenever

$$-2n \sum_{i=1}^{r}\sum_{j=1}^{s} p_{ij} \log\left(\frac{p_{i*}p_{*j}}{p_{ij}}\right) \geq 2(r-1)(s-1). \tag{6.2.24}$$

(Note that under H_0, the sampling distribution of the left-hand side of (6.2.24) is approximately chi-squared distributed with $(r-1)(s-1)$ degrees of freedom, when n is large and H_0 is true. Brown and Muentz, 1976, and Leonard, 1977a, have other justifications for comparing a chi-squared statistic with twice its degrees of freedom. This criterion seems to take account of some "practical significance" even though it can, for large $(r-1)(s-1)$, accept H_0 in situations where a standard significance test would wish to reject H_0 at any sensible significance level.)

6.3 Prior Structures, and Modeling for Nonrandomized Data

(A) *Prior Structures and the Estimation of Hyperparameters*

Now further consider the multivariate normal $N(\boldsymbol{\mu}, \mathbf{C})$ prior distributions introduced for $\boldsymbol{\gamma}$ in Sections 6.1 and 6.2. Some subjective Bayesians (e.g., Garthwaite and Dickey, 1992; Press, 1989, Ch. 4) feel able to fully specify $\boldsymbol{\mu}$ and $\mathbf{C}$, based upon prior beliefs. This can take Bayesian methods into complex areas of human psychology.

In more statistical situations, you will frequently be able to construct reasonable "prior structures," that is, specified functional forms $\boldsymbol{\mu} = \boldsymbol{\mu}(\boldsymbol{\beta})$ and $\mathbf{C} = \mathbf{C}(\boldsymbol{\zeta})$ for $\boldsymbol{\mu}$ and $\mathbf{C}$, but where the "hyperparameters" $\boldsymbol{\beta}$ and $\boldsymbol{\zeta}$ are largely unknown. A structure of the form $\boldsymbol{\mu} = \mathbf{X}\boldsymbol{\beta}$ will be reasonable in some contexts, and the possibilities described in Exercises 6.1.b and 6.1.c and Section 1.9 will sometimes be reasonable for $\mathbf{C}$, among many other possibilities. Note that the prior structure $\boldsymbol{\mu}(\boldsymbol{\beta})$ for $\boldsymbol{\mu}$ can be interpreted as a "hypothesized model" for $\boldsymbol{\gamma}$, in the spirit of Leonard and Novick (1986) and Albert (1988), in which case $\boldsymbol{\xi} = \boldsymbol{\gamma} - \boldsymbol{\mu}(\boldsymbol{\beta})$ is a vector of "parametric residuals."

As another example, statisticians often possess "equality of prior vagueness" regarding the elements of $\boldsymbol{\gamma}$ so that a vague permutable "exchangeable" prior distribution is appropriate (see Lindley and Smith, 1972; Leonard, 1972; Good, 1965), following

Laplace's 1812 principle of insufficient reason. In this case, set $\boldsymbol{\mu} = \mu \mathbf{e}_p$ and $\mathbf{C} = \sigma^2 \mathbf{I}_p$, where $\mathbf{e}_p$ is the $p \times 1$ unit vector. It is usually unnecessary to assume a common correlation, since this will quickly vanish if you then assign a uniform distribution to μ or empirically estimate μ without further assumptions. Alternatively, refer to maximum entropy and Exercise 3.4.j.

Since $\boldsymbol{\beta}$ and $\boldsymbol{\zeta}$ (e.g., μ and σ^2) are unknown, it is useful to help evaluate these quantities using the current data, by reference to their "integrated likelihood,"

$$l(\boldsymbol{\beta}, \boldsymbol{\zeta} \mid \mathbf{y}) = p(\mathbf{y} \mid \boldsymbol{\beta}, \boldsymbol{\zeta}) = \int_{R^p} l(\boldsymbol{\gamma} \mid \mathbf{y}) \pi(\boldsymbol{\gamma} \mid \boldsymbol{\beta}, \boldsymbol{\zeta})\, d\boldsymbol{\gamma}, \tag{6.3.1}$$

where $\pi(\boldsymbol{\gamma} \mid \boldsymbol{\beta}, \boldsymbol{\zeta})$ denotes our $N(\boldsymbol{\mu}, \mathbf{C})$ prior density, but with $\boldsymbol{\mu} = \boldsymbol{\mu}(\boldsymbol{\beta})$ and $\mathbf{C} = \mathbf{C}(\boldsymbol{\zeta})$. Under the likelihood approximation (6.1.2), (6.3.1) may be approximated by

$$\begin{aligned} l^*(\boldsymbol{\beta}, \boldsymbol{\zeta} \mid \mathbf{y}) \\ = (2\pi)^{\frac{1}{2}p} |\mathbf{C}|^{-\frac{1}{2}} l(\hat{\boldsymbol{\gamma}} \mid \mathbf{y}) \int_{R^p} \exp\big\{ -\tfrac{1}{2}(\boldsymbol{\gamma} - \hat{\boldsymbol{\gamma}})^T \mathbf{R} (\boldsymbol{\gamma} - \hat{\boldsymbol{\gamma}}) \\ - \tfrac{1}{2}(\boldsymbol{\gamma} - \boldsymbol{\mu})^T \mathbf{C}^{-1} (\boldsymbol{\gamma} - \boldsymbol{\mu}) \big\}\, d\boldsymbol{\gamma}. \end{aligned} \tag{6.3.2}$$

Applying Lemma 6.1.1 to the quadratic form in (6.3.2) and integrating with respect to $\boldsymbol{\gamma}$ tell us that

$$l^*(\boldsymbol{\beta}, \boldsymbol{\zeta} \mid \mathbf{y}) = \big|\mathbf{R} + \mathbf{C}^{-1}\big|^{\frac{1}{2}} |\mathbf{C}|^{-\frac{1}{2}} l(\hat{\boldsymbol{\gamma}} \mid \mathbf{y}) \exp\big\{ -\tfrac{1}{2}(\hat{\boldsymbol{\gamma}} - \boldsymbol{\mu})^T \mathbf{H} (\hat{\boldsymbol{\gamma}} - \boldsymbol{\mu}) \big\}, \tag{6.3.3}$$

where

$$\mathbf{H} = \mathbf{R}\big(\mathbf{R} + \mathbf{C}^{-1}\big)^{-1} \mathbf{C}^{-1}.$$

However, a Laplacian approximation may be obtained using the posterior mode vector $\tilde{\boldsymbol{\gamma}}$. Note that $\tilde{\boldsymbol{\gamma}}$ satisfies (6.2.2) and hence depends upon $\boldsymbol{\beta}$ and $\boldsymbol{\zeta}$, via $\boldsymbol{\mu}$ and $\mathbf{C}$. The posterior information matrix in (6.2.3) is also needed. Note that

$$l(\boldsymbol{\beta}, \boldsymbol{\zeta} \mid \mathbf{y}) = \frac{\pi(\boldsymbol{\gamma} \mid \boldsymbol{\beta}, \boldsymbol{\zeta}) l(\boldsymbol{\gamma} \mid \mathbf{y})}{\pi(\boldsymbol{\gamma} \mid \boldsymbol{\beta}, \boldsymbol{\zeta}, \mathbf{y})} \qquad (\boldsymbol{\gamma} \in R^p). \tag{6.3.4}$$

Replacing the denominator of (6.3.4) by an $N(\tilde{\boldsymbol{\gamma}}, \tilde{\mathbf{R}}^{-1})$ density and setting $\boldsymbol{\gamma} = \tilde{\boldsymbol{\gamma}}$ yield the approximation (Laird, 1975)

$$\tilde{l}(\boldsymbol{\beta}, \boldsymbol{\zeta} \mid \mathbf{y}) \propto \big|\tilde{\mathbf{R}}\big|^{-\frac{1}{2}} l(\tilde{\boldsymbol{\gamma}} \mid \mathbf{y}) \pi(\tilde{\boldsymbol{\gamma}} \mid \boldsymbol{\beta}, \boldsymbol{\zeta}) \tag{6.3.5}$$

$$\propto |\mathbf{C}|^{-\frac{1}{2}} \big|\tilde{\mathbf{R}}\big|^{\frac{1}{2}} l(\tilde{\boldsymbol{\gamma}} \mid \mathbf{y}) \exp\big\{ -\tfrac{1}{2}(\tilde{\boldsymbol{\gamma}} - \boldsymbol{\mu})^T \mathbf{C}^{-1} (\tilde{\boldsymbol{\gamma}} - \boldsymbol{\mu}) \big\}. \tag{6.3.6}$$

Racine-Poon (1985), Lindstrom and Bates (1990), and Wolfinger (1993) use related Laplacian approximations for nonlinear regression models for normally distributed data. Solomon and Cox (1992) investigate a special case of (6.3.6) and demonstrate encouraging accuracy of this approximation. The approximation contrasts with alternatives recommended in the context of multilevel modeling by Longford (1993) and Goldstein (1995).

We now have four main philosophical ways of proceeding, which can vary in technical difficulty.

(1) *Hierarchical Bayes:* Let $\boldsymbol{\beta}$ and $\boldsymbol{\zeta}$ possess joint prior density $\pi(\boldsymbol{\beta}, \boldsymbol{\zeta})$, and hence posterior density

$$\pi(\boldsymbol{\beta}, \boldsymbol{\zeta} \mid \mathbf{y}) \propto \pi(\boldsymbol{\beta}, \boldsymbol{\zeta}) l(\boldsymbol{\beta}, \boldsymbol{\zeta} \mid \mathbf{y}). \tag{6.3.7}$$

Then the unconditional posterior density of $\boldsymbol{\gamma}$ is

$$\pi(\boldsymbol{\gamma} \mid \mathbf{y}) = \int_{R^{l_1}} \int_{R^{l_2}} \pi(\boldsymbol{\gamma} \mid \boldsymbol{\beta}, \boldsymbol{\zeta}, \mathbf{y}) \pi(\boldsymbol{\beta}, \boldsymbol{\zeta} \mid \mathbf{y})\, d\boldsymbol{\beta} d\boldsymbol{\zeta}. \tag{6.3.8}$$

The contribution $\pi(\boldsymbol{\gamma} \mid \boldsymbol{\beta}, \boldsymbol{\zeta}, \mathbf{y})$ to the integrand is the posterior density of $\boldsymbol{\gamma}$, with $\boldsymbol{\mu}$ and $\mathbf{C}$ specified, which we have already considered. The exact hierarchical Bayes approach requires appropriate integrations of marginal posterior density or moments, when $\boldsymbol{\beta}$ and $\boldsymbol{\zeta}$ are specified, with respect to the posterior density of $\boldsymbol{\beta}$ and $\boldsymbol{\zeta}$. Hsu and Leonard (1997) are able to calculate all these integrations exactly, by Monte Carlo simulation with a finite standard error of simulation, for a hierarchical model for binomial regression data when two first-stage parameters, relating to global and local smoothing, are unknown. We refer the reader to Hsu and Leonard's practical example relating the mortality of mice to time and degree of exposure to nitrogen dioxide. This shows how complex models can be handled using simulation techniques. Applications of similarly complex hierarchical Bayes models to the prognostication of AIDS/HIV infection are described by Lange, Carlin, and Gelfand (1992). Kass and Steffey (1989) apply Laplacian approximations to hierarchical Bayes models, with further applications.

(2) *Parametric Empirical Bayes* (e.g., Morris, 1983): Estimate $\boldsymbol{\beta}$ and $\boldsymbol{\zeta}$ by the vectors $\hat{\boldsymbol{\beta}}$ and $\hat{\boldsymbol{\zeta}}$ maximizing the integrated likelihood (6.3.1). Then use the empirical posterior $\pi(\boldsymbol{\gamma} \mid \hat{\boldsymbol{\beta}}, \hat{\boldsymbol{\zeta}}, \mathbf{y})$ as the basis for inferences about $\boldsymbol{\gamma}$.

Note that nonparametric empirical Bayes procedures where the functional form of $\pi(\boldsymbol{\gamma})$ is left unspecified can give quite different results (Robbins, 1964; Wang and Van Ryzin, 1979; Laird and Louis, 1987; Leonard, 1984; Leonard et al., 1994b). The nonparametric empirical Bayes procedures permit the current data to play a fuller role in determining the functional form of the prior density of $\boldsymbol{\gamma}$, and hence in determining the structure of the consequent posterior estimate for $\boldsymbol{\gamma}$. However, it is less easy to make precise posterior inferences. Laird and Louis (1987) discuss the possible role of the "bootstrap" method (Efron, 1979); and hierarchical Bayes versions of these nonparametric procedures require quite informative distributions in the prior assessment. In the parametric case, Carlin and Gelfand (1991) propose a sample reuse method for empirical Bayes confidence intervals.

(3) *The Marginal Posterior Mode Compromise* (O'Hagan 1976): Estimate $\boldsymbol{\beta}$ and $\boldsymbol{\gamma}$ by their (marginal) posterior mode vectors $\tilde{\boldsymbol{\beta}}$ and $\tilde{\boldsymbol{\gamma}}$, that is, the vectors maximizing (6.3.7) under some choice of prior $\pi(\boldsymbol{\beta}, \boldsymbol{\zeta})$. Then use $\pi(\boldsymbol{\gamma} \mid \tilde{\boldsymbol{\beta}}, \tilde{\boldsymbol{\zeta}}, \mathbf{y})$ as the basis for inference about $\boldsymbol{\gamma}$.

All three preceding approaches have some individual appeal in particular cases, and all three approaches can lead to estimators for $\boldsymbol{\gamma}$ with vastly improved frequency risk properties when compared with the maximum likelihood estimator $\hat{\boldsymbol{\gamma}}$ (e.g., Exercise 6.3.j).

Except in very special cases where the joint density in (6.3.7) is approximately multivariate normal, you should not use (4), as described below. Leonard (1972, 1973a, 1975b) chose the wrong modes, a conceptual error corrected by Leonard (1976, 1977b) and Laird (1979). Frequency properties are investigated by Sun et al. (1996).

(4) *The Joint Modal Estimation of First- and Second-Stage Parameters:* If $\boldsymbol{\gamma}$, $\boldsymbol{\beta}$, and $\boldsymbol{\zeta}$ are estimated by jointly maximizing their posterior distribution

$$\pi(\boldsymbol{\gamma}, \boldsymbol{\beta}, \boldsymbol{\zeta} \mid \mathbf{y}) \propto \pi(\boldsymbol{\beta}, \boldsymbol{\zeta})\pi(\boldsymbol{\gamma} \mid \boldsymbol{\beta}, \boldsymbol{\zeta}) l(\boldsymbol{\alpha} \mid \mathbf{y}), \tag{6.3.9}$$

then this can lead to a "collapsing phenomenon," for example, with variance components overshrinking towards zero (Sun et al., 1996), and to estimators of $\boldsymbol{\gamma}$ with worse frequency properties than $\hat{\boldsymbol{\gamma}}$. Similarly, don't automatically maximize (6.3.8), for example, if it can be represented as a product of generalized multivariate t-densities, since this can lead to a similar collapsing phenomenon. (See Exercise 6.3.h (d).) For such products, the Laplacian T-approximation (see Leonard, Hsu, and Ritter, 1994) can be utilized. The collapsing phenomenon is also noticed by Miller and Fortney (1984) when developing a large regression model for insurance data. They recommend O'Hagan's marginal modes as an alternative to the joint modes.

A general computational scheme for approximations to the (marginal) posterior mode vectors of $\boldsymbol{\beta}$ and $\boldsymbol{\zeta}$ maximizing (6.3.7) is now described. This refers to the EM algorithm (Section 1.5 (E)) and often works well in practice. Regard $\boldsymbol{\gamma}$ as a set of missing observations. Setting $\log \pi(\boldsymbol{\beta}, \boldsymbol{\zeta}) = 0$ in our scheme would alternatively provide empirical Bayes estimates. If $\boldsymbol{\gamma}$ is observed, then we can estimate $\boldsymbol{\beta}$ and $\boldsymbol{\zeta}$ by maximization of

$$\begin{aligned} H(\boldsymbol{\gamma}, \boldsymbol{\beta}, \boldsymbol{\zeta}) &= \log \pi(\boldsymbol{\beta}, \boldsymbol{\zeta}) + \log \pi(\boldsymbol{\gamma} \mid \boldsymbol{\beta}, \boldsymbol{\zeta}) \\ &= \log \pi(\boldsymbol{\beta}, \boldsymbol{\zeta}) - \tfrac{1}{2} \log |\mathbf{C}| - \tfrac{1}{2}(\boldsymbol{\gamma} - \boldsymbol{\mu})^T \mathbf{C}^{-1}(\boldsymbol{\gamma} - \boldsymbol{\mu}). \end{aligned} \tag{6.3.10}$$

Since $\boldsymbol{\gamma}$ is unknown, then at the E-step, we take the posterior expectation of (6.3.10), with respect to $\boldsymbol{\gamma}$, given $\boldsymbol{\beta}$ and $\boldsymbol{\zeta}$, namely,

$$\begin{aligned} E\big[\mathbf{H}(\boldsymbol{\gamma}, \boldsymbol{\beta}, \boldsymbol{\zeta}) \mid \boldsymbol{\beta}, \boldsymbol{\zeta}, \mathbf{y}\big] &= \log \pi(\boldsymbol{\beta}, \boldsymbol{\zeta}) - \tfrac{1}{2} \log |\mathbf{C}| \\ &\quad - \tfrac{1}{2} \operatorname{trace}\big[\mathbf{C}^{-1}\big\{(\boldsymbol{\gamma} - \boldsymbol{\gamma}^*)^T(\boldsymbol{\gamma} - \boldsymbol{\gamma}^*) + \mathbf{D}^*\big\}\big], \end{aligned} \tag{6.3.11}$$

where $\boldsymbol{\gamma}^* = E[\boldsymbol{\gamma} \mid \boldsymbol{\beta}, \boldsymbol{\zeta}, \mathbf{y}]$ and $\mathbf{D}^* = \mathbf{cov}[\boldsymbol{\gamma} \mid \boldsymbol{\beta}, \boldsymbol{\zeta}, \mathbf{y}]$ should be evaluated at the latest iterates $\boldsymbol{\beta}^{(r)}$ and $\boldsymbol{\zeta}^{(r)}$ for $\boldsymbol{\beta}$ and $\boldsymbol{\zeta}$. At the M-step, we then obtain the next iterates $\boldsymbol{\beta}^{(r+1)}$ and $\boldsymbol{\zeta}^{(r+1)}$ by maximization of (6.3.11), with respect to $\boldsymbol{\beta}$ and $\boldsymbol{\zeta}$, but with $\boldsymbol{\gamma}^*$ and $\mathbf{D}^*$ fixed. This maximization can sometimes be performed analytically, but more generally the Newton–Raphson procedure may be needed. Given any set of initial vectors $\boldsymbol{\beta}^{(0)}$ and $\boldsymbol{\zeta}^{(0)}$, continued repetitions of the E-step and M-step will, according to a theorem by Wu (1983), always converge to at least a local maximum of (6.3.1) for fairly general choices of $\pi(\boldsymbol{\beta}, \boldsymbol{\zeta})$. However, there are generally problems in determining a global maximum if (6.3.1) cannot be exactly computed.

In many nonlinear problems, the posterior mean vector $\boldsymbol{\gamma}^*$ and covariance matrix $\mathbf{D}^*$ cannot be obtained analytically (they could however be computed exactly, together with (6.3.1), by importance sampling – see the next section – if you wished to repeat the importance sampling at each E-step). We therefore recommend an approximation to the E-step, also employed by Laird (1975, 1979), which replaces $\boldsymbol{\gamma}^*$ in (6.3.11) by the exact posterior mode vector $\tilde{\boldsymbol{\gamma}}$ satisfying (6.2.2), and $\mathbf{D}^*$ by the exact posterior dispersion matrix $\tilde{\mathbf{R}}^{-1}$, where $\tilde{\mathbf{R}}$ satisfies (6.2.3). The global maximum can also be approximately evaluated by using the Laplacian approximation (6.3.5) to (6.3.1) to compare local maxima obtained from different choices of the matrix vectors $\boldsymbol{\beta}^{(0)}$ and $\boldsymbol{\zeta}^{(0)}$. Furthermore, an approximate AIC $= \log l(\hat{\boldsymbol{\beta}}, \hat{\boldsymbol{\zeta}} \mid \mathbf{y}) - l_1 - l_2$ can be evaluated to compare different choices of prior structure. This is conceptually similar to methodology relating to "unbiased estimates of the risk function" (see, Efron and Morris, 1973) that also permits the comparison of different types of shrinkage estimator. There are dualities between the assumed prior structures and the types of posterior shrinkage, for example, via (6.2.2).

(B) *Sampling Models for Nonrandomized Data*

Consider the following two stages.

Stage I: Given $\boldsymbol{\gamma}$, $\mathbf{y}$ has density or probability mass function $p(\mathbf{y} \mid \boldsymbol{\gamma})$.

Stage II: $\boldsymbol{\gamma} \sim N(\boldsymbol{\mu}, \mathbf{C})$, where $\boldsymbol{\mu} = \boldsymbol{\mu}(\boldsymbol{\beta})$ and $\mathbf{C} = \mathbf{C}(\boldsymbol{\zeta})$.

So far, we have interpreted Stage II as the first stage of a prior distribution for a parameter vector $\boldsymbol{\gamma}$ appearing in a sampling model $p(\mathbf{y} \mid \boldsymbol{\gamma})$. However, the two stages can instead be combined to give the sampling model

$$
\begin{aligned}
&p(\mathbf{y} \mid \boldsymbol{\beta}, \boldsymbol{\zeta}) \\
&\quad = (2\pi)^{-\frac{1}{2}p}|\mathbf{C}|^{-\frac{1}{2}} \int_{R^p} p(\mathbf{y} \mid \boldsymbol{\gamma}) \exp\left\{ -\tfrac{1}{2}(\boldsymbol{\gamma} - \boldsymbol{\mu})^T \mathbf{C}^{-1} (\boldsymbol{\gamma} - \boldsymbol{\mu})\right\} d\boldsymbol{\gamma},
\end{aligned}
\tag{6.3.12}
$$

in which case the previous hyperparameters $\boldsymbol{\beta}$ and $\boldsymbol{\zeta}$ achieve elevated importance as the model parameters. These can be estimated by the EM algorithm procedure just described. Furthermore, the sampling distribution (6.3.12) can be approximated by either (6.3.3) or (6.3.5), so that AIC or BIC can be used to compare different choices of $\boldsymbol{\mu} = \boldsymbol{\mu}(\boldsymbol{\beta})$ and $\mathbf{C} = C(\boldsymbol{\zeta})$.

Many sampling models, such as linear regression models with truly independent errors, and multinomial distributions, can only really be justified by appropriate randomization at the design stage; and the validity of many models can be affected by problems of overdispersion, serial correlation, and spatial correlation, thus affecting the significance of many apparent conclusions (for example, Tavare and Altham, 1983; Box and Guttman, 1966; Alanko and Duffy, 1996; and Breslow, 1984, 1990b). However, the general formulation in (6.3.12) can help in numerous special cases, for example:

(a) The polychoric model in Exercise 2.5.c gives a method for modeling nonindependent binary responses.

(b) The distribution of $\mathbf{y}$, given $\boldsymbol{\mu}$ and $\mathbf{C}$, in Exercise 6.1.c gives an alternative model to the multinomial, which like the multinomial-Dirichlet distribution (Paul and Plackett, 1978) emphasizes overdispersion, but which also permits the modeling of serial dependencies between, say, the cell frequencies for a histogram. Paul and Plackett show that the associated chi-squared goodness-of-fit tests can be much less ready to reject the null hypothesis, under the overdispersion model.

(c) Consider several regressions where, given the $\boldsymbol{\beta}_i$ and $\boldsymbol{\gamma}$, the $n \times 1$ vectors $\mathbf{y}_i$ are independent $N(\mathbf{X}_i\boldsymbol{\beta}_i, \mathbf{D}(\boldsymbol{\gamma}))$, for $i = 1, \ldots, m$, with $\mathbf{D}(\boldsymbol{\gamma}) = \mathrm{diag}(e^{\gamma_1}, \ldots, e^{\gamma_n})$, and $\boldsymbol{\gamma} = (\gamma_1, \ldots, \gamma_n)^T$. Then assign a second stage to the sampling distribution by, say, taking the $\boldsymbol{\beta}_i$ to constitute a random sample from an $N(\boldsymbol{\mu}_1, \mathbf{C}_1)$ distribution and to be independent of $\boldsymbol{\gamma}$, which possesses an $N(\boldsymbol{\mu}_2, \mathbf{C}_2)$ distribution. Let $\mathbf{C}_2 = \mathbf{C}_2(\boldsymbol{\zeta})$ denote some special covariance structure for $\boldsymbol{\gamma}$. This provides a broad sampling model that can handle lack of independence of the $\mathbf{y}_i$, serial correlation for each individual regression, and overdispersion, but with a quite modest parametrization $(\boldsymbol{\mu}_1, \mathbf{C}_1, \boldsymbol{\mu}_2, \boldsymbol{\zeta})$ that can be further constrained as appropriate.

(d) The multivariate normal distribution can be substantially generalized by taking $\mathbf{y}$, given $\mathbf{D}$, to be $N(\boldsymbol{\mu}, \mathbf{D})$ and by considering the matrix logarithm (Section 1.9) $\mathbf{A} = \log \mathbf{D}$. Let $\boldsymbol{\gamma} = \mathrm{vec}(\mathbf{A})$ denote some convenient vectorization of the upper triangular elements of $\mathbf{A}$. Then take $\boldsymbol{\gamma}$ to be $N(\boldsymbol{\mu}, \mathbf{C})$, where a variety of special structures $\boldsymbol{\mu} = \boldsymbol{\mu}(\boldsymbol{\beta})$ and $\mathbf{C} = \mathbf{C}(\boldsymbol{\zeta})$ can be chosen for $\boldsymbol{\mu}$ and $\mathbf{C}$. Then the consequent marginal distribution of $\mathbf{y}$ possesses thick tails like a multivariate t-distribution, but can be used to model a variety of nonlinear dependencies between the elements of $\mathbf{y}$. (The Bayesian implications of this probability structure are investigated by Leonard and Hsu, 1992.)

Worked Example 6D: *A Poisson Random Effects Model*

Let $y_1, y_2, \ldots, y_m$ denote independent Poisson counts with respective means $\theta_1, \theta_2, \ldots, \theta_m$. Suppose that conditional on μ and σ^2, the $\gamma_i = \log \theta_i$ are independently $N(\mu, \sigma^2)$ distributed in the prior assessment. Suppose further that $\pi(\mu, \sigma^2) \propto 1$.

(a) Find equations for the exact joint posterior modes of $\gamma_1, \gamma_2, \ldots, \gamma_m$, conditional only on σ^2.

(b) Suggest a multivariate normal approximation to the posterior density of $\boldsymbol{\gamma}$, given σ^2, based upon the solutions to these equations.

(c) Consider the approximation

$$l^*(\gamma_i \mid y_i) \propto \exp\left\{ -\tfrac{1}{2} v_i^{-1} (\gamma_i - l_i)^2 \right\}$$

to the likelihood of γ_i, where

$$l_i = \log\left(y_i + \tfrac{1}{2}\right) - \frac{3}{2\left(y_i + \frac{1}{2}\right)},$$

and

$$v_i = \frac{1}{y_i + \frac{1}{2}}.$$

Use this result to obtain an approximation to the marginal posterior density of σ^2.

(d) Noting that $E[e^{\gamma_i}] = \xi$ and $\mathrm{var}\,[e^{\gamma_i}] = \xi^2(e^{\sigma^2} - 1)$, in the prior assessment, where $\xi = \exp\{\mu + \frac{1}{2}\sigma^2\}$, obtain empirical Bayes estimates for μ and σ^2, based upon the method of moments.

(e) Note that

$$p\left(\mathbf{y} \mid \sigma^2\right) = \frac{p(\mathbf{y} \mid \boldsymbol{\gamma})\pi\left(\boldsymbol{\gamma} \mid \sigma^2\right)}{\pi\left(\boldsymbol{\gamma} \mid \mathbf{y}, \sigma^2\right)}$$

for any $\boldsymbol{\gamma}$. Use the ideas in (a) to suggest a Laplacian approximation to the marginal posterior density of σ^2.

Model Answer 6D:

(a) The joint density of $\boldsymbol{\gamma} = (\gamma_1, \gamma_2, \ldots, \gamma_m)^T$ and μ, given σ^2, is

$$\begin{aligned}\pi\left(\boldsymbol{\gamma}, \mu \mid \mathbf{y}, \sigma^2\right) &\propto \exp\left\{ \sum_{i=1}^{m} \gamma_i y_i - \sum_{i=1}^{m} e^{\gamma_i} - \frac{1}{2\sigma^2} \sum_{i=1}^{m} (\gamma_i - \mu)^2 \right\} \\ &\propto \exp\left\{ \sum_{i=1}^{m} \gamma_i y_i - \sum_{i=1}^{m} e^{\gamma_i} - \frac{1}{2\sigma^2} \sum_{i=1}^{m} (\gamma_i - \bar{\gamma})^2 \right. \\ &\qquad \left. - \frac{m}{2\sigma^2} (\bar{\gamma} - \mu)^2 \right\}.\end{aligned}$$

By integrating out μ, we find that the posterior density of $\boldsymbol{\gamma}$, given σ^2, is

$$\pi\left(\boldsymbol{\gamma} \mid \mathbf{y}, \sigma^2\right) \propto \exp\left\{ \sum_{i=1}^{m} \gamma_i y_i - \sum_{i=1}^{m} e^{\gamma_i} - \frac{1}{2\sigma^2} \sum_{i=1}^{m} (\gamma_i - \bar{\gamma})^2 \right\}.$$

By differentiating with respect to the γ_i, we find that the posterior modes $\tilde{\gamma}_i$ satisfy the equation

$$e^{\tilde{\gamma}_i} = y_i - \sigma^{-2}(\tilde{\gamma}_i - \tilde{\gamma}.) \qquad (i = 1, 2, \ldots, m),$$

where $\tilde{\gamma}.$ is the average of $\tilde{\gamma}_i$. Since these are nonlinear equations, they should be solved iteratively.

(b) The posterior information matrix is

$$\tilde{\mathbf{R}} = \left[-\frac{\partial^2 \log \pi(\boldsymbol{\gamma} \mid \mathbf{y}, \sigma^2)}{\partial(\boldsymbol{\gamma}\boldsymbol{\gamma}^T)} \right]_{\boldsymbol{\gamma}=\tilde{\boldsymbol{\gamma}}},$$

where $\tilde{\boldsymbol{\gamma}} = (\tilde{\gamma}_1, \tilde{\gamma}_2, \ldots, \tilde{\gamma}_m)^T$. This matrix reduces to

$$\tilde{\mathbf{R}} = \operatorname{diag}(e^{\tilde{\gamma}_1}, e^{\tilde{\gamma}_2}, \ldots, e^{\tilde{\gamma}_m}) + \sigma^{-2}(\mathbf{I}_m - m^{-1}\mathbf{J}_m),$$

where $\mathbf{I}$ and $\mathbf{J}$, respectively, denote the $m \times m$ identity and unit matrices. Consequently, given σ^2, $\boldsymbol{\gamma}$ may be taken to be approximately $N(\tilde{\boldsymbol{\gamma}}, \tilde{\mathbf{R}}^{-1})$ distributed in the posterior assessment.

(c) Under the stated likelihood approximation, the joint posterior density of $\boldsymbol{\gamma}$, μ, and σ^2 is

$$\begin{aligned}
&\pi^*(\boldsymbol{\gamma}, \mu, \sigma^2 \mid \mathbf{y}) \\
&\quad \propto (\sigma^2)^{-\frac{m}{2}} \exp\left\{ -\frac{1}{2} \sum_{i=1}^m v_i^{-1}(\gamma_i - l_i)^2 - \frac{1}{2}\sigma^{-2} \sum_{i=1}^m (\gamma_i - \mu)^2 \right\} \\
&\quad \propto (\sigma^2)^{-\frac{m}{2}} \exp\left\{ -\frac{1}{2} \sum_{i=1}^m (v_i^{-1} + \sigma^{-2})(\gamma_i - \gamma_i^*)^2 \right. \\
&\qquad\qquad \left. - \frac{1}{2} \sum_{i=1}^m (v_i + \sigma^2)^{-1}(l_i - \mu)^2 \right\},
\end{aligned}$$

where $\gamma_i^* = (v_i^{-1} + \sigma^2)^{-1}(v_i^{-1} l_i + \sigma^{-2}\mu)$. Integrating out $\boldsymbol{\gamma}$, and then μ, gives

$$\begin{aligned}
\pi^*(\sigma^2 \mid \mathbf{y}) \propto{}& \left\{ \sum_{i=1}^m (v_i + \sigma^2)^{-1} \right\}^{-\frac{1}{2}} \prod_{i=1}^m (v_i + \sigma^2)^{-\frac{1}{2}} \\
&\times \exp\left\{ -\frac{1}{2} \sum_{i=1}^m (v_i + \sigma^2)^{-1}(l_i - \bar{l})^2 \right\},
\end{aligned}$$

where

$$\bar{l} = \left\{ \sum_{i=1}^m (v_i + \sigma^2)^{-1} \right\}^{-1} \sum_{i=1}^m (v_i + \sigma^2)^{-1} l_i.$$

This is our approximation to the marginal posterior density of σ^2.

(d) $E(y_i \mid \gamma_i) = \mathrm{var}(y_i \mid \gamma_i) = e^{\gamma_i}$. Consequently,

$$E(y_i) = \underset{\gamma_i}{E}\left(e^{\gamma_i}\right) = \xi,$$

and

$$\mathrm{var}(y_i) = \underset{\gamma_i}{E}\left(e^{\gamma_i}\right) + \underset{\gamma_i}{\mathrm{var}}\left(e^{\gamma_i}\right) = \xi + \xi^2\left(e^{\sigma^2} - 1\right).$$

Let $\bar{y}$ and s^2 denote the sample mean and variance. Typically $s^2 > \bar{y}$. Otherwise, set $\hat{\sigma}^2 = 0$ and $\hat{\mu} = \log \bar{y}$. In the typical case, our estimates should satisfy

$$\hat{\xi} = e^{\hat{\mu} + \frac{1}{2}\hat{\sigma}^2} = \bar{y}$$

and

$$\hat{\xi} + \hat{\xi}^2\left(e^{\hat{\sigma}^2} - 1\right) = s^2.$$

Consequently,

$$\bar{y} + \bar{y}^2\left(e^{\hat{\sigma}^2} - 1\right) = s^2,$$

that is,

$$\hat{\sigma}^2 = \log\left\{1 + \frac{s^2 - \bar{y}}{\bar{y}^2}\right\}$$

and

$$\hat{\mu} = \log \bar{y} - \frac{1}{2}\hat{\sigma}^2.$$

(e) Since, in the prior assessment,

$$\pi\left(\boldsymbol{\gamma}, \mu \mid \sigma^2\right) \propto \left(\sigma^2\right)^{-\frac{1}{2}m} \exp\left\{-\frac{1}{2\sigma^2}\sum_{i=1}^{m}(\gamma_i - \bar{\gamma})^2 - \frac{m}{2\sigma^2}(\bar{\gamma} - \mu)^2\right\},$$

we find, upon integrating with respect to μ, that

$$\pi\left(\boldsymbol{\gamma} \mid \sigma^2\right) \propto \left(\sigma^2\right)^{-\frac{1}{2}(m-1)} \exp\left\{-\frac{1}{2\sigma^2}\sum_{i=1}^{m}(\gamma_i - \bar{\gamma})^2\right\}.$$

This density is improper, since

$$\pi\left(\boldsymbol{\gamma} \mid \sigma^2\right) \propto \left(\sigma^2\right)^{-\frac{1}{2}(m-1)} \exp\left\{-\frac{1}{2\sigma^2}\boldsymbol{\gamma}^T \mathbf{W} \boldsymbol{\gamma}\right\},$$

where $\mathbf{W} = \mathbf{I}_m - m^{-1}\mathbf{J}_m$ is singular.

However, as a function of σ^2, and for any choice of $\boldsymbol{\gamma}$,

$$p\left(\mathbf{y} \mid \sigma^2\right) \propto \frac{\exp\left\{\sum_{i=1}^{m} \gamma_i y_i - \sum_{i=1}^{m} e^{\gamma_i}\right\} \pi\left(\boldsymbol{\gamma} \mid \sigma^2\right)}{\pi\left(\boldsymbol{\gamma} \mid \mathbf{y}, \sigma^2\right)}.$$

We take the denominator to be approximately $N(\tilde{\boldsymbol{\gamma}}, \tilde{\mathbf{R}}^{-1})$, where $\tilde{\mathbf{R}}$ is defined in part (b). Replacing $\boldsymbol{\gamma}$ by $\tilde{\boldsymbol{\gamma}}$ gives the approximation

$$p^*(\mathbf{y} \mid \sigma^2) \propto (\sigma^2)^{-\frac{1}{2}(m-1)} |\tilde{\mathbf{R}}|^{-\frac{1}{2}} \times \exp\left\{ \sum_{i=1}^m \tilde{\gamma}_i y_i - \sum_{i=1}^m e^{\tilde{\gamma}_i} - \sum_{i=1}^m \frac{1}{2\sigma^2}(\tilde{\gamma}_i - \tilde{\gamma}_\cdot)^2 \right\},$$

then $\pi(\sigma^2 \mid \mathbf{y}) \propto p^*(\mathbf{y} \mid \sigma^2)$ for $0 < \sigma^2 < \infty$. By comparing with the solution in (c), we have

$$\pi(\sigma^2 \mid \mathbf{y}) \propto \left\{ \sum_{i=1}^m (\tilde{v}_i + \sigma^2)^{-1} \right\}^{-\frac{1}{2}} \prod_{i=1}^m (\tilde{v}_i + \sigma^2)^{-\frac{1}{2}} \times \exp\left\{ \sum_{i=1}^m \tilde{\gamma}_i y_i - \sum_{i=1}^m e^{\tilde{\gamma}_i} - \sum_{i=1}^m \frac{1}{2\sigma^2}(\tilde{\gamma}_i - \tilde{\gamma}_\cdot)^2 \right\},$$

where $\tilde{v}_i = e^{-\tilde{\gamma}_i}$.

SELF-STUDY EXERCISES

6.3.a Cell frequencies $y_1, \ldots, y_m$ are independent, given $\theta_1, \ldots, \theta_m$, and possess binomial distributions with respective cell probabilities $\theta_1, \ldots, \theta_m$ and sample sizes $n_1, \ldots, n_m$, with $m \geq 3$. The θ_i are a priori exchangeable. Let the $\gamma_i = \log \theta_i - \log(1 - \theta_i)$ constitute a random sample, given μ and σ^2, from the $N(\mu, \sigma^2)$ distribution, but suppose, furthermore, that $\pi(\mu, \sigma^2) \propto 1$.

(a) Use the results of Exercise 6.2.a to write down an algebraically explicit normal approximation to the posterior distribution of γ_i, given μ and σ^2, in terms of $\hat{\gamma}_i = \log y_i - \log(n_i - y_i)$ and $n_i^{-1}\vartheta_i = y_i^{-1} + (n_i - y)^{-1}$, which is valid whenever $y_i \geq 5$ and $n_i - y_i \geq 5$.

(b) Show that the marginal posterior density of μ and σ^2 may be approximated by

$$\pi^*(\mu, \sigma^2 \mid \mathbf{y}) \propto \prod_i (\sigma^2 + n_i^{-1}\vartheta_i)^{-1} \times \exp\left\{ -\sum_{i=1}^m \frac{1}{2(\sigma^2 + n_i^{-1}\vartheta_i)} (\hat{\gamma}_i - \mu)^2 \right\}, \tag{6.3.13}$$

where $-\infty < \mu < \infty$ and $0 < \sigma^2 < \infty$.

(c) Find the corresponding approximations to the marginal posterior densities of μ and σ^2.

(d) Show that conditionally only on σ^2, the posterior distribution of γ_i is approximately $N(\gamma_i^*, \vartheta_i^*)$, where

$$\gamma_i^* = \rho_i \hat{\gamma}_i + (1 - \rho_i)\bar{\gamma} \tag{6.3.14}$$

and

$$\vartheta_i^* = (1-\rho_i)^2\bar{\vartheta} + (1-\rho_i)\sigma^2, \tag{6.3.15}$$

with

$$\rho_i = \frac{n_i\vartheta_i^{-1}}{\left(n_i\vartheta_i^{-1} + \sigma^{-2}\right)} \tag{6.3.16}$$

and for some $\bar{\gamma}$ and $\bar{\vartheta}$, which you should calculate. Hence, show how to compute approximations to the unconditional posterior mean, variance, the c.d.f. of γ_i, and the posterior c.d.f. of θ_i, using one-dimensional numerical integrations, with respect to your approximate posterior density of σ^2.

6.3.b Extending Exercise 6.3.a, show how the results in (6.3.13)–(6.3.15) can be used to obtain hierarchical Bayes shrinkage estimators for other situations involving sets of parameters that are a priori exchangeable and where the likelihood of each parameter is, after an appropriate preliminary transformation, approximately normal (e.g., several Poisson distributions, several Gamma (n, e^{γ_i}) distributions). Show how maximization of (6.3.13) with respect to μ and σ^2 can instead lead to empirical Bayes shrinkage estimators for the γ_i. If μ and σ^2 possess a proper prior density $\pi(\mu, \sigma^2)$, show how the γ_i can be estimated via posterior modal estimators for μ and σ^2, and suggest a suitable choice of the prior density $\pi(\mu, \sigma^2)$.

6.3.c In Exercise 6.3.a, suppose that some of the y_i or $n_i - y_i$ are less than 5. Use extensions of equation (6.2.7) and the general approximation (6.3.5) to demonstrate, giving full technical details, how more accurate versions of all quantities described in (6.3.14)–(6.3.16) can be computed using the Newton–Raphson procedure.

6.3.d Let $y_1, \ldots, y_m$ be independent and $N(\theta_i, \tau^2)$ distributed, with $m \geq 4$, where τ^2 is known, and suppose that $\theta_1, \ldots, \theta_m$ possess the two-stage exchangeable prior distribution described for $\gamma_1, \ldots, \gamma_m$ in Exercise 6.3.a. Show that the exact posterior mean of θ_i, given $\rho = \tau^2/(\sigma^2 + \tau^2)$, is

$$E\left[\theta_i \mid \sigma^2, \mathbf{y}\right] = \rho y_i + (1-\rho)\bar{y} \tag{6.3.17}$$

and that the exact posterior density of ρ is

$$\pi(\rho \mid \mathbf{y}) = \frac{\rho^{\frac{1}{2}(m-3)-1}}{G_{\frac{1}{2}(m-3)}(\omega/2)} \exp\left\{-\frac{1}{2}\omega\rho\right\} \qquad (0 < \rho < 1), \tag{6.3.18}$$

where $\omega = s^2/\tau^2$ with $s^2 = \sum(y_i - \bar{y})^2$ and G denotes the incomplete Gamma function (see Exercise 3.2.d). Hence show that the unconditional posterior mean of ρ is

$$E(\theta_i \mid \mathbf{y}) = y_i - \rho^*(y_i - \bar{y}), \tag{6.3.19}$$

where

$$\rho^* = \frac{2\omega^{-1}G_{\frac{1}{2}(m-1)}(\omega/2)}{G_{\frac{1}{2}(m-3)}(\omega/2)}. \tag{6.3.20}$$

Abramowitz and Stegun (1968) tell us that $G_{\nu+1}(a) = \nu G(a) - a^\nu e^{-a}$. Show that

$$\rho^* = (m-3)\omega^{-1}\{1 - h(\omega)\} = \frac{(m-3)\tau^2}{s^2}\{1 - h(\tau^2/s^2)\}, \tag{6.3.21}$$

where

$$h(\omega) = \frac{(\omega/2)^{\frac{1}{2}(m-5)}}{G_{\frac{1}{2}(m-3)}(\omega/2)} e^{-\frac{1}{2}\omega}. \tag{6.3.22}$$

Show, however, that an empirical Bayes approach, maximizing (6.3.21) with respect to ρ, would yield the Lindley–Stein estimator (Lindley, 1962), which replaces ρ^* by

$$\hat{\rho} = \max\left[\frac{(m-3)\tau^2}{s^2}, 1\right]. \tag{6.3.23}$$

For further details of this comparison, see Leonard (1976). For extensions to the replicated case where τ^2 is also unknown, see Box and Tiao (1968) and Leonard and Ord (1976). Shrinkage estimators for regression coefficients are proposed by Lindley and Smith (1972). According to the empirical study they describe, only a quarter of the sample size is needed when using these shrinkage estimators in order to obtain similar predictive accuracy, when compared with predictions based on least squares. Following the mean squared error investigation by Sun et al. (1996), we instead recommend the more conservative shrinkage estimators in Exercise 6.3.h, which generalize the Lindley–Stein empirical Bayes estimator. The substantial savings projected by Lindley and Smith are overoptimistic, for complex reasons associated with their empirical study.

6.3.e *An Autoregressive Model for Nonnormal Data:* In Exercise 6.3.a, suppose instead that given μ, σ^2, and ρ, the γ_i are $N(\mu, \sigma^2)$, that $\boldsymbol{\gamma} = (\gamma_1, \dots, \gamma_n)^T$ is multivariate normal, but that the autocorrelation between γ_i and γ_j is $\rho^{|j-k|}$ where $\rho \in (-1, 1)$, that is, $(\gamma_i - \mu_i) = \rho(\gamma_{i-1} - \mu_{i-1}) + \epsilon_i$ $(i = 1, \dots, m)$, where $\gamma_0 = \mu$, and the error terms $\epsilon_1 \sim N(0, \sigma^2)$ and $\epsilon_i \sim N(0, \sigma^2(1-\rho^2))$ $(i = 2, \dots, m)$ are independent, given ρ and σ^2. Assume further that $\pi(\mu, \sigma^2, \rho) \propto 1$.

Under the likelihood approximation $l^*(\gamma_i \mid y_i) \propto \exp\{-\frac{1}{2} n_i \vartheta_i^{-1}(\gamma_i - \hat{\gamma}_i)^2\}$, obtain Kalman filter updating relations (see Section 5.3) between the $a_i = E(\gamma_i \mid \mathbf{y}_i^* \mu, \sigma^2, \rho)$, where $\mathbf{y}_i^* = (y_1, \dots, y_i)^T$, when μ, σ^2, and ρ are known, and obtain the predictive distribution of γ_{m+j}. How would you generalize these results to the situation where μ, σ^2, and ρ are unknown?

6.3.f In the situation discussed in Exercise 5.1.f, where multinomial cell probabilities $\theta_1, \dots, \theta_p$ possess a Dirichlet distribution with parameters $\gamma\xi_1, \dots, \gamma\xi_p$, with $\boldsymbol{\xi} = (\xi_1, \dots, \xi_p)^T$ lying in the unit simplex, suppose that $\boldsymbol{\xi}$ is known and that γ is unknown. Show that γ possesses exact integrated likelihood (e.g., Good, 1965, 1967)

$$l(\gamma \mid \mathbf{y}) = \frac{K\Gamma(\gamma)}{\Gamma(\gamma + n)} \prod_{j=1}^{p} \frac{\Gamma(\gamma\xi_j + np_j)}{\Gamma(\gamma\xi_j)}$$

for some constant K, where $p_j = y_i/n$. Show that $l(\boldsymbol{\gamma} \mid \mathbf{y})$ converges to K as $\gamma \to \infty$ with n and the p_j fixed. Hence show that the posterior distribution is improper, under any distribution in the prior assessment, with improper right tail. Under the Cauchy-tail prior, $\pi(\gamma) = (1+\gamma)^{-2}$, for $0 < \gamma < \infty$, show that the posterior mean

$$E(\theta_j \mid \mathbf{y}) = \rho^* y_i + (1 - \rho^*)\xi_j$$

always exists, with $\rho^* = E[n/(n+\gamma) \mid \mathbf{y}]$.

Historical note: When $\xi_j = p^{-1}$ for $j = 1, \dots, p$, the θ_j are a priori "permutable." Irving Jack Good and Alan Turing used this assumption as the basis for their shrinkage estimates of probabilities of symbols appearing in Nazi codes during World War II. They also used Bayes factors, decision trees, and machine intelligence (Good, 1979) to solve the Nazi codes at Bletchley Park. According to Good (personal communication), this solution was unknown to Sir Winston Churchill prior to the bombing of the City of Coventry.

The dependence of ρ^* upon $y_1, \ldots, y_n$ ensures that Good's estimators for the θ_j do not satisfy "Johnson's sufficientness postulate" as developed by the philosopher William Ernest Johnson (Johnson and Braithwaite, 1932).

6.3.g Under the assumptions of Exercises 5.1.f and 6.3.f, Leonard (1977a) shows that given only γ, the marginal distribution of ρX^2 is, for large n and ρ fixed, approximately χ^2 with $p - 1$ degrees of freedom, where $\rho = n/(\gamma + n)$, and

$$X^2 = n \sum_{j=1}^{p} \frac{(p_j - \xi_j)^2}{\xi_j}.$$

Use this result to show that the integrated likelihood of ρ is approximated by

$$l^*(\rho \mid \mathbf{x}) = \rho^{\frac{1}{2}(s-1)} \exp\left\{ -\tfrac{1}{2}\rho X^2 \right\} \qquad (0 < \rho < 1),$$

so that the marginal maximum likelihood estimate $\hat{\gamma}$ of γ satisfies

$$\hat{\rho} = \frac{n}{n + \hat{\gamma}} = \min\left(\frac{s-1}{X^2}, 1 \right). \tag{6.3.24}$$

(Note that it is also exactly true that $\hat{\gamma} = \infty$ whenever $X^2 \leq s - 1$; see Levin and Reeds, 1977.)

Show that when n is large, $\pi(\gamma) = (1 + \gamma)^{-2}$ is equivalent to $\pi(\rho) \propto \rho^2$ for $0 < \rho < 1$. Under this assumption, use the incomplete Gamma function employed in Exercise 6.3.d to find an approximation, for large n, to the posterior mean of θ_j.

6.3.h *M-Group Regression:* Suppose that for $i = 1, \ldots, m$, the $n_i \times 1$ vectors $\mathbf{y}_i$ are independent $N(\mathbf{X}_i\boldsymbol{\beta}_i, \phi_i\mathbf{I}_{n_i})$, given the $\boldsymbol{\beta}_i$ and ϕ_i, where the $\boldsymbol{\beta}_i$ are a random sample from $N(\boldsymbol{\mu}, \mathbf{C})$, and the $\mathbf{X}_i$ are specified $n_i \times p$ matrices, with $\mathbf{R}_i = \mathbf{X}_i^T\mathbf{X}_i$ nonsingular.

(a) Show that, given $\boldsymbol{\mu}$, $\mathbf{C}$, and the ϕ_i, the $\boldsymbol{\beta}_i$ are a posteriori $N(\boldsymbol{\beta}_i^*, \mathbf{D}_i)$, where

$$\boldsymbol{\beta}_i^* = \left(\phi_i^{-1}\mathbf{R}_i + \mathbf{C}^{-1}\right)\left(\phi_i^{-1}\mathbf{R}_i\hat{\boldsymbol{\beta}}_i + \mathbf{C}^{-1}\boldsymbol{\mu}\right),$$

and

$$\mathbf{D}_i^{-1} = \phi_i^{-1}\mathbf{R}_i + \mathbf{C}^{-1},$$

where

$$\hat{\boldsymbol{\beta}}_i = \left(\mathbf{X}_i^T\mathbf{X}_i\right)^{-T}\mathbf{X}_i^T\mathbf{y}_i.$$

(b) Applying the EM approach of Section 6.3, the $\boldsymbol{\beta}_i$ should be regarded as missing observations. Show that the "complete data" log-likelihood $l_C(\boldsymbol{\phi}, \boldsymbol{\mu}, \mathbf{C} \mid \mathbf{y}, \boldsymbol{\beta})$ for the ϕ_i, $\boldsymbol{\mu}$, and $\mathbf{C}$, given the $\mathbf{y}_i$ and $\boldsymbol{\beta}_i$, satisfies

$$\begin{aligned} -2\log l_C(\boldsymbol{\phi}, \boldsymbol{\mu}, \mathbf{C} \mid \mathbf{y}, \boldsymbol{\beta}) &= \sum_{i=1}^{m} n_i \log\phi_i + \sum_{i=1}^{m} \phi_i^{-1} S_i^2 \\ &\quad + \sum_{i=1}^{m} n_i\phi_i^{-1}\left(\boldsymbol{\beta}_i - \hat{\boldsymbol{\beta}}_i\right)^T\mathbf{R}_i\left(\boldsymbol{\beta} - \hat{\boldsymbol{\beta}}_i\right) \\ &\quad + \sum_{i=1}^{m} (\boldsymbol{\beta}_i - \boldsymbol{\mu})^T\mathbf{C}^{-1}(\boldsymbol{\beta}_i - \boldsymbol{\mu}), \end{aligned}$$

where S_i^2 is the ith residual sum of squares.

(c) Show that the values of the ϕ_i, $\boldsymbol{\mu}$, and $\mathbf{C}$ maximizing their integrated likelihood $l^*(\boldsymbol{\phi}, \boldsymbol{\mu}, \mathbf{C} \mid \mathbf{y})$ satisfy the equations

$$\phi_i = n_i^{-1} S_i^2 + (\boldsymbol{\beta}_i^* - \hat{\boldsymbol{\beta}}_i)\mathbf{R}_i(\boldsymbol{\beta}_i^* - \hat{\boldsymbol{\beta}}_i) + \mathrm{trace}(\mathbf{R}_i\mathbf{D}_i) \qquad (i = 1, \ldots, m),$$

$$\boldsymbol{\mu} = \boldsymbol{\beta}_\cdot^* = m^{-1} \sum_{i=1}^{m} \boldsymbol{\beta}_i^*,$$

and

$$\mathbf{C} = (m-1)^{-1} \sum_i (\boldsymbol{\beta}_i^* - \boldsymbol{\beta}_\cdot^*)^T (\boldsymbol{\beta}_i^* - \boldsymbol{\beta}_\cdot^*)^T + \mathbf{D}_\cdot,$$

where

$$\mathbf{D}_\cdot = m^{-1} \sum_{i=1}^{m} \mathbf{D}_i .$$

(d) Investigate how this approach modifies Lindley and Smith (1972) and Smith (1973a and b). See also Miller and Fortney (1984) for some numerical examples comparing these procedures for insurance data.

6.3.i *Product Multinomial Model* (Leonard 1973b; Leonard and Hsu, 1994): Now generalize Exercise 6.1.c by assuming that for $i = 1, \ldots, m$, the $y_{i1}, \ldots, y_{ip}$ possess a multinomial distribution with cell probabilities $\theta_{ij}, \ldots, \theta_{ip}$ falling in the unit simplex S_U, and samples size $n_i = \sum_j y_{ij}$. All m multinomial vectors $y_i = (y_{i1}, \ldots, y_{ip})^T$ are assumed independent, given the $\boldsymbol{\theta} = (\theta_{i1}, \ldots, \theta_{ip})^T$. Consider m vectors $\boldsymbol{\gamma}_i = (\gamma_{i1}, \ldots, \gamma_{ip})^T$, of multivariate logits, satisfying

$$\theta_{ij} = \frac{e^{\gamma_{ij}}}{\sum_{g=1}^{p} e^{\gamma_{ig}}} \qquad (i = 1, \ldots, m;\ j = 1, \ldots, p).$$

Assume that, given $\boldsymbol{\mu}$ and $\mathbf{C}$, the $\boldsymbol{\gamma}_i$ are independent $N(\boldsymbol{\mu}, \mathbf{C})$, but where $\boldsymbol{\mu}$ and $\mathbf{C}$ are completely unknown, that is, the vectors $\boldsymbol{\gamma}_1, \ldots, \boldsymbol{\gamma}_m$ are exchangeable.

(a) Show that the exact posterior mode vector $\tilde{\boldsymbol{\gamma}}_i$ of $\boldsymbol{\gamma}_i$, given $\boldsymbol{\mu}$ and $\mathbf{C}$, satisfies

$$n_i \tilde{\boldsymbol{\theta}}_i = \mathbf{y}_i - \mathbf{C}^{-1}(\tilde{\boldsymbol{\gamma}}_i - \boldsymbol{\mu}) \qquad (i = 1, \ldots, m),$$

where $\tilde{\boldsymbol{\theta}}_i$ has jth element

$$\tilde{\theta}_{ij} = \frac{e^{\tilde{\gamma}_{ij}}}{\sum_{g=1}^{p} e^{\tilde{\gamma}_{ig}}} \qquad (j = 1, \ldots, p).$$

Show also that the exact posterior information matrix of $\boldsymbol{\gamma}_i$ is

$$\tilde{\mathbf{R}}_i = n_i \left[\mathrm{diag}(\tilde{\theta}_{i1}, \ldots, \tilde{\theta}_{ip}) - \tilde{\boldsymbol{\theta}}_i \tilde{\boldsymbol{\theta}}_i^T \right] + \mathbf{C}^{-1}.$$

(b) Use the EM algorithm to show that the marginal maximum likelihood estimates of $\boldsymbol{\mu}$ and $\mathbf{C}$ approximately satisfy the equations

$$\boldsymbol{\mu} = \tilde{\boldsymbol{\gamma}}_\cdot = m^{-1} \sum_{i=1}^{m} \tilde{\gamma}_i$$

and

$$\mathbf{C} = (m-1)^{-1}\sum_{i=1}^{m}(\tilde{\boldsymbol{\gamma}}_i - \tilde{\boldsymbol{\gamma}}_.)(\tilde{\boldsymbol{\gamma}}_i - \tilde{\boldsymbol{\gamma}}_.)^T + m^{-1}\sum_{i=1}^{m}\tilde{\mathbf{R}}_i^{-1}.$$

(c) Show that the integrated likelihood of $\boldsymbol{\mu}$ and $\mathbf{C}$ can be represented by the Laplacian approximation

$$\tilde{l}(\boldsymbol{\mu}, \mathbf{C} \mid \mathbf{y}) \propto |\mathbf{C}|^{-\frac{1}{2}m}\prod_{i=1}^{m} l(\tilde{\boldsymbol{\gamma}}_i \mid \mathbf{y}_i)|\mathbf{R}_i|^{-\frac{1}{2}}$$
$$\times \exp\left\{-\frac{1}{2}\sum_i(\tilde{\boldsymbol{\gamma}}_i - \boldsymbol{\mu})^T\mathbf{C}^{-1}(\tilde{\boldsymbol{\gamma}}_i - \boldsymbol{\mu})\right\}.$$

Describe how you would use this result and AIC to decide between different choices of special prior structure for $\mathbf{C}$, with unknown hyperparameters. How would you use this result and AIC to decide between (i) the current exchangeable prior distribution for the $\boldsymbol{\gamma}_i$ with unconstrained $\boldsymbol{\mu}$ and $\mathbf{C}$ and (ii) no prior information about the $\boldsymbol{\gamma}_i$?

Note: This provides an alternative to the hierarchical "Dirichlet–Dirichlet" approach described by Leonard (1977c) and Dawid and Pueschel (1998), which is applied by Dawid and Pueschel to the genetic modeling of nonhomogeneous populations.

6.3.j *Charles Stein's Integration by Parts* (Stein, 1981): Let $y_1, \ldots, y_m$ be independent, given their respective means $\theta_1, \ldots, \theta_m$ and common variance τ^2, and that each y_i is $N(\theta_i, \tau^2)$. We wish to evaluate the performance of estimators $\theta_i^* = \theta_i^*(\mathbf{y})$ of θ_i by reference to their (total) MSE,

$$r^*(\boldsymbol{\theta}) = \sum_{i=1}^{m} \underset{\mathbf{y}|\boldsymbol{\theta}}{E}\left\{(\theta_i^* - \theta_i)^2\right\} \qquad (i = 1, \ldots, m).$$

(a) Show that the MSE of the $\hat{\theta}_i = y_i$ is $\hat{r}(\boldsymbol{\theta}) = m\tau^2$.

(b) If $\theta_i^* = y_i - h_i(\mathbf{y})$, where $h_i(\mathbf{y})$ is almost surely differentable, show that the MSE $r^*(\boldsymbol{\theta})$ of the θ_i^* satisfies

$$\hat{r}(\boldsymbol{\theta}) - r^*(\boldsymbol{\theta}) = \sum_{i=1}^{m} \underset{\mathbf{y}|\boldsymbol{\theta}}{E}\left\{h_i^2(\mathbf{y})\right\} + 2\sum_{i=1}^{m} \underset{\mathbf{y}|\boldsymbol{\theta}}{E}\left\{(y_i - \theta_i)h_i(\mathbf{y})\right\}.$$

(c) Use integration by parts,

$$\int_a^b f(x)g^{(1)}(x)\,dx + \int_a^b f^{(1)}(x)g(x)\,dx = f(b)g(b) - f(a)g(a),$$

to show that

$$\underset{\mathbf{y}|\boldsymbol{\theta}}{E}\left[(y_i - \theta_i)h_i(\mathbf{y})\right] = \tau^2 \underset{\mathbf{y}|\boldsymbol{\theta}}{E} h_i^{(1)}(\mathbf{y}),$$

where $h_i^{(1)}(y_i) = \partial h_i^{(1)}(\mathbf{y})/\partial y_i$, and assuming that these integrals are finite.

Hence show that

$$\hat{r}(\boldsymbol{\theta}) - r^*(\boldsymbol{\theta}) = \underset{\mathbf{y}|\boldsymbol{\theta}}{E} \sum_{i=1}^{m} \left[2\tau^2 h_i^{(1)}(\mathbf{y}) - h_i^2(\mathbf{y})\right].$$

(d) Suppose that $h_i(\mathbf{y}) = (m-p)y_i\tau^2/u$, with $u = \sum_k y_k^2$. Show that

$$\hat{r}(\boldsymbol{\theta}) - r^*(\boldsymbol{\theta}) = (m-p)(m+p-4)\tau^4 \underset{\mathbf{y}|\boldsymbol{\theta}}{E} \left(u^{-1}\right).$$

Hence prove that the Stein estimators $\theta_i^* = y_i[1 - \{(m-2)\tau^2/u\}]$ dominate the UMVU estimator $\hat{\theta}_i = y_i$ when $m \geq 3$, so that the latter are inadmissible with respect to an additive squared error loss function. Other proofs are provided by James and Stein (1961) and Efron and Morris (1973).

6.3.k (a) In Exercise 6.3.j, note that u has expectation $Q_m(\boldsymbol{\theta}) + m\tau^2$, where $Q_m(\boldsymbol{\theta}) = \sum_{i=1}^m \theta_i^2$ and that as $m \to \infty$, $m^{-1}u$ will converge, with probability one, to $\tau^2 + \lim[m^{-1}Q_m(\boldsymbol{\theta})]$ whenever this limit exists. Use this result to show, that when m is large,

$$\frac{\hat{r}(\boldsymbol{\theta}) - r^*(\boldsymbol{\theta})}{\hat{r}(\boldsymbol{\theta})} \sim \frac{\tau^2}{\tau^2 + m^{-1}Q_m(\boldsymbol{\theta})}.$$

Does this immense potential saving in risk surprise you?

(b) *Shrinking towards* $\bar{y}$ (Lindley, 1962): Show that the empirical Bayes shrinkage estimators in (6.3.19), with ρ^* replaced by $\hat{\rho}$ in (6.3.23), also dominate the UMVU estimators with respect to an additive squared error loss function if $m > 4$. Show that the risk functions $\hat{r}(\boldsymbol{\theta})$ of the UMVU estimators and $r^*(\boldsymbol{\theta})$ of the shrinkage estimators have the approximate property, when m is large, that

$$\frac{\hat{r}(\boldsymbol{\theta}) - r^*(\boldsymbol{\theta})}{r^*(\boldsymbol{\theta})} \sim \frac{m\tau^2}{m\tau^2 + \sum(\theta_i - \theta_.)^2}.$$

By showing that the Bayes–Stein estimator (6.3.19) approximates the empirical Bayes estimator when m is large, show that the Bayes–Stein shrinkage estimator possesses similar risk properties when m is large. Show that the estimator (6.3.19), shrinking the y_i towards $\bar{y}$, is preferable to the Stein estimator of Exercise 6.3.j and of Exercise 6.3.k, which shrink the y_i either towards or beyond zero.

Note: The exact risk properties for finite m are computed by Issos (1982). These results are extended by Hudson (1978) and Ghosh, Hwang, and Tsui (1983, 1984) to justify shrinkage estimators for other members of the exponential family, and by Brown (1971) to various other choices of loss function. Fienberg and Holland (1973) investigate the risk properties of shrinkage estimators for multinomial proportions and demonstrate similarly large savings in MSE, except on the boundary of the parameter space S_U.

Following Morris (1983), it is generally meaningful to shrink the maximum likelihood estimates towards a useful null hypothesis, for example, a grand sample mean representing equality of several normal means, a hypothesized regression modelregression model, or a model representing independence of rows and columns in a two-way contingency table. Following Draper and Van Nostrand (1979), it is not usually meaningful to shrink the maximum likelihood estimates towards zero, even though savings in MSE can still be attained.

6.4 Monte Carlo Methods and Importance Sampling

(A) *Simple Monte Carlo*

First, consider cases where it is possible to obtain independent simulations $\boldsymbol{\gamma}_1, \boldsymbol{\gamma}_2, \ldots$ from the posterior distribution with density $\pi_y(\boldsymbol{\gamma})$ for a $p \times 1$ vector of parameters $\boldsymbol{\gamma}$. Then the posterior expectation, if it exists, of an arbitrary function $\eta = g(\boldsymbol{\gamma})$ of $\boldsymbol{\gamma}$ may be computed, via Monte Carlo, from

$$\mu_g = E\big[g(\boldsymbol{\gamma}) \mid y\big] = \lim_{M\to\infty} M^{-1} \sum_{j=1}^{M} g(\boldsymbol{\gamma}_j), \tag{6.4.1}$$

and the c.d.f. of $g(\boldsymbol{\gamma})$ can be computed from

$$F_g(\eta_0) = p\big(g(\boldsymbol{\gamma}) \le \eta_0\big) = \lim_{M\to\infty} M^{-1} \sum_{j=1}^{M} g_{\eta_0}(\boldsymbol{\gamma}_j), \tag{6.4.2}$$

where $g_{\eta_0}(\gamma) = I[g(\gamma) \le \eta_0]$. Note the following key issues:

(i) By the strong law of large numbers, each of the sequences in (6.4.1) and (6.4.2) will correctly converge to μ_g or $F_g(\eta_0)$ with probability one, as $M \to \infty$, whenever the limit is finite.

(ii) Let $U_M = M^{-1} \sum g(\boldsymbol{\gamma}_j)$. Then, by the central limit theorem, $M^{\frac{1}{2}}(U_M - \mu_g)$ will converge in distribution, as $M \to \infty$, to a $N(0, \sigma_g^2)$ variate whenever the posterior variance σ_g^2, of $g(\boldsymbol{\gamma})$, is finite. Therefore, for good practical convergence of (6.4.1), it is required that $\sigma_g^2 < \infty$, in which case $M^{-\frac{1}{2}}\sigma_g$ measures the standard deviation of simulation, after M simulations.

(iii) Even if $\sigma_g^2 = \infty$, the c.d.f. in (6.4.2) can still be obtained via good convergence, since the posterior variance of $g_{\eta_0}(\boldsymbol{\gamma})$ will always be finite.

(B) *Monte Carlo Procedures for Marginal Densities*

If the conditional posterior density of η, given $\boldsymbol{\gamma}$, can be specified in a functional form $\pi_y(\eta \mid \boldsymbol{\gamma})$, then the marginal posterior density of η is just the posterior expectation, with respect to $\boldsymbol{\gamma}$, for each fixed η of $\pi_y(\eta \mid \boldsymbol{\gamma})$. This can therefore be computed typically as a continuous function, using the preceding Monte Carlo method for the expectation of a function of $\boldsymbol{\gamma}$. This key point, together with issues (i), (ii), and (iii) of Section 6.4 (A), enhances the importance sampling procedures of Section 6.4 (C), together with the MCMC procedures of Sections 6.6–6.8. As an example, suppose, with $\pi_y(\boldsymbol{\gamma}) \propto \pi_y^*(\gamma_1, \gamma_2, \ldots, \gamma_p)$ specified up to a constant of proportionality, that we wish to compute the marginal posterior density $\pi_y(\gamma_1)$ of γ_1. Then for each fixed γ_1, average $\pi_y^*(\gamma_1, \gamma_2, \ldots, \gamma_p)$ across M simulations for $\boldsymbol{\gamma} = (\gamma_1, \gamma_2, \ldots, \gamma_p)^T$, but with γ_1 fixed. Then divide by the average of $\pi_y^*(\gamma_1, \gamma_2, \ldots, \gamma_p)$ across the same M simulations, but without fixing γ_1.

(C) *Importance Sampling*

If it is impossible to obtain simulations from $\pi_y(\boldsymbol{\gamma})$ because of the complicated form of this exact posterior density, it is instead possible to simulate from some simpler density $\pi_y^*(\boldsymbol{\gamma})$ with positive support on R^p, and note that

$$\mu_g = E\big(g(\boldsymbol{\gamma}) \mid \mathbf{y}\big) = E^*\big[W(\boldsymbol{\gamma})g(\boldsymbol{\gamma})\big],$$

where

$$W(\boldsymbol{\gamma}) = \frac{\pi_y(\boldsymbol{\gamma})}{\pi_y^*(\boldsymbol{\gamma})}, \tag{6.4.3}$$

and E^* denotes expectation with respect to the simpler distribution with density $\pi_y^*(\boldsymbol{\gamma})$. If $\boldsymbol{\gamma}_1, \boldsymbol{\gamma}_2, \ldots$ are independent simulations from this simpler distribution, with density, that is, importance function, $\pi_y^*(\boldsymbol{\gamma})$, then

$$\mu_g = \lim_{M\to\infty} M^{-1} \sum_{j=1}^{M} W(\boldsymbol{\gamma}_j)g(\boldsymbol{\gamma}_j). \tag{6.4.4}$$

Furthermore, $F_g(\eta_0)$ may be obtained by replacing $g(\boldsymbol{\gamma})$ in (6.4.4) by $g_{\eta_0}(\boldsymbol{\gamma}) = I[g(\boldsymbol{\gamma}) \le \eta_0]$. Note that (e.g., Geweke, 1988, 1989)

(i) the strong law of large numbers still holds for the convergence of (6.4.4) whenever μ_g is finite;

(ii) a central limit result holds, whenever μ_g is finite, as long as the variance of $W(\boldsymbol{\gamma})g(\boldsymbol{\gamma})$ for a simulated $\boldsymbol{\gamma}$ is finite, that is, if

$$E\big(W(\boldsymbol{\gamma})g^2(\boldsymbol{\gamma}) \mid \mathbf{y}\big) < \infty; \tag{6.4.5}$$

(iii) the condition in (6.4.5) will be true when computing the expectation of any bounded $g(\boldsymbol{\gamma})$, whenever

$$E^*\big(W^2(\gamma) \mid \mathbf{y}\big) = E\big(W(\boldsymbol{\gamma}) \mid \mathbf{y}\big) < \infty; \tag{6.4.6}$$

(iv) under condition (6.4.6), it is possible to obtain good convergence towards the c.d.f. $g_{\eta_0}(\boldsymbol{\gamma}) = I[g(\boldsymbol{\gamma}) \le \eta_0]$, of any function $g(\boldsymbol{\gamma})$, even if this is unbounded;

(v) a good importance function may be chosen by ensuring that $E\{W(\boldsymbol{\gamma}) \mid \mathbf{y}\}$ is as close to unity as possible, for example, by ensuring that $\pi_y^*(\boldsymbol{\gamma})$ reasonably approximates $\pi_y(\boldsymbol{\gamma})$, and that if there are differences, the extreme tails of $\pi_y^*(\boldsymbol{\gamma})$ are thicker than the extreme tails of $\pi_y(\boldsymbol{\gamma})$;

(vi) it is possible to use your simulations to calculate

$$\operatorname{var}^*\big[W(\boldsymbol{\gamma}) \mid \mathbf{y}\big] = E^*\big(W^2(\boldsymbol{\gamma}) \mid \mathbf{y}\big) - 1$$

for several different choices of the importance function, by appropriately reapplying (6.4.4). After, say, 20,000 simulations, switch to the importance function yielding the smallest simulated var $[W(\boldsymbol{\gamma}) \mid \mathbf{y}]$;

(vii) your simulated value for $M^{-1}\text{var}^*(W(\boldsymbol{\gamma}) \mid \mathbf{y})$ is your estimate variance of simulation, after M simulations;

(viii) in many applications, it is possible to know only that $\pi_y(\boldsymbol{\gamma}) = C\tilde{\pi}_y(\boldsymbol{\gamma})$, where $\tilde{\pi}_y(\mathbf{y})$ is specified and C is an unknown constant. In this case, replace (6.4.3) by

$$\tilde{W}(\boldsymbol{\gamma}) = \frac{\tilde{\pi}_y(\boldsymbol{\gamma})}{\pi_y^*(\boldsymbol{\gamma})} \tag{6.4.7}$$

and (6.4.4) by

$$\mu_g = C \lim_{M\to\infty} M^{-1} \sum_{j=1}^{M} \tilde{W}(\boldsymbol{\gamma}_j) g(\boldsymbol{\gamma}_j), \tag{6.4.8}$$

where

$$C^{-1} = \lim_{M\to\infty} M^{-1} \sum_{i=1}^{M} \tilde{W}(\boldsymbol{\gamma}_j) g(\boldsymbol{\gamma}_j). \tag{6.4.9}$$

(D) *A Generalized Multivariate t-Importance Function*

Suppose now that a $p \times 1$ vector of parameters $\boldsymbol{\gamma}$ possesses approximate multivariate normal likelihood in (6.2.1) and that the exact posterior distribution is thought to be proper under a uniform $\pi(\boldsymbol{\gamma}) \propto 1$ prior. Under this prior, the posterior distribution of $\boldsymbol{\gamma}$ is approximately $N(\hat{\boldsymbol{\gamma}}, \mathbf{R}^{-1})$. However, we recommend as importance function

$$\pi_y^*(\boldsymbol{\gamma}) = t_{\boldsymbol{\gamma}}\left(\nu, \hat{\boldsymbol{\gamma}}, \frac{\nu}{\nu + p}\mathbf{R}\right), \tag{6.4.10}$$

a generalized multivariate t-density, with ν degrees of freedom, mean vector $\hat{\boldsymbol{\gamma}}$, and precision matrix $\nu\mathbf{R}/(\nu + p)$, replacing $\lambda^{-1}\mathbf{R}_0$, as defined in Exercise 6.1.e. If the degrees of freedom are small enough, then the criterion in (6.4.5) will typically become finite. However, in practice, simply choose ν to give as fast a convergence as possible, or adjust your value of ν to reduce the estimated standard error of simulation discussed above.

(E) *Simulating from a Generalized Multivariate t-Distribution*

Note that if $\boldsymbol{\gamma}$ possesses the distribution (6.1.34), then

$$\boldsymbol{\gamma} = \boldsymbol{\mu} + \nu^{\frac{1}{2}} U_\nu^{-\frac{1}{2}} \boldsymbol{\epsilon}, \tag{6.4.11}$$

where $\boldsymbol{\epsilon}$ is $N(\mathbf{0}, \mathbf{C})$, where $\mathbf{C} = \lambda\mathbf{R}_0^{-1}$, and U_ν is independent of $\boldsymbol{\epsilon}$ and possesses a chi-squared distribution with ν degrees of freedom. Therefore, simulations from our multivariate t-distribution can be achieved by combining simulations from a multivariate normal distribution with simulations from a chi-squared distribution. The former are most readily completed by simulating p independent standard normal variates $z_1, z_2, \ldots, z_p$ and referring to a Cholesky decomposition of the multivariate normal covariance matrix. Indeed any decomposition of the form $\mathbf{C} = \mathbf{G}\mathbf{G}^T$ is appropriate, where $\mathbf{G}$ is some $p \times p$ matrix. Simply set $\boldsymbol{\epsilon} = \mathbf{G}\mathbf{z}$, where $\mathbf{z} = (z_1, z_2, \ldots, z_p)^T$.

(F) *Incorporating Prior Information*

Next suppose that subject to the likelihood approximation (6.1.2), $\boldsymbol{\gamma}$ has an $N(\boldsymbol{\mu}, \mathbf{C})$ prior distribution, with $\boldsymbol{\mu}$ and $\mathbf{C}$ specified. Then simply replace $\hat{\boldsymbol{\gamma}}$ and $\mathbf{R}$ in (6.4.10) by the explicit expressions for $\boldsymbol{\gamma}^*$ and $\mathbf{R}^*$ in (6.1.11) and (6.1.12). In situations where $\hat{\boldsymbol{\gamma}}$ and $\mathbf{R}$ are not finite, it may be possible to replace them by some ad hoc adjustments to (6.1.11) and (6.1.12). However, in some applications it might, as a last resort, be necessary to use Newton–Raphson to compute the exact posterior mode vector $\tilde{\boldsymbol{\gamma}}$ and information matrix $\tilde{\mathbf{R}}$, satisfying (6.2.2) and (6.2.3). Leonard and Hsu (1992) show how the explicit approximations can suffice.

(G) *Computing the Prior Predictive Density*

The prior predictive density

$$p(\mathbf{y} \mid \boldsymbol{\mu}, \mathbf{C}) = \underset{\boldsymbol{\gamma}|\boldsymbol{\mu},\mathbf{C}}{E}\, p(\mathbf{y} \mid \boldsymbol{\gamma}) \tag{6.4.12}$$

can be calculated by simulating our recommended importance function, in the preceding prior informative situation, to the posterior distribution of $\boldsymbol{\mu}$ and $\mathbf{C}$, and by applying (6.4.8) and (6.4.9), with $g(\boldsymbol{\gamma}) = p(\mathbf{y} \mid \boldsymbol{\gamma})$, and $\tilde{W}(\boldsymbol{\gamma}) = \pi(\boldsymbol{\gamma} \mid \boldsymbol{\mu}, \mathbf{C})/\pi_y^*(\boldsymbol{\gamma})$, where $\pi(\boldsymbol{\gamma} \mid \boldsymbol{\mu}, \mathbf{C})$ denotes our $N(\mu, \mathbf{C})$ prior density for $\boldsymbol{\gamma}$.

(H) *Hierarchical Bayesian Procedures*

It can be difficult to completely compute our hierarchical Bayesian integration (6.3.8) via importance sampling and to combine this with simulations for the posterior expectation of $g(\boldsymbol{\gamma})$, although Hsu and Leonard (1997) show that this is sometimes possible. However, importance sampling can be used to check the accuracy of the components of procedures that simulate or approximate the hierarchical Bayes integration (e.g., Leonard and Hsu, 1992; Sun et al., 1996). It is sometimes more convenient to refer to MCMC procedures relating to the Gibbs sampler (e.g., Gelfand and Smith, 1990, and our Section 6.6), which iterates between different conditional distributions and rejection sampling, which relates to the Metropolis algorithm, developed for mathematical physicists by Metropolis et al. (1953), and which will be discussed in Section 6.8. (A completely general definition of the Metropolis algorithm is described by Lee, 1997, p. 269. See also Chib and Greenberg, 1995.) However, a completely satisfactory general computational solution to the hierarchical Bayesian paradigm (6.3.8) still provides an open research question.

Table 6.4.1. *Bayesian significance probabilities* $p(\lambda_{ij}^{AB} \le 0 \mid \mathbf{y})$ *for Shopping in Oxford data.*

Car availability	Response to grocery shop question 1	2	3	4	5
None	0.9880	0.3676	0.6655	0.4160	0.0028
Some	0.0028	0.9911	0.6645	0.0847	0.1503
Full	0.4921	0.0101	0.2013	0.9507	0.9992

(I) *The Shopping in Oxford Data*

We reanalyzed the data in Table 5.2.2 based upon the independent Poisson model of Section 5.2 (B), the full-rank interaction assumption in (5.2.4), and uniform distributions for the logs of the cell means in the prior assessment. The marginal posterior densities of the interaction effects λ_{ij}^{AB} were calculated both exactly, using the Monte Carlo procedures of Section 6.4 (B), and approximately, using the Laplacian methodology paralleling Worked Example 6C. The results, however, were identical to visual accuracy. The posterior densities for the fifteen cells are reported in Figure 6.4.1.

Each posterior density should be interpreted in relation to the zero interaction value ($\lambda_{ij}^{AB} = 0$). Our graphical results support the conclusions of Section 5.2 (B). For example, there is clearly no evidence to suggest that λ_{31}^{AB} differs from zero. The exact Bayesian significance probabilities $p(\lambda_{ij}^{AB} \le 0 \mid \mathbf{y})$ are reported in Table 6.4.1.

In Figure 6.4.2, we report the posterior density of the overall measure of association

$$\eta = \frac{1}{rs} \sum_{i=1}^{r} \sum_{j=1}^{s} (\gamma_{ij} - \gamma_{i\cdot} - \gamma_{\cdot j} + \gamma_{\cdot\cdot})^2.$$

Histogram (a) represents the exact result obtained by the Monte Carlo simulations of Section 6.4 (A) and by differencing the posterior c.d.f. of η. Curve (b) is based upon Laplacian approximation (5.1.13) combined with a Gamma approximation, with exact first two moments, to the f contribution to the expression in (5.1.13). Curve (c) gives the Tierney, Kass, and Kadane saddle-point approximations, also described in Section 6.1 and referring to the 1991 corrections of Tierney, Kass, and Kadane (1989). This approximation works quite well, but approximation (5.1.13) is virtually exact, thus substantiating related conclusions by Leonard, Hsu, and Tsui (1989).

The posterior density in Figure 6.4.2 requires careful interpretation. The hypothesis $H_0 : \eta = 0$ is equivalent to the hypothesis of independence of rows and columns in our contingency table. However, $\eta = 0$ is on the boundary of the possible values for our nonnegative measure of association, so that it is not relevant to report Bayesian significance probabilities. However, $\eta^{\frac{1}{2}}$ can be interpreted as a typical value for $|\lambda_{ij}^{AB}|$ and $\exp\{\eta^{\frac{1}{2}}\}$ as a typical value for the ratio of the cell means μ_{ij} and their values

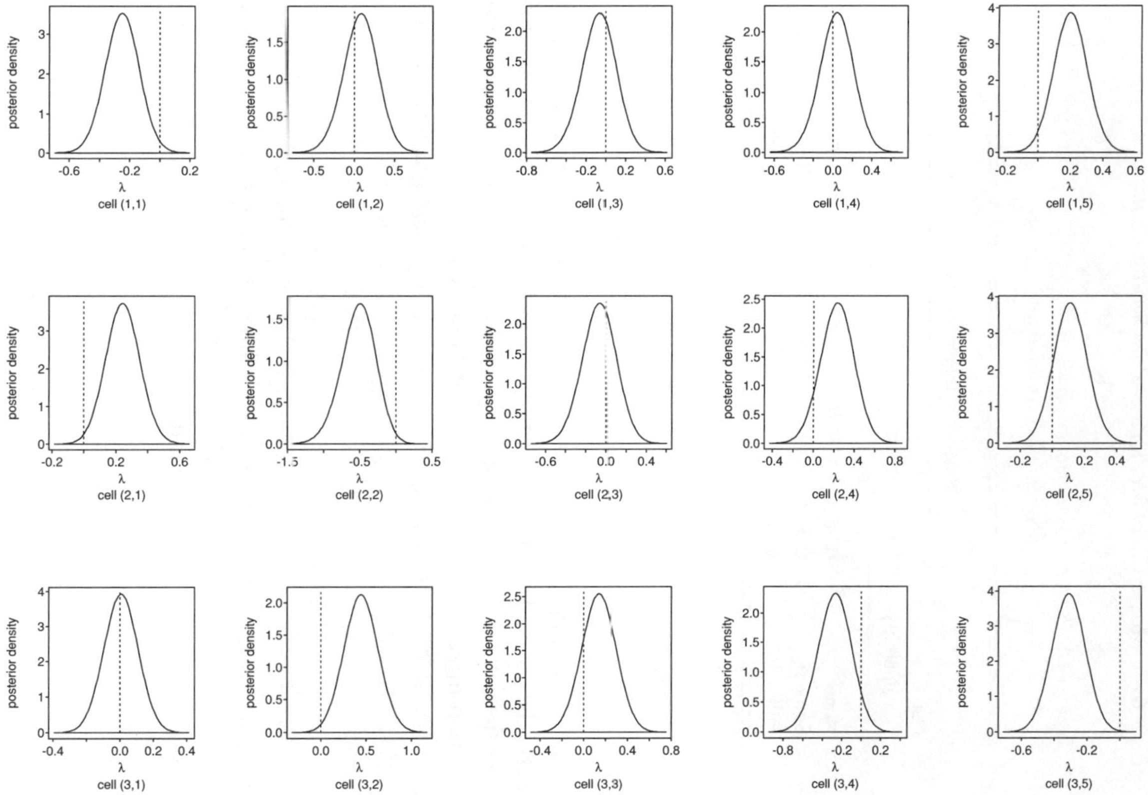

Figure 6.4.1. Posterior densities of interaction effects for Shopping in Oxford data.

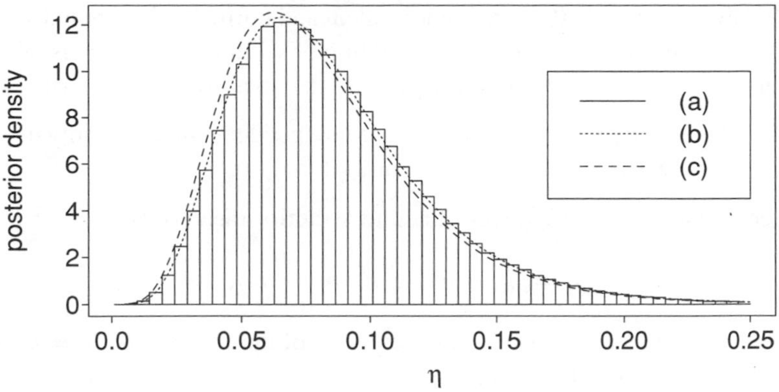

Figure 6.4.2. Posterior density of overall measure of association for Shopping in Oxford data.

$\mu_{i.}\mu_{.j}/\mu_{..}$, under the hypothesis $H_0 : \eta = 0$. The posterior mode $\tilde{\eta} = 0.053$ of η corresponds to $\tilde{\eta}^{\frac{1}{2}} = 0.23$ and $\exp\{\tilde{\eta}^{\frac{1}{2}}\} = 1.26$, indicating a substantial deviation from H_0.

SELF-STUDY EXERCISES

6.4.a Show how to perform exact computations for the analysis of failure or survival times in Section 5.1 (G). Assume a Gamma prior, with parameters κ_0 and κ_1 for γ, and an importance function that is Gamma with parameters $\kappa_0 + n$ and $\kappa_1 + t$.

6.4.b Show that the generalized multivariate t-approximation in (6.4.10) is equivalent to a second-order Taylor series (about $\boldsymbol{\gamma} = \hat{\boldsymbol{\gamma}}$) approximation to $[l(\boldsymbol{\gamma} \mid \mathbf{y}]^a$, where $a = -2/(\nu + p)$.

6.4.c Suppose that the likelihood $l(\boldsymbol{\gamma} \mid \mathbf{y})$ of a vector $\boldsymbol{\gamma}$ of unknown parameters is proportional to a density $\pi^*(\boldsymbol{\gamma})$, and let $\pi(\boldsymbol{\gamma})$ denote the prior density. Let $\boldsymbol{\gamma}_1, \boldsymbol{\gamma}_2, \ldots, \boldsymbol{\gamma}_M$ denote independent simulations from the distribution with density $\pi^*(\boldsymbol{\gamma})$. Show that the posterior expectation, if it exists, of any function $\eta = g(\boldsymbol{\gamma})$ of $\boldsymbol{\gamma}$ is

$$E(\eta \mid \mathbf{y}) = \lim_{M \to \infty} \frac{\sum_{k=1}^{M} g(\boldsymbol{\gamma}_k)\pi(\boldsymbol{\gamma}_k)}{\sum_{k=1}^{M} \pi(\boldsymbol{\gamma}_k)},$$

where this limit exists with probability one. How would you investigate the simulated estimated standard error of simulation?

6.5 Further Special Cases and Practical Examples

(A) *The Simultaneous Estimation of Treatment Means and Binomial Probabilities*

In the one-way ANOVA, observations $\{y_{ij}; i = 1, \ldots, m; j = 1, \ldots, n\}$, given $\theta_1, \ldots, \theta_m$ and τ^2, are independent and normally distributed, with common variance τ^2, and y_{ij} possesses mean θ_i. Assume the exchangeable prior distribution:

Stage I: Given μ and σ^2, the θ_i are independent and normally distributed with mean μ and (between) variance σ^2. The sampling or within variance τ^2 is taken to be independent of the θ_i and uniformly distributed over the range $(0, \infty)$.

Stage II: The hyperparameters μ and σ^2 are independent and uniformly distributed over the respective ranges $(-\infty, \infty)$ and $(0, \infty)$.

Then Leonard and Ord (1976) show that the posterior mean of θ_i is

$$\theta_i^* = y_{i.} - F^{-1} g(F)(y_{i.} - y_{..}), \tag{6.5.1}$$

where F denotes the usual test statistic for equality of the means, on $\nu_B = m - 1$ and $\nu_W = m(n-1)$ degrees of freedom, and

$$g(F) = 1 - 2\nu_B^{-1} - 2\nu_B^{-1} \frac{A^{\frac{1}{2}\nu_B - 1}(1-A)^{\frac{1}{2}\nu_W}}{B_A\{\frac{1}{2}\nu_B - 1, \frac{1}{2}\nu_W + 1\}}, \tag{6.5.2}$$

with

$$A = \frac{\nu_B F}{\nu_B F + \nu_W} \tag{6.5.3}$$

and

$$B_A\{u, v\} = \int_0^A \phi^{\frac{1}{2}u - 1}(1-\phi)^{\frac{1}{2}v - 1} d\phi. \tag{6.5.4}$$

The shrinkage estimator in (6.5.1) slightly amends a suggestion by Box and Tiao (1968), who instead assume a prior distribution that is dependent upon the sample size. It generalizes the estimator in (6.3.19), which takes τ^2 to be known. Consider the data reported by Hald (1952, p. 434) relating to the tensile strengths of $n = 12$ wires contained in each of $m = 9$ cables. The group sample means of the $m = 9$ cables are described in the first column of Table 6.5.1 (Hald has subtracted 340 from each observation).

Table 6.5.1. *Estimates of group means (Hald's tensile strength data).*

Group	Group mean	Posterior mean	Standard deviation	Marginal mode	Joint mode
1	−4.083	−3.796	1.463	−3.791	−3.696
2	−7.000	−6.463	1.488	−6.461	−6.275
3	−6.083	−5.625	1.479	−5.621	−5.465
4	−2.667	−2.501	1.456	−2.497	−2.442
5	1.917	1.690	1.459	−1.685	1.611
6	0.833	0.699	1.455	0.696	0.652
7	0.917	0.776	1.456	0.772	0.726
8	3.333	2.985	1.468	2.980	2.863
9	6.250	5.652	1.497	5.653	5.443

In the second column, we describe the corresponding posterior means (6.5.1). These provide shrinkage estimators for the group means θ_i, which substantially smooth the sample means. The corresponding posterior standard deviations are in the third column. In the fourth column are the modes of the marginal posterior distributions of the θ_i. These may be contrasted with the joint posterior modes of the θ_i (with σ^2) in the fifth column.

Sun et al. (1996, p. 750) provide frequency simulations that suggest that the posterior means and marginal modes can be considerably superior, under a variety of loss functions, when compared with the group means, and quite superior when compared with the joint modes.

As a further numerical example, suppose that $n = 6$, $m = 10$, and $F = 1.34$. Then the estimator (6.5.1) becomes

$$\phi_i^* = \tfrac{1}{2}\bar{y}_{i.} + \tfrac{1}{2}\bar{y}_{..},$$

so that our estimate for the ith treatment is exactly the average of the ith sample mean $\bar{y}_{i.}$ and the grand mean $\bar{y}_{..}$. Leonard and Ord use this result to argue that when $n = 6$ and $m = 10$, the null hypothesis of equality of the means can be refuted whenever F exceeds the critical value 1.34. As m becomes large, the estimator (6.5.1) approaches

$$\phi_i^* = \bar{y}_{i.} - \min\left(F^{-1}, 1\right)(\bar{y}_{i.} - \bar{y}_{..}). \tag{6.5.5}$$

Leonard and Ord use this result to argue that for large m, the null hypothesis of equality of the means should be refuted whenever $F^{-1} \leq 0.5$, that is, $F \geq 2$, a result that parallels Akaike's information criterion (1.1.5).

Note that the simple shrinkages in (6.5.1) are consequences of the parametric form assumed for the mixing distribution introduced at Stage I of the model. Laird (1982) and Leonard (1984) show that if this mixing distribution is instead estimated nonparametrically from the data, then quite different estimates for the θ's can be obtained. These comments hold also for nonnormal situations. The data in Table 6.5.2 were extracted from the occupational mobility data (Table 5.2.7) and denote the numbers of sons y_i out of n_i following their fathers' occupation, but where the occupations have been rearranged according to the magnitude of the observed proportions. Leonard (1984) assumed the y_i to be independent and to possess binomial distributions that are independent, given their respective probabilities θ_i and sample sizes n_i. The θ_i were assumed to be a priori exchangeable and, given G, to constitute a random sample from a distribution with c.d.f. G. The common c.d.f. was then estimated by a constrained marginal maximum likelihood procedure that produced a discrete distribution assigning probabilities $\frac{1}{14}$, $\frac{4}{14}$, $\frac{4}{14}$, $\frac{4}{14}$, and $\frac{1}{14}$ to the points 0.020, 0.103, 0.257, 0.480, and 0.823.

The estimates in the last column of Table 6.5.2 are the Bayesian posterior means of the θ_i, under this discrete choice of prior distribution. As the latter is empirically estimated from the data, our estimates for the θ_i are *nonparametric empirical Bayes*. Note that they do not shrink the observed proportions in a simple manner. Eleven of the smoothed proportions cluster into three groups. The first and fourteenth proportions are regarded as outliers, and hence are only slightly smoothed. The ninth proportion

Table 6.5.2. *Proportions of sons following their fathers' occupation.*

Occupation (i)	y_i	n_i	Observed proportion	Smoothed proportion
1	0	26	0.000	0.020
2	6	88	0.068	0.103
3	11	106	0.104	0.103
4	7	54	0.130	0.115
5	6	44	0.137	0.127
6	4	19	0.211	0.221
7	18	69	0.261	0.257
8	9	32	0.281	0.270
9	6	18	0.333	0.334
10	23	51	0.451	0.477
11	54	115	0.470	0.480
12	20	41	0.488	0.480
13	28	50	0.560	0.480
14	51	62	0.823	0.823

is not appreciably smoothed, since it is not obvious whether this proportion should be smoothed upwards or downwards.

Murray (1992) applies these ideas to the analysis of health care performance measures for an inhomogeneous population of physicians, obtains related empirical Bayes confidence intervals, and justifies these procedures via a variety of simulations.

(B) *Mean Squared Error Properties of Empirical Bayes Estimators*

Reinsel (1985) considers the one-way ANOVA model described in Section 6.5 (A) and compares the least squares estimators $\hat{\theta}_i = y_{i.}$ of the θ_i with empirical Bayes shrinkage estimators $\hat{\theta}_i$. While his shrinkage estimators are more complex than (6.5.1), they are somewhat similar in spirit. In his simulations, he takes $m = 1$ and $n = 4$ and generates the θ_i from

$$\theta_i = i + \lambda_i,$$

where the ϵ_i are independent and normally distributed with zero mean and common variance $\sigma_\lambda^2 = 1$. He then generated the observations from our normal sampling model, with $\tau^2 = 4$. Reinsel reports substantial savings in simulated mean squared error.

(C) *The Impact of Chlorofluoromethanes on Stratospheric Ozone*

Reinsel and Tiao (1987) analyze monthly averages of data on stratospheric ozone and vertical distribution ozone profiles from a global collection of ground-based recording stations and detect a trend in ozone that may be associated with the possible effects of release of chlorofluoromethanes (CFMs).

They develop an interesting random effects model by assuming that the estimated trend y_i, in ozone data at an individual station j within a given geographical region i, satisfies

$$y_{ij} = \omega + \alpha_i + \beta_{ij} + \epsilon_{ij} \qquad (i = 1, \ldots, m;\ j = 1, \ldots, n_i),$$

where ω is the overall true global trend, α_i is a random component associated with station j, and the ϵ_{ij} are random error terms. Under one set of assumptions, they take the α_i, β_{ij}, and ϵ_{ij} to be normally distributed with zero means, the α_i and β_{ij} to possess respective common unknown variances σ_α^2 and σ_β^2, and the ϵ_{ij} to possess known variance v_{ij}. In this case, these are just three unknown parameters ω, σ_A^2, and σ_β^2, which Reinsel and Tiao estimate by maximum likelihood.

Noting that y_{ij}, given α_i, is normally distributed with mean $\omega + \alpha_i$ and variance $\sigma_\beta^2 + v_{ij}$, it is straightforward to obtain Bayes estimates for ω and α_i, when σ_α^2 and σ_β^2 are known. The joint density of the y_{ij} and α_i, given ω, σ_α^2, and σ_β^2, is

$$\begin{aligned} p(\mathbf{y}, \boldsymbol{\alpha} \mid \omega, \sigma_\alpha^2, \sigma_\beta^2) &\propto (\sigma_\alpha^2)^{-\frac{1}{2}m} \\ &= \prod_{i=1}^{m} \prod_{j=1}^{n_i} (\sigma_\beta^2 + v_{ij})^{-1} \\ &\quad \times \exp\left\{ -\frac{1}{2}(\sigma_\beta^2 + v_{ij})^{-1} \sum_{i=1}^{m} (y_{ij} - \omega\alpha_i)^2 - \frac{1}{2}\sigma_\alpha^{-2} \sum_{i=1}^{m} \alpha_i^2 \right\}. \end{aligned}$$

From a Bayesian perspective, the choice of prior $\pi(\omega, \sigma_\alpha^2, \sigma_\beta^2) \propto 1$ might be plausible. In this case, the hierarchical Bayes analysis indicated in Section 6.3 (A), is easily performed and can be used to exactly compute the posterior means, variances, and distribution of all parameters of interest. The reader is referred to the paper by Reinsel and Tiao for some numerical examples of their approach.

(D) *Smoothing Histograms*

The data in Table 6.5.3 describes the observed frequencies for a grouped histogram relating to the distribution of the gains in weights of 511 pigs (see Snedecor and Cochran, 1967, p. 71). We applied the method described in Exercises 6.1.c and 6.2.c for smoothing the $p = 21$ cell probabilities in the histogram. The smoothed frequencies in Table 6.5.3 are the values of the $100\hat{\theta}_j$, where the $\hat{\theta}_j$ denote the posterior modes of the θ_j, which satisfy (6.2.13). The prior means μ_j of the logits were all set equal to zero, corresponding to a uniform prior estimate, over the interval (18.5, 39.5), for the underlying distribution of the observations. The choices $\sigma^2 = 0.3$ and $\rho = 0.7$ were made for the prior parameters. These choices were based upon approximate empirical Bayes procedures.

Note that the smoothed frequencies in Table 6.5.3 smooth the observed frequencies in an autoregressive fashion. The smoothing is therefore of a different type than the simple shrinkages recommended in Section 6.5 (A), and this is created by the prior assumptions.

Table 6.5.3. *Frequency distribution of pig gains.*

Interval midpoint (lbs)	Observed frequency (x_i)	Smoothed frequency ($100\hat{\theta}_i$)
19	1	2.82
20	1	3.10
21	0	3.43
22	7	5.68
23	5	6.69
24	10	10.62
25	30	28.32
26	30	28.32
27	41	39.11
28	48	47.59
29	66	62.99
30	72	68.15
31	56	54.98
32	46	45.60
33	45	40.92
34	22	22.41
35	24	20.64
36	12	11.81
37	5	6.95
38	0	4.58
39	1	4.29

(E) *Smoothing Mortality Tables: An Edinburgh Tradition*

Hickman and Miller (1977, 1981) developed Bayesian graduation procedures for mortality tables in actuarial science, based on a 1972 collaboration with Leonard (see Leonard, 1973a) and the histogram smoothing methods of Exercise 6.1.c. They instead assumed autoregressive priors for arc–sine transformations of binomial probabilities and are hence able to smooth both univariate and bivariate mortality tables.

Note that an earlier graduation method by Whittaker (1923) refers to a smoothing criterion corresponding to the quadratic form (6.1.33), which appears in the prior density (6.1.32). Whittaker's criterion suggested that a second-order autoregressive prior for the parameters (or a first-order process for successive differences, as suggested by Leonard, 1973a) might be more appropriate.

Maclaurin (see Martin, 1764) employed European mortality tables in Scotland, in relation to pensions for widows of clergy and academics. Both Colin Maclaurin (1724–1746) and Sir Edmund Whittaker (1912–1946) held the Chair of Mathematics at the University of Edinburgh, and Whittaker's successor, Alexander Aitken, published his D.Sc. dissertation on mortality tables (Aitken, 1925), effectively discovering polynomial splines. The theory of cardinal splines was first developed by Schoenberg (1946), also in the context of actuarial graduation. Consequently, the University of Edinburgh's tradition has been continued at Wisconsin (Miller, Hickman, Schoenberg). Some of this information has been provided by John Searle and Derek Arthur, who have maintained further archives and correspondence at Edinburgh.

(F) *The London High School Data*

Leonard and Hsu (1994) apply the method described in Exercise 6.3.i to a data set involving $m = 40$ London high schools. At each high school, the numbers of students fully in $p = 6$ grade categories are observed, and the data were assumed to follow a "product multinomial" model, with 40 independent multinomial distributions, each with 6 cells. Our exchangeable prior distribution was assumed, with a common mean vector $\boldsymbol{\mu}$ and covariance matrix $\mathbf{C}$, for the logits of each multinomial distribution.

The approximate EM algorithm, described in Exercise 6.3.i, was applied, yielding data-based estimates $\tilde{\boldsymbol{\mu}}$ and $\tilde{\mathbf{C}}$, for $\boldsymbol{\mu}$ and $\mathbf{C}$, satisfying

$$\begin{aligned}\tilde{\xi} &= \left(\sum_{j=1}^{b} e^{\tilde{\mu}_j}\right)^{-1} \left(e^{\tilde{\mu}_1}, \ldots, e^{\tilde{\mu}_b}\right)^T \\ &= (0.087 \quad 0.217 \quad 0.255 \quad 0.262 \quad 0.064 \quad 0.115)^T, \\ \operatorname{diag}(\tilde{\mathbf{C}}) &= (1.08, 0.39, 0.25, 0.26, 0.46, 0.92),\end{aligned}$$

and

$$\mathbf{B} = \begin{pmatrix} 1 & 0.79 & 0.57 & -0.07 & 0.38 & -0.57 \\ 0.79 & 1 & 0.79 & 0.29 & -0.14 & -0.31 \\ 0.57 & 0.79 & 1 & 0.61 & 0.21 & -0.01 \\ 0.07 & 0.25 & 0.61 & 1 & 0.68 & 0.62 \\ -0.38 & -0.14 & 0.21 & 0.68 & 1 & 0.83 \\ -0.57 & -0.31 & -0.01 & 0.62 & 0.83 & 1 \end{pmatrix},$$

where $\mathbf{B}$ denotes the corresponding estimated prior correlation matrix. The observed and corresponding smoothed properties $\tilde{\theta}_{15,1}, \ldots, \tilde{\theta}_{15,6}$ for the $n_{15} = 22$ students at School 15 are described in Table 6.5.4, and similar results are available for the other schools. The smoothing is now highly complex. For example, for School 15, the smoothed proportions (a) take account of proportions for adjacent and other grades at the

Table 6.5.4. *Observed and smoothed percentages for School 15.*

	Grade					
	1	2	3	4	5	6
Observed	0.0	18.2	13.6	54.5	9.1	4.5
Smoothed	4.2	17.2	23.2	34.7	7.4	13.4
Overall (40 schools)	8.7	21.7	25.5	26.2	6.4	11.5

same school, (b) take account of collateral information from all schools (as represented by $\tilde{\boldsymbol{\xi}}$), and (c) refer to the complicated structure of the estimated covariance matrix $\tilde{\mathbf{C}}$.

Note that following the concepts of Section 6.3 (B), the above analysis could instead be interpreted as non-Bayesian and relating to a sampling distribution for the cell frequencies generalizing the product multinomial and taking allowance (e.g., overdispersion/serial correlation) of the nonrandomized nature of the high school data. In this case, $\boldsymbol{\mu}$ and $\mathbf{C}$ should be interpreted as the parameters of our sampling model, and these are hence estimated by approximate maximum likelihood.

(G) *Shrinkage Estimators for Several Poisson Parameters, and the Oilwell and Audit Sampling Data*

Now consider the situation where observations $y_1, \ldots, y_m$ are independent and Poisson distributed with respective means $\theta_1, \ldots, \theta_m$. Then Leonard (1976) considers shrinkage estimators for $\theta_1, \ldots, \theta_m$ taking the form

$$\theta_i^* = y_i - \min\left[\frac{(m-1)\bar{y}}{s^2}, 1\right](y_i - \bar{y}) \qquad (i = 1, \ldots, m). \tag{6.5.6}$$

In particular, Leonard shows that as $m \to \infty$, the average mean squared error

$$r^*(\boldsymbol{\theta}) = m^{-1} \underset{\mathbf{y}|\boldsymbol{\theta}}{E} \sum_{i=1}^{m} \left(\theta_i^* - \theta_i\right)^2 \tag{6.5.7}$$

multiplies the average mean squared error $\bar{\theta}$ of the unbiased estimators y_i by a factor approximately equal to

$$\frac{\sum_{i=1}^m \left(\theta_i - \bar{\theta}\right)^2}{\sum_{i=1}^m \left(\theta_i - \bar{\theta}\right)^2 + m\bar{\theta}}. \tag{6.5.8}$$

Other estimators for Poisson means, under squared error loss, are recorded by Tsui (1981). Clevenson and Zidek (1975) instead proposed estimators shrinking the y_i towards zero, but based upon a different loss function. In their oilwell example, they report the number of discoveries of oil wells in $m = 36$ different areas during a specified time period. The numbers of discoveries in the $m = 36$ areas was equal to 0 exactly nineteen times, equal to 1 ten times, equal to 2 four times, equal to 3 two times, and equal to 5 one time. The Clevenson–Zidek estimates corresponding to the observed frequencies (0, 1, 2, 3, and 5) were (0, 0.45, 0.89, 1.34, and 2.23), respectively. However, (6.5.6) gives the quite different estimates (0.50, 0.88, 1.26, 1.64, and 2.41), since they shrink the frequencies towards $\bar{y} = 0.806$. In particular, the zero frequency is increased to the data-based value of 0.50.

Clevenson and Zidek empirically justify their estimates by predicting "true values" calculated by using data over a much larger time period. The smoothed estimate of 0.50 based upon zero observed frequencies was closer to the true value, when compared with the Clevenson–Zidek estimate of 0, seventeen out of nineteen times. The smoothed estimate 0.88, based upon unit cell frequencies, was closer to the true value, five out of ten times. The smoothed 1.26, based upon all frequencies equal to 2, was closer three out of four times, and the smooth estimates 1.64 and 2.41 were both closer once. This empirical justification of the equation (6.5.6), shrinking towards $\bar{y}$ rather than zero, is of course free from a choice of loss function.

Ijiri and Leitch (1980) and Matsumura and Tsui (1982) discuss applications in audit sampling. Other types of shrinkage estimators are reported by Peng (1975) and Hudson and Tsui (1981). These possess uniformly smaller average mean squared error, for any finite $m \geq 3$, when compared with the unbiased estimates y_i. For example, Hudson and Tsui propose

$$\tilde{\theta}_i = y_i - \left(m - \sum_{j=0}^{\lambda} m_j^*\right) \frac{\left[h(y_i) - h(\lambda)\right]}{\sum_{j=1}^{m}\left[h(y_j) - h(\lambda)\right]^2}, \tag{6.5.9}$$

where $m_j^* = \max\{(m_j - 3), 0\}$, with m_j denoting the number of frequencies equal to j, λ denoting a common prior estimate for the θ_j, and

$$h(y) = \sum_{j=1}^{y} \frac{1}{j}. \tag{6.5.10}$$

Matsumura and Tsui consider the numbers of auditing errors $y_1, y_2, \ldots, y_9$ from $m = 9$ different accounts. The observations were 0, 1, 0, 1, 1, 2, 2, 3, and 6, yielding $\bar{y} = 1.667$ with sample variance $s^2/(m-1) = 30/8 = 3.75$. Since $(m-1)\bar{y}/s^2 = 0.444$, the shrinkage estimates in (6.5.5), with $\lambda = \bar{y}$, therefore shrink the observed y_i as proportion 0.444 of the distance towards $\bar{y}$, giving the estimates 0.74, 0.74, 0.74, 1.30, 1.30, 1.85, 1.85, 2.41, 4.07. Hudson and Tsui instead propose setting $\lambda = 1$ in (6.5.9), yielding the estimators 0.16, 0.16, 0.16, 1.00, 1.00, 1.92, 1.92, 2.87, and 5.77. However, a higher choice of λ may yield more appealing estimates. George, Makov, and Smith (1994) discuss a hierarchical Bayesian approach to the Poisson shrinkage problems, which they solve using an excellent application of MCMC procedures. In this Poisson case, there is considerable debate regarding an appropriate direction for the shrinkages of the y_i. For example, should the shrinkages be taken towards zero, towards $\bar{y}$, or even towards the smallest observation? For the best average mean squared error properties, there is little doubt that when m is large, the shrinkages should be taken towards $\bar{y}$. However, when m is small to moderate, some compromize between $\bar{y}$ and zero is appropriate, according to unpublished calculations at Edinburgh, by George Streftaris.

(H) *The Linear Model with Unequal Variances, and the Breaking Strengths of Fabrics*

Consider n vectors of observations $\mathbf{y}_1, \ldots, \mathbf{y}_n$, which are independent, given a vector $\boldsymbol{\beta}$ of parameters, and possess m-dimensional multivariate normal distributions, with respective mean vectors $\mathbf{X}_1\boldsymbol{\beta}, \ldots, \mathbf{X}_n\boldsymbol{\beta}$ and common covariance matrix $\mathbf{C} = \mathrm{diag}(\phi_1, \ldots, \phi_m)^T$. Let $\gamma_j = \log \phi_j$ and $\boldsymbol{\gamma} = (\gamma_1, \ldots, \gamma_m)^T$. Rather than assuming conjugate inverted Gamma or chi-squared distributions for the ϕ_j in the prior assessment, Leonard (1975a) instead suggests taking $\boldsymbol{\beta}$ and $\boldsymbol{\gamma}$ to be independent and each to possess a multivariate normal distribution. In a numerical example, he suggests an exchangeable hierarchical distribution for the elements γ_j of $\boldsymbol{\gamma}$, which, given unknown hyperparameters μ and σ^2, takes the γ_j to be normally distributed with mean μ and variance σ^2.

Consider the data previously analyzed by Leonard (1975a) and concerning the breaking strengths of six different fabrics. There are six independent samples, with ten observations in each sample, yielding maximum likelihood estimates 0.72, 14.26, 3.39, 5.79, 1.93, and 0.81 of the population variances. The authors show that Bartlett's test for equality of the population variances $\phi_1, \ldots, \phi_6$ suggests inequality at any sensible significance level. This test is also based upon the log-variances.

Leonard (1975a) uses the preceding exchangeable prior distribution and recommends the estimates 0.91, 10.61, 3.24, 5.07, 2.04, and 1.00 for the population variances and an estimate of $\tilde{\sigma}^2 = 0.91$ for the between variance σ^2 of the log-variances. The estimates for the population variances smooth the maximum likelihood estimates of the population variances towards the value 2.61, their geometric mean. For applications of Leonard's more general formulation to stochastic volatility in time series, see Section 5.7.

(I) *Bayesian Inference for a Covariance Matrix, and the Project* TALENT *American High School Data*

Let $\mathbf{y}_1, \ldots, \mathbf{y}_n$ denote a random sample from a multivariate normal distribution with $p \times 1$ mean vector $\boldsymbol{\theta}$ and $p \times p$ covariance matrix $\mathbf{C}$. Then Evans (1965) proposed the conjugate prior formulation, where, given $\mathbf{C}, \boldsymbol{\theta}$ possesses a multivariate normal distribution with mean vector $\boldsymbol{\mu}$ and covariance matrix $\zeta^{-1}\mathbf{C}$, where $\boldsymbol{\mu}$ and ζ are specified. Furthermore, $\mathbf{C}^{-1}$ has a Wishart prior distribution with ν degrees of freedom and mean matrix $\mathbf{R}_0$. See Bernardo and Smith (1994, p. 435) for a full definition.

Under the conjugate prior, the posterior distribution of $\mathbf{C}^{-1}$ is Wishart with $\nu + n$ degrees of freedom, and the inverse of the posterior mean matrix of $\mathbf{C}^{-1}$ is

$$\mathbf{C}^* = (\nu + n)^{-1}\left(\nu R_0^{-1} + \mathbf{U}^*\right), \tag{6.5.11}$$

where

$$\mathbf{U}^* = (1 - \rho)\mathbf{U}(\bar{\mathbf{y}}) + \rho\mathbf{U}(\boldsymbol{\mu}), \tag{6.5.12}$$

$\rho = \zeta/(\zeta + n)$, and

$$\mathbf{U}(\boldsymbol{\mu}) = \sum_i (\mathbf{y}_i - \boldsymbol{\mu})(\mathbf{y}_i - \boldsymbol{\mu})^T. \tag{6.5.13}$$

Furthermore, the posterior distribution of $\boldsymbol{\theta}$, given $\mathbf{C}$, is multivariate normal with covariance matrix $(\zeta + n)^{-1}\mathbf{C}$ and mean vector

$$\boldsymbol{\theta}^* = (1 - \rho)\bar{\mathbf{y}} + \rho\boldsymbol{\mu}_0. \tag{6.5.14}$$

Empirical and hierarchical Bayesian procedures are described by Chen (1979) and Dickey, Lindley, and Press (1985).

An inverted Wishart prior for $\mathbf{C}$ is quite restrictive. The matrix $\mathbf{R}_0^{-1}$ can be based upon prior estimates for $\mathbf{C}$. However, there is only a single further prior parameter ν. It is therefore impossible to express different degrees of belief regarding prior information for different elements of $\mathbf{C}$ or to flexibly represent prior dependencies between elements of $\mathbf{C}$. Consequently, the posterior estimate (6.5.11) is very specialized. There is no posterior smoothing of elements of $\mathbf{C}$, and each diagonal and nondiagonal element is shrunk the same proportionate distance towards the corresponding prior estimate. Leonard and Hsu (1992) instead consider $\mathbf{A} = \log \mathbf{C}$, the matrix logarithm of the covariance matrix, as defined in Section 1.9. Let $\boldsymbol{\alpha} = \text{vec}(\mathbf{A})$ denote some convenient $q \times 1$ vectorization of the upper triangular elements of $\mathbf{A}$, so that $q = \frac{1}{2}p(p+1)$. Then take $\boldsymbol{\alpha}$ to possess a multivariate normal prior distribution, say, with $q \times 1$ mean vector $\boldsymbol{\xi}$ and covariance matrix $\boldsymbol{\Delta}$. This prior distribution contains $q + \frac{1}{2}q(q+1)$ distinct prior parameters and could, in principle, be assessed subjectively. See, for example, the prior assessment procedures of Garthwaite and Dickey (1992).

Consider, however, the Project TALENT American high school data reported by Flanagan et al. (1964). Leonard and Hsu report the observed correlation matrix for a sample of $n = 78$ eighteen-year-old female twelfth-grade students and their results on $p = 8$ tests. The sample mean vector of scores was $\bar{\mathbf{y}} = (0.514, 0.496, 0.784, 0.677, 0.555, 0.446, 0.602, 0.410)^T$ with respective sample standard deviations $(0.118, 0.136, 0.085, 0.181, 0.183, 0.168, 0.190, 0.165)^T$.

For these data, Leonard and Hsu assume an "exchangeable" prior distribution for the positive definite matrix $\mathbf{C}$, that is, the distribution is invariant under any permutation of the rows of $\mathbf{C}$, together with the same permutation of the columns. In particular, the first p elements of $\boldsymbol{\xi}$, denoting the prior means of the diagonal elements of $\mathbf{A}$, were set equal to a common value $\xi - 1$, and the remaining $q - p$ elements, denoting the prior means of the nondiagonal elements of $\mathbf{A}$, were set equal to a common value ξ_2. Also $\boldsymbol{\Delta}$ was taken to be diagonal, with the first p elements equal to σ_1^2, and the last $q - p$ elements equal to σ_2^2. Therefore, different common prior variances were chosen for the diagonal and nondiagonal elements of $\mathbf{A}$. A hierarchical stage was then added by assuming an improperly uniform distribution for (ξ_1, ξ_2, σ_1^2, and σ_2^2). Further justifications of this prior formulation are described by Leonard and Hsu.

Leonard and Hsu use importance sampling to calculate an empirical posterior mean matrix for $\mathbf{C}$, and they report the elements of the corresponding correlation matrix. There is substantial smoothing of the raw correlations. While Leonard and Hsu are able to compute exact posterior means, they report an approximation, under the above exchangeable prior, to the posterior mean vector of $\boldsymbol{\alpha} = \text{vec}\,(\mathbf{A})$, conditional only upon σ_1^2 and σ_2^2 and when the mean vector $\boldsymbol{\theta}$ is zero. This takes the form

$$\boldsymbol{\alpha}^* = \left(\mathbf{Q} + \boldsymbol{\Delta}^{-1}\right)^{-1}\left(\mathbf{Q}\boldsymbol{\lambda} + \boldsymbol{\Delta}^{-1}\mathbf{X}\boldsymbol{\mu}^*\right), \tag{6.5.15}$$

where

$$\boldsymbol{\mu}^* = \left[\mathbf{X}^T\left(\mathbf{Q}^{-1} + \boldsymbol{\Delta}\right)^{-1}\mathbf{X}\right]^{-1}\mathbf{X}^T\left(\mathbf{Q}^{-1} + \boldsymbol{\Delta}\right)^{-1}\boldsymbol{\lambda}; \tag{6.5.16}$$

$\boldsymbol{\lambda}$ and $\mathbf{Q}$, respectively, denote the maximum likelihood vector and the likelihood information matrix for $\boldsymbol{\alpha}$; $\boldsymbol{\Delta}$ assumes the special diagonal form described above for our exchangeable prior; and the prior mean vector of $\boldsymbol{\alpha}$ is the hypothesized linear model $\mathbf{X}\boldsymbol{\mu}$, where $\boldsymbol{\mu} = (\mu_1, \mu_2)^T$ and $\mathbf{X}$ denotes a $q \times 2$ matrix with unit entries in the first p rows of the first column and the last $q - p$ rows of the second column, and zeros elsewhere.

A variety of simulation procedures are described by Leonard and Hsu (1992). For example, if the observation vectors are generated from a multivariate normal distribution with zero mean vector and an intraclass covariance matrix, both the estimator in (6.5.15) and the corresponding estimator for $\mathbf{C}$ possess superior mean squared error properties when compared with any estimators based upon multiples of $\mathbf{C}$, and for many choices of σ_1^2 and σ_2^2. However, the estimators shrinking towards intraclass form can possess superior mean squared error, even if the true matrix does not assume intraclass form.

For example, we performed 10,000 simulations, each generating $n = 50$ vectors from a $(p = 6)$-dimensional multivariate normal distribution with zero mean vector. The diagonal elements of the true covariance matrix $\mathbf{C}$ were equal to 1, 4, 9, 16, 25, and 26, and all true applications were equal to 0.7.

The total mean squared error for diagonal components of the maximum likelihood estimator $\Lambda = \log(\mathbf{U}/50)$, where $\mathbf{U} = \sum \mathbf{y}_i \mathbf{y}_i^T$, for $\mathbf{A}$ was 0.274, and the total mean squared error for the upper triangular elements of $\mathbf{\Lambda}$ was 0.236. However, the estimator $\mathbf{\Lambda} = \log(\mathbf{U}/47)$ adjusted these two components of mean squared error to 0.242 and 0.236. Our approximate Bayesian estimator (6.5.15) performed as well or better for many values of $\kappa_1 = \log \sigma_1^2$ and $\kappa_2 = \log \sigma_2^2$. For example, $\kappa_1, = 2$ and $\kappa_2 = -2$ adjusted the components to 0.269 and 0.192.

Our estimators seem to perform better with respect to mean squared error on the original matrix $\mathbf{C}$. The diagonal and nondiagonal components of mean squared error for $\hat{\mathbf{C}} = \mathbf{U}/50$ were 92.8 and 92.2, whereas the multiple $\tilde{\mathbf{C}} = \mathbf{U}/52$ reduced these to 89.0 and 81.2. However, the exponential transformation of our approximate Bayes estimator for $\mathbf{A}$ reduced both components of mean squared error even further, to 84.5 and 81.5.

(J) *Smoothing Probability Densities and the Chondrite Meteor Data*

We now describe a practical example of the semi-parametric smoothing methods discussed in Exercise 6.1.b. Leonard (1978) reports data relating to silica assays for $n = 22$ chondrite meteors. The data have been renormalized to fall in the range (0,1). They were previously analyzed by two physicists, Burch and Parsons (1976), who developed a "squeeze" significance test to refute an assumption of normality of the observations. Note that the observations appear to lie in three groups, corresponding to different types of meteors. However, Leonard (1978) describes a normal density, fit by maximum likelihood.

With $n = 22$, assume that the observations $y_1, \ldots, y_n$ are a random sample from a distribution with density $f(t)$, for $t \in (a, b)$, where $a = 0$ and $b = 1$. Then Leonard (1978) and Lenk (1988, 1991) consider the logistic density transform $g(t)$ satisfying

$$f(t) = \frac{e^{g(t)}}{\int_a^b e^{g(s)} ds} \tag{6.5.17}$$

and the related influence function

$$g^{(1)}(t) = \frac{f^{(1)}(t)}{f(t)}. \tag{6.5.18}$$

In the prior assessment, suppose that $g^{(1)}(t)$ follows a Gaussian process with mean value function $\mu^{(1)}(t)$, for $t \in (a, b)$, and covariance kernel

$$C(s, t) = \sigma^2 \exp\{-\beta|s - t|\} \qquad (0 < s, t < \infty), \tag{6.5.19}$$

which expresses elasticity properties from the Ornstein–Uhlenbeck stochastic process and represents prior information relating to the regular behavior of f. Leonard (1978) takes $\mu(t)$ to be the logarithm of a normal curve, representing a null hypothesis of normality of the observations. He made the choices $\sigma^2 = 25$ and $\beta = 50$ subjectively, but by reference to the behavior of the posterior estimate of f, under different choices of σ^2 and β.

Leonard (1978) discusses a "prior likelihood" approach and shows that the preceding prior formulation justifies estimation of g by maximization of the penalized log-likelihood

$$L_1(g) = L(g) + L_0(g), \tag{6.5.20}$$

where

$$L(g) = \sum_{i=1}^{n} g(y_i) - n \log \int_a^b e^{g(t)}\,dt \tag{6.5.21}$$

and

$$\begin{aligned} L_0(y) = {} & \frac{1}{2\beta\sigma^2}\int_a^b \left(g^{(2)}(t) - \mu^{(2)}(t)\right)^2 dt + \frac{\beta}{2\sigma^2}\int_a^b \left(g^{(1)}(t) - \mu^{(1)}(t)\right)^2 dt \\ & + \frac{1}{\sigma^2}\left\{\left(g^{(1)}(a) - \mu^{(1)}(a)\right)^2\right\} + \frac{1}{\sigma^2}\left\{\left(g^{(2)}(a) - \mu^{(2)}(a)\right)^2\right\}. \end{aligned} \tag{6.5.22}$$

Hence $-2L_2(g)$ provides a roughness penalty, in the spirit of Good and Gaskins (1971), Silverman (1978, 1986), and Green and Silverman (1994). In response to a comment by Whittle, Leonard indicates that if more generally, $\operatorname{cov}(g(s), g(t)) = K(s, t)$, then the penalized maximum likelihood estimator for g satisfies the nonlinear Fredholm equation

$$\tilde{g}(t) = \mu(t) + \sum_{i=1}^{n} K(y_i, t) - \int_a^b \exp\{\tilde{g}(s)\} K(s, t)\,ds. \tag{6.5.23}$$

There are many ways to solve this equation in special cases, for example, via Gu's (1993) representation in terms of nonlinear smoothing splines. Approximate solutions are discussed by Thorburn (1986). The consequent exact estimate of $f(t)$ is described by Leonard (1978). This adjusts the hypothesized normal curve and possesses three bulges, corresponding to the different types of meteors. Obviously, if a different prior estimate $\mu(t)$ is used, then a quite different posterior estimate would be obtained. For any semi-parametric procedure, the greatest attention should be paid to the prior estimate of the unknown function. It always seems necessary to state an appropriate "prior hypothesis" for g in order to obtain a meaningful posterior estimate.

Good and Gaskins (1980) reanalyze the chondrite meteor data, under different choices of roughness penalty, and recommend a trimodal estimate. They appear to strongly refute an initial assumption of normality, based upon $n = 22$ observations; their results should be contrasted with those of Burch and Parsons. Lenk (1988, 1991) substantially extends the Bayesian approach described in this section. He discusses several further numerical examples. Frequency properties of the density estimate are developed by Cox and O'Sullivan (1990). Stone (1986) further discusses the relationship between a linear model for our logistic density transform and the exponential family of sampling distributions.

(K) *The Pelican Crossing Data and Nonhomogeneous Poisson Processes*

Griffiths and Cresswell (1976) developed a mathematical model for pelican crossings during the debate in Britain concerning the potential modernization of zebra crossings (black and white walkways) to pelican crossings (with green man). One assumption underlying their model is that pedestrians arrive in a Poisson process with constant rate λ.

We investigated this assumption by instead supposing that the arrival times $y_1, \ldots, y_n$, as described by Leonard (1978), with $n = 45$, occur in a nonhomogeneous Poisson process with time-varying rate or intensity function $\lambda(t)$, for $t \in (a, b)$. Then the number $N(t_1, t_2)$ of arrivals in any interval (t_1, t_2) is Poisson distributed with mean $\int_{t_1}^{t_2} \lambda(t) dt$, and the likelihood of λ is

$$l(\lambda \mid \mathbf{y}) = \left\{ \prod_{i=1}^{n} \lambda(y_i) \right\} \exp\left\{ -\int_a^b \lambda(t)\, dt \right\}. \tag{6.5.24}$$

Let $g(t) = \log \lambda(t)$ and make the same prior assumptions as in Section 6.5 (J), but with $\mu^{(1)}(t) \equiv 0$, corresponding to the null hypothesis that the arrivals follow a homogeneous Poisson process. Note that conditional upon n, the arrival times possess the same distribution as the order statistics of a random sample from a distribution, with density f satisfying (6.5.17). Consequently, Leonard (1978) applies the procedures of Section 6.5 (J) to estimate f and then multiplies by n to estimate λ.

Leonard (1978) reports our estimates for λ under three different choices of the shrinkage parameter σ^2 and the smoothing parameter β. Note that in principle it is possible to use the techniques of Section 6.3 to empirically estimate σ^2 and β from the current data. However, the choice $K(s, t) = \sigma^2 \exp\{-\beta(s - t)^2\}$ of covariance kernel may lead to more sensible results.

Simpler analyses of nonhomogeneous Poisson processes are provided by linear assumptions of the form

$$g(t) = \log \lambda(t) = \beta_0 + \beta_1 \phi_1(t) + \cdots + \beta_p \phi_p(t), \tag{6.5.25}$$

where the ϕ functions are specified and p can be chosen by AIC or BIC. For example, Steinijans (1976) fits a quadratic model when analyzing the years of arrivals of major freezes on Lake Constance. However, the prior assumptions of this section can also be regarded as providing formulations of "doubly stochastic Poisson processes," which are quite important in electrical engineering (e.g., Snyder, 1975). These are also developed by Clevenson and Zidek (1977).

6.6 Markov Chain Monte Carlo (MCMC) Methods: The Gibbs Sampler

Let $\boldsymbol{\theta}_1, \boldsymbol{\theta}_2, \ldots, \boldsymbol{\theta}_p$ possess joint posterior density $\pi(\boldsymbol{\theta}_1, \boldsymbol{\theta}_2, \ldots, \boldsymbol{\theta}_p)$, but where it is difficult to simulate directly from this joint density and where no convenient importance function (see Section 6.4) is available. Let $\pi_k^*(\boldsymbol{\theta}_k)$ denote the conditional posterior density of $\boldsymbol{\theta}_k$, given all the remaining $\boldsymbol{\theta}_j$. Assume that it is possible to simulate directly from each $\pi_k^*(\boldsymbol{\theta}_k)$ for $k = 1, 2, \ldots, p$.

Geman and Geman (1984) advise us to start with initial vectors $\boldsymbol{\theta}_1, \boldsymbol{\theta}_2, \ldots, \boldsymbol{\theta}_p$. Then take repeated simulations from $\pi_1^*(\boldsymbol{\theta}_1), \pi_2^*(\boldsymbol{\theta}_2), \ldots, \pi_p^*(\boldsymbol{\theta}_p)$, where, for each k, we should always condition $\pi_k^*(\boldsymbol{\theta}_k)$ on the latest vectors for $\boldsymbol{\theta}_j$ $(j \neq k)$. This gives repeated sets of simulated vectors for $\boldsymbol{\theta}_1, \boldsymbol{\theta}_2, \ldots, \boldsymbol{\theta}_p$, which will not, in general, be independent. However, if we ignore this problem and apply the simple Monte Carlo methodology of Section 6.4, we can still calculate the posterior expectation for any parameter η of interest, which is some function of $\boldsymbol{\theta}_1, \boldsymbol{\theta}_2, \ldots, \boldsymbol{\theta}_p$. Under fairly general regularity conditions, the posterior expectation of η will equal the almost sure limit of the successive values of η calculated from the simulated $\boldsymbol{\theta}_1, \boldsymbol{\theta}_2, \ldots, \boldsymbol{\theta}_p$. Precise regularity conditions are described by Kipnis and Varadhan (1986) and Geyer (1992). Variations on the iterative scheme are described by Gelfand and Smith (1990). Geyer and Thompson (1992) and Tanner and Wong (1987) describe a variety of scientific and medical applications. An extensive review of later work is provided by Cowles and Carlin (1996).

Note that it is difficult to prove strong convergence of the estimated standard error of simulation, a disadvantage when compared with simple Monte Carlo. In practice, convergence needs to be judged with care. For further explanation of Gibbs sampling, see Casella and George (1992).

6.7 Modeling Sampling Distributions, Using MCMC

A key remaining problem for the applied statistician is how to model a multivariate sampling density $f(\mathbf{y})$ for a $p \times 1$ vector $\mathbf{y}$ of observations when a simple parametric model, with just a few parameters, is unavailable. Following Titterington, Smith, and Makov (1985), McLachlan and Basford (1988), Lavine and West (1992), Escobar and

West (1995), Green (1995), Leonard et al. (1994b), Richardson and Green (1997), and Roeder and Wasserman (1997), we believe that this problem is frequently handled well via mixtures, for example, the mixture

$$f(\mathbf{y}) = \sum_{j=1}^{q} \xi_j \Psi_{\mathbf{y}}(\boldsymbol{\theta}_j, \mathbf{C}_j) \qquad \left(\sum_{j=1}^{q} \xi_j = 1\right), \tag{6.7.1}$$

of q multivariate normal densities with respective mean vectors $\boldsymbol{\theta}_1, \ldots, \boldsymbol{\theta}_q$ and covariance matrices $\mathbf{C}_1, \mathbf{C}_2, \ldots, \mathbf{C}_q$. As an illustration, suppose that $y_1, \ldots, y_n$ are a random sample from a univariate distribution, with density

$$f(y) = \xi \Psi_y(\theta_1, \phi_1) + (1 - \xi)\Psi_y(\theta_2, \phi_2), \tag{6.7.2}$$

where $\Psi_y(\theta_i, \phi_i)$ denotes a normal density with mean θ_i and variance ϕ_i. Then both the EM algorithm and the Lavine–West Gibbs-sampling procedure introduce hypothetical (i.e., missing) binary variables $z_1, \ldots, z_n$, where the (y_i, z_i) are a random sample from the distribution with (a) $z_i = 1$ with probability ξ and $z_i = 0$ with probability $1 - \xi$; (b) given $z_i = 1$, y_i has density $\Psi_y(\theta_1, \phi_1)$; and (c) given $z_i = 0$, y_i has density $\Psi_y(\theta_2, \phi_2)$.

Suppose that in the prior assessment, ξ possesses a beta distribution with parameters α and β and is independent of $\theta_1, \theta_2, \phi_1$, and ϕ_2. Furthermore, given ϕ_1 and ϕ_2, θ_1 and θ_2 are independent and normally distributed with respective means μ_1 and μ_2 and variances ϕ_1/κ_1 and ϕ_2/κ_2. Finally, $\nu_1\lambda_1/\phi_1$ and $\nu_2\lambda_2/\phi_2$ possess independent chi-squared distributions with ν_1 and ν_2 degrees of freedom. Then (e.g., Exercise 5.1.c), in the posterior assessment,

(i) conditional on $\theta_1, \theta_2, \xi, \phi_1$, and ϕ_2, the z_i are independent and binary, with

$$p(z_i = 1 \mid \theta_1, \theta_2, \xi, \phi_1, \phi_2) = \frac{\xi \Psi_{y_i}(\theta_1, \phi_1)}{\xi \Psi_{y_i}(\theta_1, \phi_1) + (1 - \xi)\Psi_{y_i}(\theta_2, \phi_2)} \qquad (i = 1, \ldots, n);$$

(ii) given the z's, θ_2, ϕ_1, ϕ_2, and ξ, θ_1 is normally distributed with mean $(n_1\bar{y}_1 + \kappa_1\mu_1)/(n_1 + \kappa_1)$ and variance $\phi/(n_1 + \kappa_1)$, with $\bar{y}_1$ denoting the sample mean of those n_1 observations with $z_i = 1$;

(iii) given the z's, $\theta_1, \theta_2, \phi_2$, and ξ, the quantity $[\nu_1\lambda_1 + s_1^2(\theta_1)]/\phi_1$ possesses a chi-squared distribution with $\nu_1 + n_1$ degrees of freedom, where

$$s_1^2(\theta_1) = \sum_{i:z_i=1} (y_i - \theta_1)^2;$$

(iv) given the z's, θ_1, ϕ_1, ϕ_2, and ξ, θ_2 is normally distributed with mean $(n_2\bar{y}_2 + \kappa_2\mu_2)/(n_2 + \kappa_2)$ and variance $\phi/(n_2 + \kappa_2)$, with $\bar{y}_2$ denoting the sample mean of these n_2 observations with $z_i = 0$;

(v) given the z's, $\theta_1, \theta_2, \phi_1$, and ξ, the quantity $[\nu_2\lambda_2 + s_2^2(\theta_2)]/\phi_2$ possesses a chi-squared distribution with $\nu_2 + n_2$ degrees of freedom, where

$$s_2^2(\theta_2) = \sum_{i:z_i=0} (y_i - \theta_2)^2;$$

(vi) given the z's, θ's, and ϕ's, ξ possesses a beta distribution with parameters $\alpha + n_1$ and $\beta + n_2$.

Simulations for $z_1, \ldots, z_n, \theta_1, \phi_1, \theta_2, \phi_2$, and ξ may be obtained by iterating between these n binary distributions, two normal distributions, two inverted chi-squared distributions, and one beta distribution. Then the Gibbs sampling scheme of Section 6.6 may be used to simulate the posterior expectation, or distribution function, of any parameter of interest.

Note that the preceding procedure is open to improvement by replacing stages (iii) and (v) by

(iiia) given the z's, θ_2, and ϕ_2, $\phi_1^{-1}[\nu_1\lambda_1 + s_1^2(\bar{y}_1) + (\kappa_1^{-1} + n_1^{-1})^{-1}(\mu_1 - \bar{y}_1)^2]$ possesses a chi-squared distribution with $\nu_1 + n_1$ degrees of freedom;

(va) given the z's, θ_1, and ϕ_1, $\phi_2^{-1}[\nu_2\lambda_2 + s_2^2(\bar{y}_2) + (\kappa_2^{-1} + n_2^{-1})^{-1}(\mu_2 - \bar{y}_2)^2]$ possesses a chi-squared distribution with $\nu_2 + n_2$ degrees of freedom.

Then (ii) and (iii) together permit exact simulations from the posterior density of (θ_1, ϕ_1), given the z's, and (iv) and (v) together permit exact simulations from the posterior density of (θ_2, ϕ_2), given the z's if stages (iii) and (v) are performed before stages (ii) and (iv). These procedures will work only if the prior distribution is proper. Indeed, when analyzing mixtures, improper prior distributions will almost always provide improper posterior distributions. When specifying the prior distribution, it might be useful to set all the prior sample sizes, $\gamma = \alpha + \beta, \kappa_1, \kappa_2, \nu_1$, and ν_2, equal and positive. Then the choices of $\xi = \alpha/(\alpha + \beta)$, μ_1, μ_2, λ_1, and λ_2 may be based upon a prior estimate of the mixture of two normal distributions.

6.8 Equally Weighted Mixtures and Survivor Functions

Leonard (1984) and Leonard et al. (1994b) suggest that it will frequently suffice to consider equally weighted mixtures. This parallels the theory of kernel estimators (e.g., Silverman, 1986; Chan and Aitken, 1989). Consider, for example, n survival times $y_1, y_2, \ldots, y_n$ and censoring times $c_1, c_2, \ldots, c_n$. Let

$$d_i = \begin{cases} 1 & \text{if } y_i \le c_i \\ 0 & \text{if } y_i > c_i \quad (i = 1, 2, \ldots, n), \end{cases} \tag{6.8.1}$$

and $x_i = \min(y_i, c_i)$. If $y_1, y_2, \ldots, y_n$ are assumed to constitute a random sample from a distribution with density $f(t)$ and c.d.f. $F(t) = \int_0^t f(u)du$, then the likelihood function of f, given the x_i and d_i, is

$$l(f \mid \mathbf{x}, \mathbf{d}) = \prod_{i:d_i=1} f(x_i) \prod_{i:d_i=0} G(x_i), \tag{6.8.2}$$

where $G(t) = 1 - F(t) = \int_t^\infty f(u)du$ denotes the survivor function. Assume that

$$G(t) = m^{-1} \sum_{k=1}^{m} \exp\{-\xi_k t^b\} \qquad (0 < t < \infty), \tag{6.8.3}$$

where b and m are specified. Then the sampling distribution is an equally weighted mixture of Weibull distributions, and the power transformations y_i^b constitute a random sample from a mixture of exponential distributions. Conditional on the realizations of random variables z_i, each y_i has density

$$p(y_i \mid z_i = k) = b\xi_k y_i^{b-1} \exp\{-\xi_k y_i^b\} \qquad (0 < y_i < \infty), \qquad k = 1, 2, \ldots, m, \tag{6.8.4}$$

where the z_i are independent and uniformly distributed on the integers $\{1, 2, \ldots, k\}$. It is straightforward to evaluate the posterior expectation of the survivor function (6.8.3) by using MCMC and a convenient prior distribution. We find, from (6.8.2), that conditional on the z_i, the posterior distribution of the ξ_k, given the $u_i = x_i^b$ and d_i, is

$$\pi(\boldsymbol{\xi} \mid \mathbf{u}, \mathbf{d}, \mathbf{z}) \propto \prod_{k=1}^{m} \xi_k^{l_k} \exp\{-\xi_k g_k\}, \tag{6.8.5}$$

where

$$l_k = \sum_{i:z_i=k} d_i \tag{6.8.6}$$

and

$$g_k = \sum_{i:z_i=k} u_i. \tag{6.8.7}$$

Assume the two-stage (permutable) prior distribution:

Stage I: Given β, the ξ_k are a random sample from the Gamma distribution, with parameters α and β, and

$$\pi(\xi_k \mid \beta) = \frac{\beta^\alpha}{\Gamma(\alpha)} \xi_k^{\alpha-1} \exp\{-\xi_k \beta\} \qquad (0 < \xi_k < \infty),$$

where α is specified.

Stage II: β possesses the Gamma distribution, with specified parameters a_0 and b_0, and density

$$\pi(\beta) \propto \beta^{a_0-1} \exp\{-b_0 \beta\} \qquad (0 < \beta < \infty). \tag{6.8.8}$$

Combining these two stages, we find that the marginal prior density of ξ_k is

$$\pi(\xi_k) \propto \xi_k^{\alpha-1} (b_0 + \xi_k)^{\alpha+a_0} \qquad (0 < \xi_k < \infty), \tag{6.8.9}$$

so that $\xi_k/(b_0 + \xi_k)$ has a beta distribution with parameters α and α_0. At the second stage of the model, note that β has mean $\beta_0 = a_0/b_0$ and variance β_0/b_0.

The current section will be continued after the following exercises.

SELF-STUDY EXERCISES

6.8.a Show that the prior mean, given β, of $G(t)$ is

$$G_0(t) = \frac{\beta^\alpha}{(t^b + \beta)^\alpha}.$$

6.8.b Show that the prior density (6.8.9) of ξ_k relates to the F-distribution (see Section 5.2 (A)), and discuss how the prior parameters α, a_0, and b_0 might be specified.

6.8.c Show that in the posterior assessment,

(a) given β and the z_i, ξ_k possesses a Gamma distribution with parameters $\alpha + l_k$ and $\beta + g_k$;

(b) given the ξ_k and z_i, β possesses a beta distribution with parameters $a_0 + m\alpha$ and $b_0 + \sum_{i=1}^m \xi_k$;

(c) given β and the ξ_k, z_i is equal to k with probability

$$p_{ik} = \frac{A_{ik}}{\sum_{l=1}^m A_{il}},$$

where

$$A_{ik} = \xi_k^{d_i} \exp\{-\xi_k u_i\} \qquad (i = 1, 2, \ldots, n;\, k = 1, 2, \ldots, m).$$

Using the results of Exercise 6.8.c, it is straightforward to calculate the posterior expectation of the survivor function (6.8.3) by using MCMC based upon successive simulations from these distributions. However, it is essential, when drawing inferences based upon mixture models, either to input substantial prior information or to employ some sensible empirical Bayesian procedure.

Curve (a) in Figure 6.8.1 describes the Kaplan–Meier estimate (see Susarla and Van Ryzin, 1976) of the survivor function for a set of $n = 30$ survivor times in days, reported by Bull and Spiegelhalter (1997). Curve (b) describes a single Weibull curve fitted by maximum likelihood, which yields the survivor function

$$G(t) = \exp\left\{-\left(\frac{t}{4887.02}\right)^{0.4608}\right\} \qquad (0 < t < \infty). \tag{6.8.10}$$

With $m = 20$, we made the empirical choices $b = 0.4608$ and

$$E\left(\xi_k^{-1}\right) = \frac{a_0}{b_0(\alpha - 1)} = (4887.02)^{0.4608}.$$

Curve (c) in Figure 6.8.1 reports the posterior mean value function of $G(t)$, using the additional, subjective choices $b_0 = 15$ and $\alpha = 2$. This provides a semi-parametric smoothing procedure contrasting with Leonard (1978). Curve (d) instead plots the posterior probability that $G(t) \geq 0.5$. In all cases, MCMC appeared to converge to four-decimal-place accuracy in under 200,000 simulations.

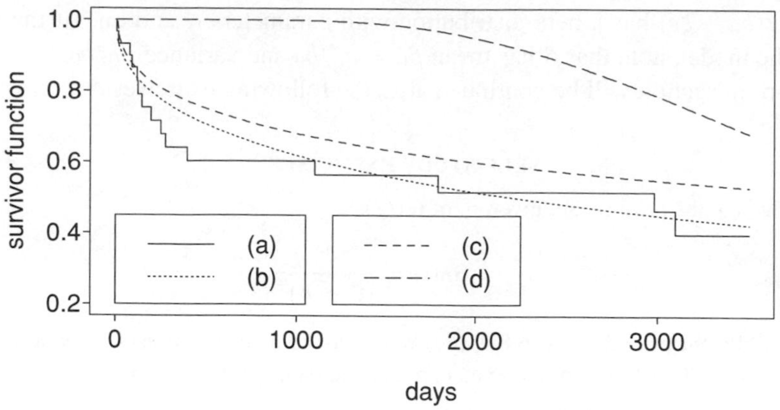

Figure 6.8.1. Estimated survivor functions and posterior probability curve (d).

6.9 A Hierarchical Bayes Analysis

When performing MCMC, and in situations where it is impossible to simulate from one of the relevant conditional distributions, it is better to employ rejection sampling (Ripley, 1987) rather than to attempt importance sampling from an approximate conditional distribution, since it is then unclear how to weight your simulations.

Consider, for example, a generalized linear model with random effects. Let $y_1, y_2, \ldots, y_m$ denote Poisson observations with respective means $\mu_1, \mu_2, \ldots, \mu_m$ and let $\gamma_i = \log \mu_i$ for $i = 1, 2, \ldots, m$. Given $\boldsymbol{\beta}$ and σ^2, take the γ_i to be independent and normally $N(\mathbf{x}_i^T\boldsymbol{\beta}, \sigma^2)$ distributed, for $i = 1, 2, \ldots, m$, where $\mathbf{x}_1, \mathbf{x}_2, \ldots, \mathbf{x}_m$ are specified design vectors and $\boldsymbol{\beta}$ is an identifiable $p \times 1$ vector of unknown parameters. At the second stage of our hierarchical model, take $(\boldsymbol{\beta}, \sigma^2)$ to be uniformly distributed. Then the joint posterior density of the γ_i, $\boldsymbol{\beta}$, and σ^2 is

$$\pi(\boldsymbol{\gamma}, \boldsymbol{\beta}, \sigma^2 \mid \mathbf{y}) \propto (\sigma^2)^{-\frac{1}{2}m} \exp\left\{\sum_{i=1}^{m} y_i\gamma_i - \sum_{i=1}^{m} e^{\gamma_i} - \frac{1}{2}\sigma^{-2}\sum_{i=1}^{m}(\gamma_i - \mathbf{x}_i^T\boldsymbol{\beta})^2\right\}. \tag{6.9.1}$$

If, instead, $\nu\lambda/\sigma^2$ is taken to possess a chi-squared distribution with ν degrees of freedom, in the prior assessment, then all marginal posterior densities may be computed with a finite standard error of simulation by importance sampling. An importance-sampling approach is available by simulating each γ_i independently as $\gamma_i = \log \theta_i$, where $\theta_i = e^{\gamma_i}$ has a Gamma distribution with mean and variance both equal to $y_i + 1$. See Hsu and Leonard (1997) for similar results in a binomial situation. If, however, an MCMC approach is preferred, note that when $(\boldsymbol{\beta}, \sigma^2)$ is uniformly distributed,

(a) given $\gamma_1, \gamma_2, \ldots, \gamma_m$, and σ^2, $\boldsymbol{\beta}$ possesses the conditional distribution

$$\boldsymbol{\beta} \sim N\left(\left(\sum_{i=1}^{m}\mathbf{x}_i\mathbf{x}_i^T\right)^{-1}\sum_{i=1}^{m}\mathbf{x}_i\gamma_i, \left(\sum_{i=1}^{m}\mathbf{x}_i\mathbf{x}_i^T\right)^{-1}\right);$$

(b) given $\gamma_1, \gamma_2, \ldots, \gamma_m$, and $\boldsymbol{\beta}$, σ^2 possesses an inverted chi-squared distribution, such that

$$\frac{1}{\sigma^2}\sum_{i=1}^{m}\left(\gamma_i - \mathbf{x}_i^T\boldsymbol{\beta}\right)^2 \sim \chi^2_{m-2};$$

(c) given $\boldsymbol{\beta}$ and σ^2, the γ_i are independent and have posterior densities respectively proportional to

$$b_i(\gamma_i) = \exp\left\{y_i\gamma_i - e^{\gamma_i} - \frac{1}{2\sigma^2}\left(\gamma_i - \mathbf{x}_i^T\boldsymbol{\beta}\right)^2\right\} \qquad (i = 1, \ldots, m).$$

It is straightforward to simulate from the distributions in (a) and (b), but not from the distribution in (c). However, Zeger and Karim (1991) follow one of the major themes of our current volume by recommending simulation from some normal $N(a_i, v_i)$ distribution, with density $\pi_i^*(\gamma_i)$ that approximates our exact conditional density. To perform rejection sampling, they need to restrict γ_i to some range where there exists a constant c_i such that $c_i\pi_i^*(\gamma_i) \geq b_i(\gamma_i)$ for all γ_i in this range. Then, following Ripley (1987), the following steps result in the simulation of a γ_i from its exact conditional distribution:

(1) Generate a γ_i as a $N(a_i, v_i)$ variate.
(2) Generate a uniformly random number u_i on (0, 1).
(3) If $b_i(\gamma_i)/c_i\pi_i^*(\gamma_i) < u_i$, γ_i is appropriately generated. Otherwise, return to step (1).

It would be possible to simulate γ_i over an unrestricted range by replacing $\pi_i^*(\gamma_i)$ by a generalized t-density, with appropriately low degrees of freedom. In either case, it is necessary to avoid a high rejection rate and to use a precise computational procedure in order to fully identify efficient choices of $\pi^*(\gamma_i)$. This can increase the computer time needed. Modifications to this procedure are described by Carlin and Gelfand (1991) and Tierney (1994). Applications of MCMC to image processing are described by Johnson et al. (1991) and many others.

References

Abramowitz, M. and Stegun, I. A. (eds.) (1968). *Handbook of Mathematical Functions*, Dover, New York.

Aitchison, J. (1986). *The Statistical Analysis of Compositional Data*, Chapman and Hall, London.

Aitchison, J. and Dunsmore, I. R. (1975). *Statistical Prediction Analysis*, Cambridge University Press, Cambridge.

Aitken, A. C. (1925). *The Graduation of Observational Data*, Ph.D. thesis, University of Edinburgh.

Aitken, A. C. (1944). *Statistical Mathematics*, Oliver and Boyd, Edinburgh, London.

Aitken, C. G. G. (1995). *Statistics and the Evaluation of Evidence for Forensic Scientists*, Wiley, New York.

Aitken, C. G. G., Bring, J., Leonard, T., and Papasouliotis, O. (1996). Estimation of drugs handled and burden of proof, *Journal of Royal Statistical Society, Ser. A*, **159**: 333–51.

Akaike, H. (1978). A Bayesian analysis of the minimum AIC procedure, *Annals of the Institute of Statistical Mathematics* **30(A)**: 9–14.

Alanko, T. and Duffy, J. C. (1996). Compound binomial distributions for modelling consumption data, *The Statistician* **45**: 268–86.

Albert, J. H. (1983). A Bayesian treatment of nonresponse when sampling from a dichotomous population, *ASA Proc. of Survey Res. Methods Sect.*, pp. 314–7.

Albert, J. H. (1988). Computational methods using a Bayesian hierarchical generalized linear model, *Journal of the American Statistical Association* **83**: 1037–44.

Albert, J. H. (1992). Bayesian estimation of the polychoric correlation coefficient, *Comput. and Simul.* **10**: 1261–8.

Albert, J. H. and Pepple, P. A. (1989). A Bayesian approach to some overdispersion models, *Canadian Journal of Statistics* **17**: 333–44.

Allais, M. (1953). Le comportement de l'homme rationnel devant le risque: Critique des postulats et axiomes de l'école américaine, *Econometrica* **21**: 503–46.

Allais, M. and Hagen, O. (eds.) (1979). *Expected Utility Hypotheses and the Allais Paradox*, Reidel, Dordrecht, Holland.

Altham, P. M. E. (1969). Exact Bayesian analysis of a 2×2 contingency table, and Fisher's 'exact' significance test, *Journal of Royal Statistical Society, Ser. B*, **31**: 261–9.

Anderson, D. A. and Aitkin, M. (1985). Variance component models with binary responses: interviewer variability, *Journal of Royal Statistical Society, Ser. B*, **47**: 203–10.

Anderson, J. A. (1975). Quadratic logistic discriminination, *Biometrika* **62**: 149–54.

Anderson, T. W. (1974). *An Introduction to Multivariate Statistical Analysis*, Wiley, New York.

Antoniak, C. E. (1974). Mixtures of Dirichlet process with applications to Bayesian nonparametric problems, *Annals of Statistics* **6**: 1152–74.

Aoki, M. (1975). *Optimal Control and System Theory in Dynamic Economic Analysis*, North-Holland, Amsterdam.

Arjas, E. and Gasbarra, D. (1994). Non-parametric Bayesian inference from right censored survival data, using the Gibbs sampler, *Statistica Sinica* **4**: 505–24.

Arjas, E. and Heikkinen, J. (1997). An algorithm for nonparametric Bayesian estimation of a Poisson intensity, *Computational Statistics* **12**: 385–402.

Arnold, B. (1991). Dependence in conditionally specified distributions, *in* H. W. Block, A. R. Sampson, and T. H. Savits (eds.), *Topics in Statistical Dependence*, IMS, Hayward, California, pp. 13–8.

Ash, R. B. (1972). *Real Analysis and Probability*, Academic Press, New York.

Astrom, K. J. (1970). *Introduction to Stochastic Control Theory*, Academic Press, New York.

Atilgan, T. (1983). *Parameter Parsimony, Model Selection, and Smooth Density Estimation*, Ph.D. thesis, University of Wisconsin–Madison.

Atilgan, T. (1990). On derivation and application of AIC as a data-based criterion for histograms, *Communications in Statistics, Ser. A*, **19**: 885–903.

Atilgan, T. and Bozdogan, H. (1990). Selecting the number of knots in fitting cardinal B-splines for density estimation using AIC, *J. Japan Statist. Soc.* **20**: 179–90.

Atilgan, T. and Leonard, T. (1987). On the application of AIC to bivariate density estimation, non-parametric regression, and discrimination, *in* H. Bozadogan and A. K. Gupta (eds.), *Multivariate Statistical Modeling and Data Analysis*, Reidel, Dordrecht, Holland, pp. 1–16.

Barlow, R. E. and Singpurwalla, N. (1985). Assessing the reliability of computer network: An opportunity for partnership with computer scientists, *American Statistician* **39**: 88–94.

Barlow, R. E. and Zhang, X. (1987). Bayesian analysis of inspection sampling procedures discussed by Deming, *Journal of Statistical Planning and Inference* **16**: 285–96.

Barnard, G. A. (1958). Thomas Bayes' essay towards solving a problem in the doctrine of chances, *Biometrika* **45**: 293–315.

Barndorff-Nielsen, O. E. (1983). On a formula for the distribution of the maximum likelihood estimator, *Biometrika* **70**: 343–65.

Barndorff-Nielsen, O. E. and Cox, D. R. (1979). Edgeworth and saddle-point approximations with statistical applications (with discussion), *Journal of Royal Statistical Society, Ser. B*, **41**: 279–312.

Barnett, V. and Lewis, T. (1978). *Outliers in Statistical Data*, Wiley, New York.

Barlett, M. S. and Kendall, D. G. (1946). The statistical analysis of variance – heterogeneity and the logarithmic transformation, *Journal of Royal Statistical Society, Ser. B*, **8**: 128–38.

Bates, D. M. and Watts, D. G. (1988). *Nonlinear Regression Analysis and Its Applications*, Wiley, New York.

Bayes, T. (1763). An essay towards solving a problem in the doctrine of chances, *Philos. Trans. Roy. Soc.* **53**: 370–418.

Bell, D. E. (1982). Regret in decision-making under uncertainty, *Operations Research* **30**: 961–81.

Bellman, R. (1971). *Introduction to the Mathematical Theory of Control Processes*, Academic Press, New York.

Bennett, J. H. (ed.) (1971–4). *Collected Papers of R. A. Fisher* (5 volumes), University of Adelaide Press, Adelaide.

Berger, J. O. (1985). *Statistical decision theory and Bayesian analysis*, Springer-Verlag, New York.

Berger, J. O. and Wolpert, R. L. (1984). *The Likelihood Principle*, Institute of Mathematical Statistics Monograph Series, Hayward, California.

Bernardo, J. M. (1979). Reference posterior distributions for Bayesian inference (with discussion), *Journal of Royal Statistical Society, Ser. B*, **41**: 113–47.

Bernardo, J. M. and Ramon, J. M. (1998). An introduction to Bayesian reference analysis: inference on the ratio of multinomial parameters, *The Statistician* **47**: 101–35.

Bernardo, J. M. and Smith, A. F. M. (1994). *Bayesian Theory*, Wiley, New York.

Besag, J. and Green, P. J. (1993). Spatial statistics and Bayesian computation (with discussion), *Journal of Royal Statistical Society, Ser. B*, **55**: 25–37.

Birnbaum, A. (1962). On the foundations of statistical inference (with discussion), *Journal of the American Statistical Association* **57**: 269–326.

Birnbaum, A. (1969). Statistical theory for logistic mental test models with a prior distribution of ability, *J. Math. Psych.* **6**: 258–76.

Bishop, C. M. (1995). *Neural Networks for Pattern Recognition*, Oxford University Press, Oxford.

Blattberg, R. C. and George, E. I. (1991). Shrinkage estimation of price and promotional elasticities: Seemingly unrelated equations, *Journal of the American Statistical Association* **86**: 304–15.

Blattberg, R. C. and George, E. I. (1992). Estimation under profit-driven loss functions, *J. Busn. Econ. Statist.* **10**: 437–44.

Blight, B. J. N. (1974). Recursive solutions for the estimation of a stochastic parameter, *Journal of the American Statistical Association* **69**: 477–81.

Blight, B. J. N. and Ott, L. (1975). A Bayesian approach to model inadequacy for polynomial regression, *Biometrika* **62**: 79–88.

Bloch, D. A. and Watson, G. S. (1967). A Bayesian study of the multinomial distribution, *Annals of Mathematical Statistics* **38**: 1423–35.

Bock, R. D. and Aitkin, M. (1981). Marginal maximum likelihood estimation of item parameters: An application to an EM algorithm, *Psychometrika* **35**: 179–97.

Bonsall, C., Lennon, R., and McSweeney, K. (1997). Mesolithic and early neolithic in the iron gates: A palaeodietary perspective, *Journal of European Archaeology* **5**: 50–92.

Bowlby, S. and Silk, J. (1982). Analysis of qualitative data using GLIM: two examples based on shopping survey data, *Professional Geographer* **34**: 80–90.

Box, G. E. P. (1980). Sampling and Bayes inference in scientific modeling and roubustness (with discussion), *Journal of Royal Statistical Society, Ser. A*, **143**: 383–430.

Box, G. E. P. (1983). An apology for ecumenism in statistics, *in* G. E. P. Box, T. Leonard, and C.-F. Wu (eds.), *Scientific Inference, Data Analysis, and Robustness*, Academic Press, London, pp. 51–84.

Box, G. E. P. and Guttman, I. (1966). Some aspects of randomization, *Journal of Royal Statistical Society, Ser. B*, **28**: 543–58.

Box, G. E. P. and Jenkins, G. M. (1976). *Time Series Analysis: Forecasting and Control*, Holden-Day, San Francisco.

Box, G. E. P. and Tiao, G. C. (1968). Bayesian estimation of means for the random effect model, *Journal of the American Statistical Association* **63**: 174–81.

Box, G. E. P. and Tiao, G. C. (1973). *Bayesian Inference in Statistical Analysis*, Wiley, New York.

Box, G. E. P., Hunter, W. G., and Hunter, J. S. (1978). *Statistics for Experimenters: An Introduction to Design, Data Analysis, and Model Building*, Wiley, New York.

Breslow, N. E. (1984). Extra-Poisson variation in log-linear models, *Applied Statistics* **33**: 38–44.

Breslow, N. E. (1990a). Biostatistics and Bayes (with discussion), *Statistical Science* **5**: 269–98.

Breslow, N. E. (1990b). Tests of hypotheses in overdispersed Poisson regression, and other quasi-likelihood models, *Journal of the American Statistical Association* **85**: 565–71.

Bretthorst, G. L. (1988). *Bayesian Spectrum Analysis and Parameter Estimation*, Springer-Verlag, Berlin.

Brown, C. C. and Muentz, L. (1976). Reduced mean squared error estimation in contingency tables, *Journal of the American Statistical Association* **71**: 176–82.

Brown, L. D. (1971). Admissible estimators, recurrent diffusions, and insoluble boundary value problems, *Annals of Mathematical Statistics* **42**: 855–903.

Brown, L. D., Chow, M., and Fong, D. K. H. (1992). On the admissibility of maximum likelihood estimator of the binomial variance, *Canadian Journal of Statistics* **20**: 253–8.

Brown, R. L., Leonard, T., Rounds, L. A., and Papasouliotis, O. (1997). A two item screening test for alcohol and other drug problems, *Journal Family Practice* **44**: 151–60.

Bull, K. and Spiegelhalter, D. J. (1997). Tutorial in biostatistics, *Statistics in Medicine* **10**: 1041–74.

Burch, C. R. and Parsons, I. T. (1976). "Squeeze" significance tests, *Applied Statistics* **25**: 287–91.

Butler, R. W. (1986). Predictive likelihood inference with applications (with discussion), *Journal of Royal Statistical Society, Ser. B*, **41**: 1–38.

Byar, D. P., Simon, R. M., Friedewald, M. D., Schlesselman, J. J., DeMets, D. L., Ellenberg, J. H., Gail, M. H., and Ware, J. H. (1976). Randomized clinical trials, perspectives on some recent ideas, *New England J. Med.* **295**: 74–80.

Carlin, B. P. and Gelfand, A. E. (1991). A sample re-use method for accurate parametric empirical Bayes confidence intervals, *Journal of Royal Statistical Society, Ser. B*, **53**: 189–200.

Carlin, B. P. and Louis, T. A. (1996). *Bayes and Empirical Bayes Methods for Data Analysis*, Chapman and Hall, London.

Casella, G. and George, E. I. (1992). Explaining the Gibbs Sampler, *American Statistician* **46**: 160–74.

Cercignani, C. (1988). *The Boltzman Equation and Its Applications*, Springer-Verlag, Berlin.

Chaloner, K. M. and Duncan, G. T. (1983). Assessment of a beta prior distribution: PM elicitation, *The Statistician* **32**: 174–80.

Chaloner, K. M. and Larntz, G. T. (1989). Optimal Bayesian design applied to logistic regression experiments, *Journal of Statistical Planning and Inference* **21**: 191–208.

Chamberlain, G. and Leamer, E. E. (1976). Matrix weighted averages and posterior bounds, *Journal of Royal Statistical Society, Ser. B*, **38**: 73–84.

Chan, K. P. S. and Aitken, C. G. G. (1989). Estimation of the Bayes' factor in a forensic science problem, *JSCS* **33**: 249–64.

Chen, C. (1979). Bayesian inference for a normal dispersion matrix and its application to stochastic multiple regression analysis, *Journal of Royal Statistical Society, Ser. B*, **41**: 235–48.

Chew, S. H. (1983). A generalization of the quasilinear mean with applications to the measurement of income inequality and decision theory resolving the Allais paradox, *Econometrica* **51**: 1065–92.

Chib, S. and Greenberg, E. (1995). Understanding the Metropolis-Hastings algorithm, *American Statistician* **49**: 327–35.

Chiu, T. Y. M., Leonard, T., and Tsui, K. (1996). The matrix-logarithmic covariance model, *Journal of the American Statistical Association* **91**: 198–210.

Clayton, M. K. and Berry, D. A. (1985). Bayesian nonparametric bandits, *Annals of Statistics* **13**: 1523–34.

Clevenson, M. L. and Zidek, J. V. (1975). Simultaneous estimation of the means of independent Poisson laws, *Journal of the American Statistical Association* **70**: 698–705.

Clevenson, M. L. and Zidek, J. V. (1977). Bayes linear estimators of the intensity function of the unstationary Poisson process, *Journal of the American Statistical Association* **69**: 50–7.

Cohen, M. (1992). Security level, potential level, expected utility: a three-criteria decision model under risk, *Theory and Decision* **33**: 101–34.

Cooke, A. J., Espey, B., and Carswell, R. (1996). The evolution of ionising flux at high redshifts, *Technical Report*, Institute of Astronomy, University of Edinburgh.

Copas, J. M. and Li, H. G. (1997). Inference for non-random samples (with discussion), *Journal of Royal Statistical Society, Ser. B*, **59**: 55–95.

Cowles, M. K. and Carlin, B. P. (1996). Markov Chain Monte Carlo convergence diagnostics: a comparative review, *Journal of the American Statistical Association* **91**: 883–904.

Cox, D. D. and O'Sullivan, F. (1990). Asymptotic analysis of penalized likelihood and related estimators, *Annals of Statistics* **18**: 1676–95.

Cox, D. R. (1958). Some problems connected with statistical inference, *Annals of Mathematical Statistics* **29**: 357–72.

Cox, D. R. and Snell, E. J. (1981). *Applied Statistics: Principles and Examples*, Chapman and Hall, London.

Craig, P. S., Goldstein, M., Seheult, A. H., and Smith, J. A. (1998). Constructing partial prior specifications for models of complex physical systems (with discussion), *The Statistician* **47**: 37–53.

Crook, J. F. and Good, I. J. (1982). The powers and strengths of tests for multinomials and contingency tables, *Journal of the American Statistical Association* **77**: 793–802.

Cyert, R. M. and DeGroot, M. H. (1987). *Bayesian Analysis and Uncertainty in Economic Theory*, Rowman & Littlefield, Totowa, New Jersey.

Dale, A. I. (1982). Bayes or Laplace? An examination of the origin and early application of Bayes' theorem, *Arch. for History of Exact Sciences* **27**: 23–47.

Dale, A. I. (1991). *A History of Inverse Probability from Thomas Bayes to Karl Pearson*, Springer-Verlag, Berlin.

Davison, A. C. (1986). Approximate predictive likelihood, *Biometrika* **73**: 323–32.

Dawid, A. P. (1973). Posterior expectations for large observations, *Biometrika* **60**: 664–6.

Dawid, A. P. and Dickey, J. M. (1977). Likehood and Bayesian inference from selective reported data, *Journal of the American Statistical Association* **72**: 845–50.

Dawid, A. P. and Pueschel, J. (1999). Hierarchical models for DNA profiling, using heterogeneous data bases, *in* J. O. Berger, J. M. Bernardo, A. P. Dawid, and A. F. M. Smith (eds.), *Bayesian Statistics VI*, Clarendon Press, Oxford, pp. 103–20.

Dawid, A. P., Stone, M., and Zidek, J. V. (1973). Marginalization paradoxes in Bayesian and structural inference (with discussion), *Journal of Royal Statistical Society, Ser. B*, **35**: 189–233.

Dayan, R. E. (1987). Intensity parameter estimation of inhomogeneous Poisson process in the almost smooth case, *Izvestija Akademii Nauk Armjanskoi SSR Serija Matematika* **22**: 200–7.

de Finetti, B. (1937). Le Prévision: ses lois logiques, ses sources subjectives, *Ann. Inst. Henri Poincaré,* tome **VII**, Fasc. **1**: 1–68.

DeGroot, M. H. (1970). *Optimal Statistical Decisions*, McGraw-Hill, New York.

DeGroot, M. H. (1974). Reaching a consensus, *Journal of the American Statistical Association* **69**: 118–21.

Dempster, A. P., Laird, N. M., and Rubin, D. B. (1977). Maximum likelihood from incomplete data via the EM algorithm (with discussion), *Journal of Royal Statistical Society, Ser. B*, **39**: 1–38.

Devore, A. J. (1991). *Probability and Statistics for Engineering and Sciences*, Brooks/Core, Pacific Grove, California.

Dickey, J. M. (1973). Scientific reporting, *Journal of Royal Statistical Society, Ser. B*, **35**: 285–305.

Dickey, J. M. (1976). Approximate posterior distributions, *Journal of the American Statistical Association* **71**: 680–9.

Dickey, J. M. and Chen, C. (1985). Direct subjective-probability modeling using ellipsoidal distributions, *in* J. M. Bernardo, M. H. DeGroot, D. V. Lindley, and A. F. M. Smith (eds.), *Bayesian Statistics II*, North-Holland, Amsterdam, pp. 157–82.

Dickey, J. M., Lindley, D. V., and Press, S. J. (1985). Bayesian estimation of the dispersion matrix of a multivariate normal distribution, *Communications in Statistics, Ser. A*, **14**: 1019–34.

Doksum, K. A. and Lo, A. Y. (1990). Consistent and robust Bayes procedures for location based on partial information, *Annals of Statistics* **18**: 443–53.

Donner, A. (1984). Approaches to sample size estimation in the design of clinical trials – a review, *Statistics in Medicine* **3**: 199–214.

Draper, N. R. and Van Nostrand, R. C. (1979). Ridge regression and James-Stein estimation: Review and comments, *Technometrics* **21**: 451–66.

Duncan, G. T. (1974). An empirical Bayes approach to scoring multiple-choice tests in the misinformation model, *Journal of the American Statistical Association* **69**: 50–7.

Edgeworth, F. (1899). On the representation of statistics by mathematical formulae. Parts 111 and 1v, *Journal of Royal Statistical Society, Ser. B*, **52**: 373–385 and 534–555.

Edwards, A. W. F. (1972). *Likelihood*, Cambridge University Press, Cambridge.

Edwards, A. W. F. (1976). Fiducial probability, *The Statistician* **25**: 15–35.

Efron, B. (1979). Bootstrap methods: Another look at the jacknife, *Annals of Statistics* **7**: 1–26.

Efron, B. and Hinkley, D. V. (1978). Assessing the accuracy of the maximum likelihood estimator: Observed versus expected Fisher information (with discussion), *Biometrika* **65**: 457–87.

Efron, B. and Morris, C. N. (1973). Stein's estimation rule and its competitors – An empirical Bayes approach, *Journal of the American Statistical Association* **68**: 117–30.

Ellsberg, D. (1961). Risk, ambiguity and the Savage axioms, *Quarterly J. Econ.* **75**: 643–99.

Emerson, P. A. (1974). Decision theory in the prevention of thrombo-embolism, *Annals of Statistics* **7**: 1–26.

Ericson, W. A. (1970). On the posterior mean and variance of a population mean, *Journal of the American Statistical Association* **65**: 649–52.

Escobar, M. D. and West, M. (1995). Bayesian density estimation and inference using mixtures, *Journal of the American Statistical Association* **90**: 577–88.

Essen-Möller, E. (1938). Die Beweiskraft der Ähnlichkeit in Vaterschaftsnachweis; theoretische Grundlagen, *Mitt. Anthr. Ges. (Wien)* **68**: 9–53.

Evans, I. G. (1965). Bayesian estimation of parameters of a multivariate normal distribution, *Journal of Royal Statistical Society, Ser. B*, **27**: 279–83.

Ferguson, T. S. (1967). *Mathematical Statistics: A Decision-Theoretic Approach*, Academic Press, New York.

Ferguson, T. S. (1973). A Bayesian analysis of some nonparametric problems, *Annals of Statistics* **1**: 209–30.

Fernandez, C. and Steel, M. F. (1998). On Bayesian modelling of fat tails and skewness, *Journal of the American Statistical Association* **93**: 359–67.
Fienberg, S. E. (1987). *The Analysis of Cross-Classified Categorical Data*, The MIT Press, Cambridge, Massachusetts.
Fienberg, S. E. and Holland, P. W. (1973). Simultaneous estimation of multinomial cell probabilities, *Journal of the American Statistical Association* **68**: 683–91.
Fingleton, B. (1984). *Models of Category Counts*, Cambridge University Press, Cambridge.
Fishburn, P. C. (1965). Independence in utility theory with whole product sets, *Operations Research* **13**: 28–45.
Fishburn, P. C. (1974). von Neumann-Morgenstern utility functions on two attributes, *Operations Research* **22**: 35–45.
Fishburn, P. C. and La Valle, I. H. (1992). Multiattribute expected utility without the Archimedean axiom, *J. Math. Psych.* **36**: 573–91.
Fisher, R. A. (1925). *Statistical Methods for Research Workers*, Oliver and Boyd, Edinburgh.
Fisher, R. A. (1935). The fiducial argument in statistical inference, *Annals of Eugenics* **6**: 391–8.
Fisher, R. A. (1959). *Statistical Methods and Scientific Inference*, 2d ed., Oliver and Boyd, Edinburgh.
Flanagan, J. C., Davis, F. B., Dailey, J. T., Shaycroft, M. F., Orr, D. B., Goldberg, I., and Neyman, C. A. (1964). Project TALENT, *Technical Report*, University of Pittsburgh.
Foulley, J. L., Gianola, D., and Im, S. (1990). Genetic evaluation for discrete polygenic trials in animal breeding, *in* D. Gianola and K. Hammond (eds.), *Advances in Statistical Methods for Genetic Improvement of Livestock*, Springer-Verlag, Berlin, pp. 361–411.
Fraser, D. A. S. (1966). Structural probability and a generalization, *Biometrika* **53**: 1–9.
Freedman, D., Pisani, R., and Purves, R. (1991). *Statistics*, 2d ed., Norton, New York.
Friedman, M. and Savage, L. J. (1948). The utility analysis of choices involving risk, *J. Political Economy* **56**: 279–304.
Garthwaite, P. H. and Dickey, J. M. (1992). Elicitation of prior distributions for variable-selection problems in regression, *Annals of Statistics* **20**: 1697–719.
Geisel, M. S. (1975). Bayesian comparisons of simple macroeconomic models, *in* S. E. Fienberg and A. Zellner (eds.), *Studies in Bayesian Econometrics and Statistics in Honor of Leonard J. Savage*, North-Holland, Amsterdam, pp. 227–56.
Geisser, S. (1974). A predictive approach to the random effect model, *Biometrika* **61**: 101–7.
Geisser, S. (1987). Comment on "The statistical precision of medical screening procedures: Application to polygraph and AIDS antibodies test data," *Statistical Science* **2**: 231–2.
Geisser, S. (1992). Some statistical issues in medicine and forensics, *JASA* **87**: 607–14.
Geisser, S. (1993). *Predictive Inference: An Introduction*, Chapman and Hall, London.
Geisser, S. and Johnson, W. (1992). Optimal administration of dual screening tests for detecting a characteristic with special reference to low prevalence diseases, *Biometrics* **48**: 839–52.
Gelfand, A. E. and Smith, A. F. M. (1990). Sampling-based approaches to calculating marginal densities, *JASA* **85**: 398–409.
Gelman, A., Carlin, J. B., Stern, H. S., and Rubin, D. B. (1995). *Bayesian Data Analysis*, Chapman and Hall, New York.
Geman, S. and Geman, D. (1984). Stochastic relaxation, Gibbs distributions, and the Bayesian restoration of images, *IEEE Transactions on Pattern Analysis and Machine Intelligence* **6**: 721–41.
George, E. I. (1984). An empirical Bayes approach using mixtures of models, *ASA Proc. of Busn. and Econ. Sect.*, pp. 32–41.
George, E. I. and Wecker, W. E. (1985). Estimating damages in a class action litigation, *J. Busn. Econ. Statist.* **3**: 132–9.
George, E. I., Makov, U. E., and Smith, A. F. M. (1994). Fully Bayesian hierarchical analysis for exponential families, via Monte Carlo computation, *in* P. R. Freeman and A. F. M. Smith (eds.), *Aspects of Uncertainty: A Tribute to D. V. Lindley*, Wiley, Chichester, pp. 181–99.
Geweke, J. (1988). Antithetic acceleration of Monte-Carlo integration in Bayesian inference, *Journal of Econometrics* **38**: 73–89.
Geweke, J. (1989). Exact predictive density for linear models with arch distributions, *Journal of Econometrics* **40**: 63–86.

Geyer, C. J. (1992). Practical Markov Chain Monte Carlo (with discussion), *Statistical Science* **7**: 473–503.

Geyer, C. J. and Thompson, E. A. (1992). Constrained Monte Carlo maximum likelihood for dependent data (with discussion), *Journal of Royal Statistical Society, Ser. B*, **54**: 657–99.

Ghosh, M., Hwang, J., and Tsui, K. (1983). Construction of improved estimators in multiparameter estimation for discrete exponential families (with discussion), *Annals of Statistics* **11**: 351–76.

Ghosh, M., Hwang, J., and Tsui, K. (1984). Construction of improved estimators in multiparameter estimation for continuous exponential families, *Journal of Multivariate Analysis* **14**: 212–20.

Gianola, D. and Hammond, K. (eds.) (1990). *Advances in Statistical Methods for Genetic Improvement of Livestock*, Springer-Verlag, New York.

Gilardoni, G. L. (1989). *Combining Prior Opinions*, Ph.D. thesis, University of Wisconsin–Madison.

Gilboa, I. (1988). A combination of expected utility theory and maxmin decision criteria, *J. Math. Psych.* **32**: 405–20.

Goldstein, H. (1995). *Multilevel Statistical Models*, Edward Arnold, London.

Good, I. J. (1965). *The Estimation of Probabilities*, M.I.T. Press, Boston.

Good, I. J. (1967). A Bayesian significance test for multinomial distributions (with discussion) (corrected in vol. 36, pp. 109–110), *Journal of Royal Statistical Society, Ser. B*, **29**: 399–431.

Good, I. J. (1979). Studies in the History of Probability and Statistics. Part XXXVll: A.M. Turing's work in World War II, *Biometrika* **66**: 393–6.

Good, I. J. (1991). Weight of evidence and the likelihood ratio, *in* C. G. G. Aitken and D. A. Stoney (eds.), *The Use of Statistics in Forensic Science*, Ellis Horwood, Chichester, pp. 85–106.

Good, I. J. and Gaskins, R. A. (1971). Nonparametric roughness penalties for probability densities, *Biometrika* **58**: 255–71.

Good, I. J. and Gaskins, R. A. (1980). Density estimation and bump-hunting by the penalized likelihood method exemplified by scattering and meteorite data (with discussion), *Journal of the American Statistical Association* **75**: 42–73.

Goodman, L. A. (1964). Interactions in multidimensional contingency tables, *Annals of Mathematical Statistics* **35**: 632–46.

Goodman, L. A. (1968). The analysis of cross-classified data: Independence, quasi-independence, and interactions in contingency tables with or without missing entries, *Journal of the American Statistical Association* **63**: 1091–131.

Green, P. J. (1995). Reversible jump Markov Chain Monte Carlo computation and Bayesian model determination, *Biometrika* **82**: 711–32.

Green, P. J. and Silverman, B. W. (1994). *Non-Parametric Regression and Generalised Linear Models: A Roughness Penalty Approach*, Chapman and Hall, London.

Griffiths, J. D. and Cresswell, C. (1976). A mathematical model of a pelican crossing, *J. Inst. Math. Applics.* **18**: 381–94.

Gu, C. (1993). Smoothing spline density estimation: A dimensionless automatic algorithm, *Journal of the American Statistical Association* **88**: 495–504.

Hald, A. (1952). *Statistical Theory with Engineering Applications*, Wiley, New York.

Harrison, P. J. and Stevens, C. F. (1976). Bayesian forecasting (with discussion), *Journal of Royal Statistical Society, Ser. B*, **38**: 205–47.

Harrison, P. J., Leonard, T., and Gazard, T. N. (1983). An application of multivariate hierarchical forecasting, *Technical Report*, University of Warrick.

Heikkinen, J. and Arjas, E. (1998). Non-parametric Bayesian estimation of a spatial Poisson intensity, *Scandinavian Journal of Statistics* **25**: 435–450.

Hickman, J. and Miller, R. B. (1977). Notes on Bayesian graduation (with discussion), *Trans. Soc. Actuaries* **29**: 7–49.

Hickman, J. C. and Miller, R. B. (1981). Bayesian bivariate graduation and forecasting, *Scandinavian Actuarial Journal*, pp. 129–50.

Highfield, R. A. (1986). *Forecasting with Bayesian State Space Models*, Ph.D. thesis, University of Chicago.

Hill, B. M. (1986). Some subjective Bayesian consideration in the selection of models, *Econometric Review* **4**: 191–246.

Hill, B. M. (1987). The validity of the likelihood principle, *American Statistician* **41**: 95–100.
Hinkley, D. V. (1979). Predictive likelihood, *Annals of Statistics* **7**: 718–28.
Hoadley, B. (1971). Asymptotic properties of maximum likelihood estimators for the independent not identically distributed case, *Annals of Mathematical Statistics* **42**: 1977–91.
Hogarth, R. (1980). *Judgement and Choice*, Wiley, New York.
Hong, C. (1989). *Forecasting Real Output Growth Rate and Cyclical Properties of Models: A Bayesian Approach*, Ph.D. thesis, University of Chicago.
Hsu, J. S. J. (1990). *Bayesian Inference and Marginalization*, Ph.D. thesis, University of Wisconsin–Madison.
Hsu, J. S. J. (1995). Generalized Laplacian approximations in Bayesian inference, *Canadian Journal of Statistics* **23**: 399–410.
Hsu, J. S. J. and Leonard, T. (1997). Bayesian semi-parametric procedures for logistic regression, *Biometrika* **84**: 85–93.
Hsu, J. S. J., Leonard, T., and Tsui, K. (1991). Statistical inference for multiple choice tests, *Psychometrika* **56**: 327–48.
Hudson, H. M. (1978). A natural identity for exponential families with applications in multiparameter estimation, *Annals of Statistics* **6**: 473–84.
Hudson, H. M. and Tsui, K. (1981). Simultaneous Poisson estimators for a priori hypotheses about means, *Journal of the American Statistical Association* **76**: 182–7.
Ijiri, Y. and Leitch, R. A. (1980). Stein's paradox and audit sampling, *J. Accounting Research* **18**: 91–108.
Issos, J. N. (1982). *Risks of Stein-type Estimators and Applications to Forest Inventory*, Ph.D. thesis, University of Wisconsin–Madison.
Izenman, A. J. and Zabell, S. L. (1976). Babies and the blackout: a time series analysis of New York City birth data, *ASA Proceedings of Social Statistics Section*, pp. 427–33.
Izenman, A. J., Papasouliotis, O., Leonard, T., and Aitken, C. G. G. (1998). Bayesian predictive evaluation of measurement error with application to the assessment of illicit drug quantity, *Technical Report 3*, Statistical Laboratory, University of Edinburgh.
Jaffray, J. Y. (1988). Choice under risk and the security factor, *Theory and Decision* **24**: 169–200.
James, W. and Stein, C. M. (1961). Estimation with quadratic loss, *Proc. Fourth Berkeley Sympos.*, vol. 1, pp. 361–79.
Jeffreys, H. (1961). *Theory of Probability*, 3d ed., Wiley, New York.
Johnson, B. M. (1971). On the admissible estimators for certain fixed sample binomial problems, *Annals of Mathematical Statistics* **42**: 1579–87.
Johnson, R. A. (1967). An asymptotic expansion for posterior distributions, *Annals of Mathematical Statistics* **38**: 1899–906.
Johnson, R. A. (1970). Asymptotic expansions associated with posterior distributions, *Annals of Mathematical Statistics* **41**: 851–64.
Johnson, R. A. and Klotz, J. (1993). Estimating hot numbers and testing uniformity for the lottery, *Journal of the American Statistical Association* **88**: 662–8.
Johnson, R. A. and Ladalla, J. N. (1979). The large sample behavior of posterior distributions which sample from multiparameter exponential models, and allied results, *Sankyha, Ser. B*, **41**: 196–215.
Johnson, V. E., Wong, W. H., Hu, X., and Chen, C. T. (1991). Image restoration using Gibbs priors: Boundary modeling, treatment of blurring and selection of hyperparameter, *IEEE Trans. on Pattern Analysis and Machine Intelligence* **13**: 413–25.
Johnson, W. E. and Braithwaite, R. B. (1932). Probability: Deductive and inductive problems, *Mind* **41**: 421–3.
Kadane, J. B. (1992). Healthy scepticism as an expected-utility explanation of the phenomena of Allais and Ellsberg, *Theory and Decision* **32**: 57–64.
Kadane, J. B. and Wolfson, L. J. (1998). Experiences in elicitation (with discussion), *The Statistician* **46**: 3–19.
Kahneman, D. and Tversky, A. (1979). Prospect theory: An analysis of decisions under risk, *Econometrika* **47**: 263–91.

Kahneman, D., Slovic, P., and Tversky, A. (1982). *Judgement under Uncertainty: Heuristics and Biases*, Cambridge University Press, Cambridge.

Kalbfleisch, J. D. and Sprott, D. A. (1970). Application of likelihood methods to models involving large number of parameters, *Journal of Royal Statistical Society, Ser. B*, **32**: 175–208.

Kalman, R. E. (1960). A new approach to linear filtering and prediction problems, *Trans. ASME, D* **82**: 35–45.

Kaplan, S. (1983). On a "two stage" Bayesian procedure for determining failure rates from experimental data, *IEEE Trans. on Power Apparatus and Systems* **102**: 195–202.

Kaplan, S., Garrick, B. J., and Bieniarz, P. P. (1981). On the use of Bayes' theorem in assessing the frequency of anticipated transients, *Nuclear Engineering and Design* **65**: 23–31.

Karmarkar, U. S. (1978). Subjectively weighted utility: A descriptive extension of the expected utility model, *Organizational Behavior and Human Performance* **21**: 61–72.

Kass, R. E. and Steffey, D. (1989). Approximate Bayesian inference in conditionally independent hierarchical models (parametric empirical Bayes models), *Journal of the American Statistical Association* **84**: 717–26.

Katz, R. W. (1981). On some criteria for estimating the order of a Markov chain, *Technometrics* **23**: 243–9.

Kerr, D. R. (1990). *Sedimentology and Stratigraph of Pennsylvanian and Lower Permian Strata (Upper Amsden Formation and Tensleep Sandstone) in North Wyoming*, Ph.D. thesis, University of Wisconsin–Madison.

Kim, S. H., Cohen, A. S., Baker, F. B., Suboviak, M. J., and Leonard, T. (1994). An investigation of hierarchical Bayes procedures in item response theory, *Psychometrika* **59**: 405–21.

Kimeldorf, G. S. and Wahba, G. (1970). A correspondence between Bayesian estimation on stochastic processes and smoothing by splines, *Annals of Mathematical Statistics* **41**: 495–502.

Kimeldorf, G. S. and Wahba, G. (1971). Some results on Tchebycheffian spline functions, *Journal of Mathematical Analysis and Applications* **33**: 82–95.

Kipnis, C. and Varadhan, S. R. S. (1986). Central Limit Theorem of additive functionals of reversible Markov processes and applications to simple exclusions, *Comm. Math. Phys.* **104**: 1–19.

Kloek, T. and van Dijk, H. K. (1978). Bayesian estimates of equation system parameters: An application of intergration by Monte Carlo, *Econometrica* **46**: 1–19.

Knill-Jones, R. P. (1974). The diagnosis of jaundice by the computation of probabilities, *J. Roy. Coll. Phys. Lond.* **9**: 205–10.

Koehler, A. B. and Murphree, E. S. (1988). A comparison of the Akaike and Schwarz criteria for selecting model order, *Applied Statistics* **37**: 187–95.

Ladalla, J. N. (1976). *The Large Sample Behavior of Posterior Distributions when Sampling from Multiparameter Exponential Family Models, and Allied Results*, Ph.D. thesis, University of Wisconsin–Madison.

Laird, N. M. (1975). *Empirical Bayes Methods for Two-Way Contingency Tables*, Ph.D. thesis, Harvard University.

Laird, N. M. (1979). Empirical Bayes methods for two-way contingency tables, *Biometrika* **65**: 581–90.

Laird, N. M. (1982). Empirical Bayes estimation using the nonparametric maximum likelihood estimate for the prior, *Journal of Statistical Computing and Simulation* **15**: 211–20.

Laird, N. M. and Louis, T. A. (1987). Empirical Bayes confidence intervals based on bootstrap samples (with discussion), *Journal of the American Statistical Association* **82**: 739–57.

Lange, N., Carlin, B. P., and Gelfand, A. E. (1992). Hierarchical Bayes models for the progression of HIV infection using longitudinal CD4 T-cell numbers (with discussion), *Journal of the American Statistical Association* **87**: 615–32.

Laplace, P. S. (1812). *Théorie Analytique des Probabilitées*, Courcier, Paris.

Lavalle, I. H. and Fishburn, P. C. (1991). Lexicographic state-dependent subjective expected utility, *J. Risk and Uncertainty* **4**: 251–69.

Lavine, M. (1991). Sensitivity in Bayesian statistics: The prior and the likelihood, *Journal of the American Statistical Association* **86**: 396–9.

Lavine, M. and West, M. (1992). A Bayesian method for classification and discrimination, *Canadian Journal of Statistics* **20**: 451–61.

Lee, P. (1997). *Bayesian Statistics*, 2d ed., Arnold, London.

Lehmann, E. (1991). *Theory of Point Estimation*, Wadsworth, Belmont, California.

Lejeune, M. and Faulkenberry, G. D. (1982). A simple predictive density function (with discussion), *Journal of the American Statistical Association* **77**: 654–9.

Lenk, P. J. (1988). The logistic normal distribution for Bayesian, nonparametric, predictive densities, *Journal of the American Statistical Association* **83**: 509–16.

Lenk, P. J. (1991). Towards a practicable Bayesian nonparametric density estimator, *Biometrika* **78**: 531–43.

Leonard, T. (1972). Bayesian methods for binomial data, *Biometrika* **59**: 581–9.

Leonard, T. (1973a). A Bayesian method for histograms, *Biometrika* **60**: 297–308.

Leonard, T. (1973b). *Bayesian Methods for the Simultaneous Estimation of Several Parameters*, Ph.D. thesis, University of London.

Leonard, T. (1974). A modification to the Bayes estimate for the mean of a Normal distribution, *Biometrika* **61**: 627–8.

Leonard, T. (1975a). A Bayesian approach to the linear model with unequal variances, *Technometrics* **17**: 95–102.

Leonard, T. (1975b). Bayesian estimation methods for two-way contingency tables, *Journal of Royal Statistical Society, Ser. B*, **37**: 23–37.

Leonard, T. (1976). Some alternative approaches to multiparameter estimation, *Biometrika* **63**: 69–76.

Leonard, T. (1977a). A Bayesian approach to some multinomial estimation and pretesting problems, *Journal of the American Statistical Association* **72**: 865–8.

Leonard, T. (1977b). An alternative Bayesian approach to the Bradley-Terry model for paired comparisons, *Biometrics* **33**: 121–30.

Leonard, T. (1977c). Bayesian simultaneous estimation for several multinomial distributions (describing the Dirichlet-Dirichlet approach), *Communications in Statistics, Ser. A*, **6**: 610–30.

Leonard, T. (1978). Density estimation, stochastic processes and prior information (with discussion), *Journal of Royal Statistical Society, Ser. B*, **40**: 113–46.

Leonard, T. (1980). The roles of inductive modeling and coherence in Bayesian statistics (with discussion), *in* J. M. Bernardo, M. H. DeGroot, D. V. Lindley, and A. F. M. Smith (eds.), *Bayesian Statistics I*, University Press, Valencia, pp. 537–55.

Leonard, T. (1982a). Comment on "A simple predictive density function," *Journal of the American Statistical Association* **77**: 657–8.

Leonard, T. (1982b). A Bayesian approach to Markovian models for normal and Poisson data, *Technical Report 2337*, Mathematics Research Center, University of Wisconsin–Madison.

Leonard, T. (1982c). An empirical Bayesian approach to the smooth estimation of unknown functions, *Technical Report 2339*, Mathematics Research Center, University of Wisconsin–Madison.

Leonard, T. (1983). Some philosophies of inference and modeling, *in* G. E. P. Box, T. Leonard, and C.-F. Wu (eds.), *Scientific Inference, Data Analysis, and Robustness*, Academic Press, London, pp. 51–84.

Leonard, T. (1984). Some data-analytic modifications to Bayes-Stein estimation, *Annals of the Institute of Statistical Mathematics* **36**: 11–21.

Leonard, T. (1996). On exchangeable sampling distributions for uncontrolled data, *Statistics and Probability Letters* **26**: 1–6.

Leonard, T. (1999). On the application of Bayes' theorem to the evaluation of legal evidence, *Statistical Science*, invited revision.

Leonard, T. and Hsu, J. S. J. (1992). Bayesian inference for a covariance matrix, *Annals of Statistics* **20**: 1669–96.

Leonard, T. and Hsu, J. S. J. (1994). The Bayesian analysis of categorical data – a selective review, *in* P. R. Freeman and A. F. M. Smith (eds.), *Aspects of Uncertainty: A Tribute to D. V. Lindley*, Wiley, Chichester.

Leonard, T. and Hsu, J. S. J. (1996). On small sample Bayesian inference and sequential design for quantal response, *Modeling and Prediction: Honoring Seymour Geisser*, Springer-Verlag, New York, pp. 169–75.

Leonard, T. and Novick, M. R. (1986). Bayesian full rank marginalization for two-way contingency tables, *Journal of Educational Statistics* **11**: 33–56.

Leonard, T. and Ord, K. (1976). An investigation of the F test procedure as an estimation short-cut, *Journal of Royal Statistical Society, Ser. B*, **38**: 95–8.

Leonard, T., Hsu, J. S. J., and Ritter, C. (1994a). The Laplacian t-approximation in Bayesian inference, *Statistica Sinica* **4**: 127–42.

Leonard, T., Hsu, J. S. J., and Tsui, K. (1989). Bayesian marginal inference, *Journal of the American Statistical Association* **84**: 1051–8.

Leonard, T., Hsu, J. S. J., Tsui, K., and Murray, J. (1994b). Bayesian and likelihood inference from equally weighted mixtures, *Annals of the Institute of Statistical Mathematics* **46**: 203–20.

Leonard, T., Tsui, K., and Hsu, J. S. J. (1990). Comment on "Predictive likelihood – a review," *Statistical Science* **98**: 39–54.

Lesage, J. P. and Magura, M. (1990). Using Bayesian techniques for data pooling in regional payroll forecasting, *J. Busn. Econ. Statist.* **8**: 127–35.

Levin, B. and Reeds, J. (1977). Compound multinomial likelihood functions: proof of a conjecture by I. J. Good, *Annals of Statistics* **5**: 79–87.

Lindley, D. V. (1956). On a measure of the information provided by an experiment, *Annals of Mathematical Statistics* **27**: 986–1005.

Lindley, D. V. (1957). A statistical paradox, *Biometrika* **44**: 187–92.

Lindley, D. V. (1958). Fiducial distributions and Bayes' theorem, *Journal of Royal Statistical Society, Ser. B*, **20**: 102–7.

Lindley, D. V. (1962). Discussion of paper by Stein, *Journal of Royal Statistical Society, Ser. B*, **24**: 285–7.

Lindley, D. V. (1964). The Bayesian analysis of contingency tables, *Annals of Mathematical Statistics* **35**: 1622–43.

Lindley, D. V. (1965). *Introduction to probability and statistics from a Bayesian viewpoint*, Cambridge University Press, Cambridge.

Lindley, D. V. (1977). A problem in forensic science, *Biometrika* **64**: 207–23.

Lindley, D. V. (1983). Comment on "Parametric empirical Bayes inference: Theory and applications," *Journal of the American Statistical Association* **78**: 61–2.

Lindley, D. V. (1985). *Making Decisions*, 2d ed., Wiley, Chichester.

Lindley, D. V. (1991). Subjective probability, decision analysis and their legal consequences, *Journal of Royal Statistical Society, Ser. A*, **154**: 83–92.

Lindley, D. V. and Novick, M. R. (1981). The role of exchangeability in inference, *Annals of Statistics* **9**: 45–58.

Lindley, D. V. and Smith, A. F. M. (1972). Bayes estimates for the linear model (with discussion), *Journal of Royal Statistical Society, Ser. B*, **34**: 1–41.

Lindley, D. V., Tversky, A., and Brown, R. V. (1979). On the reconciliation of probability assessments (with discussion), *Journal of Royal Statistical Society, Ser. A*, **142**: 146–80.

Lindstrom, M. J. and Bates, D. M. (1990). Non-linear mixed effects models for repeated measures data, *Biometrics* **46**: 673–87.

Longford, N. T. (1993). *Random Coefficient Models*, Oxford University Press, Oxford.

Lord, F. M. and Novick, M. R. (1968). *Statistical theories of mental test scores (with contributions by Allan Birnbaum)*, Addison-Wesley, Reading, Mass.

Low, J. A., Karchmar, J., Broekhoven, L., Leonard, T., McGrath, M. J., Pancham, S. R., and Piercy, W. N. (1981). The probability of fetal metabolic acidosis during labor in a population at risk as determined by clinical factors (with discussion), *Amer. J. of Obstetrics and Gynecology* **141**: 941–51.

Machina, M. J. (1982). Expected utility analysis with the independence axiom, *Econometrika* **50**: 277–323.

Machina, M. J. (1983). Generalized expected utility analysis and the nature of observed violations of the independence axiom, *in* B. Stigum and F. Wenstop (eds.), *Foundations of Utility and Risk Theory with Applications*, Reidel, Dordrecht, Holland.

Machina, M. J. (1984). Temporal risk and the nature of induced axiom, *J. Econometric Theory* **33**: 199–231.

Machina, M. J. (1987). Choice under uncertainty: Problems solved and unsolved, *Economic Perspectives* **1**: 121–54.

Mangasarian, O. L. (1968). Multi-surface method of pattern separation, *IEEE Trans. on Information Theory* **14**: 801–7.

Mangasarian, O. L. (1992). Mathematical programming in neural networks, *Technical Report 1129*, Computer Science Department, University of Wisconsin–Madison.

Mangasarian, O. L., Setiono, R., and Wolberg, W. H. (1989). Pattern recognition via linear programming: Theory and applications to medical diagnosis, *in* T. F. Coleman and Y. Li (eds.), *The Workshop on Large-scale Numerical Optimerical Optimization, Cornell University, Ithaca, New York*, SIAM, Philadephia, pp. 314–17.

Martin, B. (1764). *The Life of Colin Maclaurin (1698–1746)*, Biographia Philiosophica, National Library of Scotland, Edinburgh.

Matsumura, E. and Tsui, K. (1982). Stein-type Poisson estimators in audit sampling, *J. Accounting Research* **20**: 162–70.

McCullagh, P. and Nelder, J. A. (1989). *Generalized Linear Models*, Chapman and Hall, London.

McCulloch, R. E. and Rossi, P. E. (1992). Bayes factors for nonlinear hypotheses and likelihood distributions, *Biometrika* **79**: 663–76.

McLachlan, G. J. and Basford, K. E. (1988). *Mixture models: Inference and applications to clustering*, Marcel Dekker, New York.

Meginniss, J. R. (1976). *Alternatives to the Expected Utility Rule*, Ph.D. thesis, University of Chicago.

Metropolis, N., Rosenbluth, A. W., Rosenbluth, M. N., Teller, A. H., and Teller, E. (1953). Equations of state calculations by fast computing machines, *J. Chemical Physics* **21**: 1087–91.

Miller, R. B. and Fortney, W. G. (1984). Industry-wide expense standards using random coefficient regression, *Insurance: Mathematics and Econometrics* **3**: 19–33.

Min, C. and Zellner, A. (1993). Bayesian and non-Bayesian methods for combining models and forecasts with applications to forecasting international growth rates, *Journal of Econometrics* **56**: 89–118.

Moore, D. S. and McCabe, G. P. (1989). *Introduction to the Practice of Statistics*, W. H. Freeman, New York.

Morgan, J. P., Chaganty, N. R., Dahiya, R. C., and Doviak, M. J. (1991). Let's make a deal: The player's dilemma (with discussion), *American Statistician* **45**: 284–9.

Morris, C. N. (1983). Parametric empirical Bayes inference: Theory and applications (with discussion), *Journal of the American Statistical Association* **78**: 47–65.

Morris, M. D., Mitchell, T. J., and Ylvisaker, D. (1993). Bayesian design and analysis of computer experiments: Use of derivatives in surface prediction, *Technometrics* **35**: 243–55.

Morrison, D. G. and Brockway, G. (1979). A modified beta binomial model with applications to multiple choice and taste tests, *Psychometrika* **44**: 427–42.

Murray, J. F. (1992). *An Evaluation of an Empirical bayes Approach for Analyzing Health Care Performance Measures*, Ph.D. thesis, University of Wisconsin–Madison.

Nazaret, W. A. (1987). Bayesian log-linear estimates for three-way contingency tables, *Biometrika* **74**: 401–10.

Nyquist, J. E. (1986). *Thermal and Mechanical Models of the Midcontinent Rift*, Ph.D. thesis, University of Wisconsin–Madison.

Nyquist, J. E. and Wang, H. F. (1988). Flexural modeling of the Midcontinent Rift, *Geophysical Research* **93**: 8852–68.

O'Hagan, A. (1976). On posterior joint and marginal modes, *Biometrika* **63**: 329–33.

O'Hagan, A. (1978). Curve fitting and optimal design for prediction, *Journal of Royal Statistical Society, Ser. B*, **40**: 1–24.

O'Hagan, A. (1988). *Probability: Methods and Measurement*, Chapman and Hall, London.

O'Hagan, A. (1994). *Kendall's Advanced Theory of Statistics, Volume 2B: Bayesian Inference*, Arnold, London.

O'Hagan, A. (1998). Eliciting expert beliefs in substantial practical applications (with discussion), *The Statistician* **46**: 1–15.

O'Hagan, A. and Leonard, T. (1976). Bayes estimation subject to uncertainty about parameter constraints, *Biometrika* **63**: 201–3.

O'Sullivan, F., Yandell, B. S., and Raynor, W. J. (1986). Automatic smoothing of regression functions in generalized linear models, *Journal of the American Statistical Association* **81**: 96–103.

Parton, R. M., Fischer, S., Malho, O., Papasouliotis, O., Jelitto, T. C., Leonard, T., and Read, N. D. (1997). Pronounced cytoplasmic ph gradients are not required for tip growth in plant and fungal cells, *Journal of Cell Science* **11**: 1187–98.

Paul, S. R. and Plackett, R. L. (1978). Inference sensitivity for Poisson mixtures, *Biometrika* **65**: 591–602.

Pearson, K. P. (1904). On the theory of contingency and its relation to association and normal correlation, *Technical Report*, Drapers Co. Res. Mem. Biometrics Series, London.

Peng, J. C. M. (1975). Simultaneous estimation of the parameters of independent Poisson distribution, *Technical Report 78*, Department of Statistics, Stanford University.

Poundstone, W. (1992). *Prisoner's Dilemma*, Doubleday, New York.

Press, S. J. (1982). *Applied Multivariate Analysis: Using Bayesian and Frequency Measures of Inference*, 2d ed., Krieger, Melbourne, Florida.

Press, S. J. (1989). *Bayesian Statistics: Principles, Models, and Applications*, Wiley, New York.

Quesenberry, C. P. and Kent, J. (1982). Selecting among probability distributions used in reliability, *Technometrics* **24**: 59–65.

Quiggin, J. (1982). A theory of anticipated utility, *Journal of Economic Behavior and Organization* **3**: 323–43.

Racine-Poon, A. (1985). A Bayesian approach to nonlinear random effects models, *Biometrics* **51**: 1015–23.

Radelet, M. (1981). Racial characteristics and imposition of the death penalty, *Amer. Sociol. Rev.* **46**: 918–27.

Raftery, A. E. (1996). Approximate Bayes factors and accounting for model uncertainty in generalised linear models, *Biometrika* **83**: 251–66.

Ratkowsky, D. A. (1983). *Nonlinear Regression Modeling: A Unified Practical Approach*, Marcel Dekker, New York.

Reid, N. (1988). Saddlepoint methods and statistical inference (with discussion), *Statistical Science* **3**: 213–38.

Reinsel, G. C. (1985). Mean squared error properties of empirical Bayes estimators in a multivariate random effects general linear model, *Journal of the American Statistical Association* **80**: 642–50.

Reinsel, G. C. and Tiao, G. C. (1987). Impact of chlorofluoromethanes on stratospheric ozone: A statistical analysis of ozone data for trends, *Journal of the American Statistical Association* **82**: 20–30.

Rényi, A. (1970). *Probability Theory*, American Elsevier, New York.

Richardson, S. and Green, P. J. (1997). On Bayesian analysis of mixtures with an unknown number of components, *Journal of Royal Statistical Society, Ser. B*, **59**: 731–8.

Ripley, B. (1987). *Stochastic Simulation*, Wiley, New York.

Ritter, C. (1992). *Modern Inference in Non-linear Least Square Regression*, Ph.D. thesis, University of Wisconsin–Madison.

Ritter, C. (1994). Statistical analysis of spectra from electron spectroscopy for chemical analysis, *The Statistician* **43**: 111–27.

Robbins, H. (1964). The empirical Bayes approach to statistical decision problems, *Annals of Mathematical Statistics* **35**: 1–20.

Roberts, H. V. (1967). Informative stopping rules and inference about population size, *Journal of the American Statistical Association* **62**: 763–75.

Robinson, G. K. (1975). Some counterexamples to the theory of confidence intervals, *Biometrika* **62**: 155–62.

Roeder, K. and Wasserman, L. (1997). Practical Bayesian density estimation using mixtures of normals, *Journal of the American Statistical Association* **92**: 894–902.

Rosenkrantz, R. D. (1989). *E. T. Jaynes: Papers on Probability, Statistics and Statistical Physics*, Kluwer Academic, Norwell, Massachusetts.

Rubin, D. B. (1981). The Bayesian bootstrap, *Annals of Statistics* **9**: 130–4.

Rubin, D. B. (1984). Bayesianly justifiable and relevant frequency calculations for the applied statistician, *Annals of Statistics* **12**: 1151–72.

Rubinstein, R. Y. (1981). *Simulation and the Monte Carlo Method*, Wiley, New York.

Savage, L. J. (1972). *The Foundations of Statistics*, 2d ed., Dover, New York.

Scales, L. E. (1985). *Introduction to Non-linear Optimization*, Springer-Verlag, New York.

Schoenberg, I. J. (1946). Contributions to the problem of approximation of equidistant data by analytic functions, part a – On the problem of smoothing or graduation: a first class of analytic approximation formulae, *Quarterly Applied Mathematics* **4**: 45–99.

Schwarz, G. (1978). Estimating the dimension of a model, *Annals of Statistics* **6**: 461–4.

Sen, P. K. and Singer, J. M. (1993). *Large Sample Methods in Statistics: An Introduction with Applications*, Chapman and Hall, London.

Severini, T. A. and Wong, W. H. (1992). Profile likelihood and conditionally parametric models, *Annals of Statistics* **20**: 1768–802.

Shepherd, N. and Pitt, M. K. (1998). Analysis of time varying covariances: a factor stochastic volatility approach, *in* J. O. Berger, J. M. Bernardo, A. P. Dawid, and A. Smith (eds.), *Bayesian Statistics VI*, Clarendon Press, Oxford, pp. 433–63.

Shiau, J. J. (1985). Smoothing spline estimation of functions with discontinuities, *Technical Report 768*, Department of Statistics, University of Wisconsin–Madison.

Shiau, J. J., Wahba, G., and Johnson, D. R. (1987). Partial spline models for the inclusion of tropause and frontal boundary information in otherwise smooth two- and three-dimensional objective analysis, *Journal of Atmospheric and Oceanic Technology* **3**: 714–24.

Shun, Z. M. and McCullagh, P. (1995). Laplace approximation of high dimensional integrals, *Journal of Royal Statistical Society, Ser. B*, **57**: 749–60.

Silver, R. N., Martz, H. F., and Wallstrom, T. (1993). Quantum statistical inference for density estimation, *ASA Proceedings of the Section on Bayesian Statistical Science*, pp. 131–9.

Silverman, B. W. (1978). Density ratios, empirical likelihood and cot death, *Applied Statistics* **27**: 26–33.

Silverman, B. W. (1986). *Density Estimation for Statistics and Data Analysis*, Chapman and Hall, London.

Simpson, E. H. (1951). The interpretation of interactions in contingency tables, *Journal of Royal Statistical Society, Ser. B*, **13**: 238–41.

Smith, A. F. M. (1973a). Bayes estimates in one-way and two-way models, *Biometrika* **60**: 319–29.

Smith, A. F. M. (1973b). A general Bayesian linear model, *Journal of Royal Statistical Society, Ser. B*, **35**: 67–75.

Smith, J. Q. (1978). *Problems in Bayesian Statistics Relating to Discontinuous Phenomena, Catastrophe Theory and Forecasting*, Ph.D. thesis, University of Warwick.

Smith, J. Q. (1980). Bayes estimates under bounded loss, *Biometrika* **67**: 629–38.

Smith, J. Q. (1988). *Decision Analysis: A Bayesian Approach*, Chapman and Hall, London.

Smith, P. J. and Sedransk, J. (1982). Bayesian optimization of the estimation of the age composition of a fish population, *Journal of the American Statistical Association* **77**: 707–13.

Snedecor, G. W. and Cochran, W. G. (1967). *Statistical Methods*, 6th ed., Iowa State University Press, Ames.

Snyder, D. L. (1981). *Random Point Processes*, Wiley, Chichester.

Solomon, P. J. and Cox, D. R. (1992). Non-linear component of variance models, *Biometrika* **79**: 1–11.

Soofi, E. S. (1992). A generalized formulation of conditional logits with diagnostics, *Journal of the American Statistical Association* **87**: 812–6.

Sprott, D. A. and Kalbfleisch, J. D. (1969). Examples of likelihoods and large sample approximations, *Journal of the American Statistical Association* **64**: 468–84.

Stein, C. (1956). Inadmissibility of the usual estimator for the mean of a multivariate normal distribution, *Proc. Third Berkeley Symp. Math. Statist. Probab. 1*, University of California Press, Berkeley, pp. 197–206.

Stein, C. (1964). Inadmissibility of the usual estimator for the variance of a normal distribution with unknown mean, *Annals of the Institute of Statistical Mathematics* **16**: 155–60.

Stein, C. M. (1981). Estimation of the mean of a multivariate normal distribution, *Annals of Statistics* **9**: 1135–151.

Steinijans, V. W. (1976). A stochastic point-process model for the occurrence of major freezes in Lake Constance, *Applied Statistics* **25**: 58–61.

Stigler, S. M. (1983). Who discovered Bayes' theorem? *American Statistician* **77**: 290–6.

Stiratelli, R., Laird, N. M., and Ware, J. H. (1984). Random-effects models for serial observations with binary response, *Biometrics* **40**: 961–71.

Stone, C. J. (1986). The dimensionality reduction principle for generalized additive models, *Annals of Statistics* **14**: 590–606.

Stone, C. J. (1991). Asymptotics for doubly flexible logspline response models, *Annals of Statistics* **19**: 1832–54.

Stone, M. (1974). Crossvalidatory choice and assessment of statistical predictions (with discussion), *Journal of Royal Statistical Society, Ser. B*, **36**: 111–47.

Stone, M. (1977). An asymptotic equivalence of choice of model by crossvalidation and Akaike's criterion, *Journal of Royal Statistical Society, Ser. B*, **39**: 44–7.

Stone, M. (1979). Comments on model selection criteria of Akaike and Schwarz, *Journal of Royal Statistical Society, Ser. B*, **41**: 276–8.

Sun, L., Hsu, J. S. J., Guttman, I., and Leonard, T. (1996). Bayesian methods for variance component models, *Journal of the American Statistical Association* **91**: 743–52.

Susarla, V. and Van Ryzin, J. (1976). Nonparametric Bayesian estimation of survival curves from incomplete observations, *Journal of the American Statistical Association* **71**: 897–902.

Tanner, M. and Wong, W. (1987). The calculation of posterior distributions by data augmentation (with discussion), *Journal of the American Statistical Association* **82**: 528–50.

Tavare, S. and Altham, P. M. E. (1983). Serial dependence of observations leading to contingency tables, and corrections to chi-squared statistics, *Biometrika* **70**: 139–44.

Taylor, S. J. (1994). Modelling stochastic volatility, *Mathematical Finance* **4**: 183–204.

Tempelman, R. J. (1993). *Poisson Mixed Models for the Analysis of Counts with an Application to Dairy Cattle Breeding*, Ph.D. thesis, University of Wisconsin–Madison.

Thorburn, D. (1986). A Bayesian approach to density estimation, *Biometrika* **73**: 65–75.

Tierney, L. (1994). Markov chains for exploring posterior distributions, *Annals of Statistics* **22**: 1701–28.

Tierney, L. and Kadane, J. B. (1986). Accurate approximations for posterior moments and marginal densities, *Journal of the American Statistical Association* **81**: 82–6.

Tierney, L., Kass, R. E., and Kadane, J. B. (1989). Approximate marginal densities of non-linear functions, *Biometrika* **76**: 425–33. Corrections (1991) **78**, 233–4.

Titterington, D. M., Smith, A. F. M., and Makov, U. E. (1985). *Statistical Analysis of Finite Mixture Distributions*, Wiley, New York.

Tsui, K. (1981). Simultaneous estimation of several Poisson parameters under squared error loss, *Annals of the Institute of Statistical Mathematics* **33**: 215–23.

Tukey, J. W. (1977). *Exploratory Data Analysis*, Addison-Wesley, Reading, Massachusetts.

Tversky, A. and Kahneman, D. (1974). Judgment under uncertainty: Heuristics and biases, *Science* **185**: 1124–31.

Tversky, A. and Kahneman, D. (1992). Advances in prospect theory: Cumulative representation of uncertainty, *Journal of Risk and Uncertainty* **5**: 297–323.

Tversky, A. and Wakker, P. P. (1995). Risk attitudes and decision weight, *Econometrica* **63**: 1255–80.

van der Linde, A. (1995). Splines from a Bayesian point of view, *Test* **4**: 63–81.

van der Linde, A. (1996). The invariance of statistical analyses with smoothing splines with respect to inner product in reproducing kernel Hilbert space, *in* W. Hardle and M. Schimek (eds.), *Statistical Theory and Computational Aspects of Smoothing*, Physica Verlag, Berlin, pp. 149–64.

van der Linde, A. (1997). Variance estimation and smoothing parameter determination for spline-based regression, *in* M. Schimek (ed.), *Smoothing and Regression: Approaches, Computation and Applications*, Wiley, New York.

Vaughan, A. (1979). *Incredible Coincidence*, J. M. Lippincott, New York.

Viscusi, W. K. (1989). Prospective reference theory: toward an explanation of the paradoxes, *J. Risk and Uncertainty* **2**: 235–64.

von Neumann, J. and Morgenstern, O. (1953). *Theory of Games and Economic Behavior*, 3d ed., Princeton University Press, Princeton, New Jersey.

Wahba, G. (1978). Improper priors, spline smoothing and the problem of guarding against model errors in regression, *Journal of Royal Statistical Society, Ser. B*, **40**: 364–72.

Wahba, G. (1985a). A comparison of GCV and GML for choosing the smoothing parameter in the generalized spline smoothing problem, *Annals of Statistics* **12**: 1378–402.

Wahba, G. (1985b). Design criteria and eigensequence plots for satellite-computed tomography, *J. Atmosphere and Oceanic Tech.* **2**: 125–32.

Wahba, G. (1990). *Spline Models for Observational Data*, SIAM, Philadelphia.

Wahba, G. and Wendelberger, J. (1980). Some new mathematical methods for variational objective analysis using splines and cross validation, *Monthly Weather Review* **108**: 1122–43.

Walker, A. M. (1969). On the asymptotic behaviour of posterior distributions, *Journal of Royal Statistical Society, Ser. B*, **31**: 80–8.

Wang, M. C. and Van Ryzin, J. (1979). Discrete density smoothing applied to the empirical Bayes estimation of a Poisson mean, *Jour. Stat. Comp. and Sim.* **8**: 207–26.

Weber, N. C. (1980). A martingale approach to central limit theorems for exchangeable random variables, *Journal of Applied Probability* **17**: 662–73.

Weisberg, S. (1980). *Applied Linear Regression*, Wiley, New York.

West, M. (1984). Outlier model and prior distributions in Bayesian linear regression, *Journal of Royal Statistical Society, Ser. B*, **46**: 431–9.

West, M. (1988). Modeling expert opinion, *in* J. M. Bernardo, M. H. DeGroot, D. V. Lindley, and A. F. M. Smith (eds.), *Bayesian Statistics III*, Clarendon Press, Oxford, pp. 493–508.

West, M. and Harrison, J. (1990). *Bayesian Forecasting and Dynamic Models*, Springer-Verlag, Berlin.

Wetzler, P. (1983). *Provisioning the Sambre and Meuse Army: Military-Civilian Relations and the Occupation of Belgium and the Northern Rhineland in 1794*, Ph.D. thesis, University of Wisconsin–Madison.

White, D. J. (1975). *Decision Methodology*, Wiley, Chichester.

Whittaker, E. (1923). A method of graduation based upon probability, *Proc. Edin. Math. Soc.* **41**: 63–75.

Wijsman, R. A. (1973). On the attainment of the Cramer-Rao lower bound, *Annals of Statistics* **1**: 538–42.

Wilkinson, G. N. (1977). On resolving the controversy in statistical inference, *Journal of Royal Statistical Society, Ser. B*, **39**: 119–43.

Williams, D. A. (1975). The analysis of binary responses from toxilogical experiments involving reproduction and teratogenicity, *Biometrics* **31**: 949–52.

Williams, D. A. (1982). Extra-binomial variation in logistic linear models, *Applied Statistics* **31**: 144–8.

Wolfinger, R. (1993). Laplace's approximation for nonlinear mixed models, *Biometrika* **80**: 791–5.

Wong, W. H. (1986). Theory of partial likelihood, *Annals of Statistics* **14**: 88–123.

Wong, W. H. and Li, B. (1992). Laplace expansions for posterior densities of non-linear functions of parameters, *Biometrika* **79**: 393–8.

Wu, C. F. J. (1981). Asymptotic theory of nonlinear least squares estimation, *Annals of Statistics* **9**: 501–13.

Wu, C. F. J. (1983). On the convergence properties of the EM algorithm, *Annals of Statistics* **11**: 95–103.

Yandell, B. S. and Green, P. J. (1986). Semi-parametric generalized linear model diagnostics, *ASA Proceedings of Statistical Computing Section*, pp. 48–53.

Zeger, S. L. and Karim, M. R. (1991). Generalised linear models with random effects, a Gibbs sampling approach, *Journal of the American Statistical Association* **86**: 79–86.

Zellner, A. (1971). *An Introduction to Bayesian inference in Econometrics*, Wiley, New York.

Zellner, A. (1983). *Applied Time Series Analysis of Economic Data*, U.S. Commerce, Washington, D. C.

Zellner, A. (1988). Optimal information processing and Bayes' theorem (with discussion), *American Statistician* **42**: 278–84.

Zellner, A. and Hong, C. (1989). Forecasting internal growth rates using Bayesian shrinkage and other procedures, *Journal of Econometrics* **40**: 183–202.

Zellner, A. and Rossi, P. E. (1984). Bayesian analysis of dichotomous quantal response models, *Journal of Econometrics* **25**: 365–93.

Zellner, A., Hong, C., and Min, C. (1991). Forecasting turning points in international output growth rates using Bayesian exponentially weighted autoregression, time-varying parameter and pooling techniques, *Journal of Econometrics* **49**: 275–304.

Author Index

Subject Index